Scattering of Waves by Wedges and Cones with Impedance Boundary Conditions

The Mario Boella Series on Electromagnetism in Information & Communication

The Mario Boella series offers textbooks and monographs in all areas of radio science, with a special emphasis on the applications of electromagnetism to information and communication technologies. The series is scientifically and financially sponsored by the Istituto Superiore Mario Boella affiliated with the Politecnico di Torino, Italy, and is scientifically co-sponsored by the International Union of Radio Science (URSI). It is named to honor the memory of Professor Mario Boella of the Politecnico di Torino, who was a pioneer in the development of electronics and telecommunications in Italy for half a century, and a Vice President of URSI from 1966 to 1969.

Scattering of Waves by Wedges and Cones with Impedance Boundary Conditions, ISMB Series

Mikhail A. Lyalinov and Ning Yan Zhu

This monograph on scattering of electromagnetic waves by impedance wedges and cones is the most comprehensive treatise in existence on a specialized topic that is of great interest to electrical engineers and mathematical physicists. Written by two international experts on scattering and diffraction, it constitutes a fundamental and lasting contribution to the field of applied mathematics. Even though its obvious applications are to radar and to mobile communications, the mathematical methods employed herein are also applicable to problems involving scattering of acoustic and elastic waves. Thus, this monograph will be of interest to all researchers on wave propagation and will remain a standard reference for the foreseeable future.

Piergiorgio L. E. Uslenghi – Series Editor
Chicago, July 2012

Other titles in this series

Fundamentals of Wave Phenomena, Second Edition
by Akira Hirose and Karl E. Lonngren

Forthcoming titles in this series

The Wiener-Hopf Method in Electromagnetics
by Vito G. Daniele and Rodolfo S. Zich

Higher Order Numerical Solution Techniques in Electromagnetics
by Roberto D. Graglia and Andrew F. Peterson

Scattering of Waves by Wedges and Cones with Impedance Boundary Conditions

ISMB Series

Mikhail A. Lyalinov

Institute of Physics
St. Petersburg University
St. Petersburg, Russia

Ning Yan Zhu

Institut für Hochfrequenztechnik
Universität Stuttgart
Stuttgart, Germany

SciTech Publishing
an imprint of the IET

Edison, NJ
scitechpub.com

SciTech Publishing
an imprint of the IET

Published by SciTech Publishing, an imprint of the IET.
www.scitechpub.com
www.theiet.org

10 9 8 7 6 5 4 3 2 1

ISBN 978-1-61353-003-0 (hardback)
ISBN 978-1-61353-025-2 (PDF)

Typeset in India by MPS Ltd
Printed in the USA by Sheridan Books, Inc.

Contents

Preface

In view of the widespread applications of the geometrical theory of diffraction (GTD)[1] to scattering of high-frequency waves by natural or artificial obstacles, the wave diffraction on bodies with singular lines or points on their boundaries like wedges or cones attracts traditionally much attention from researchers. The GTD is a short-wavelength asymptotic theory, the application of which is based on the commonly accepted localization principle. The localization principle, which V. A. Fock apparently formulated in its modern form in the 1940s,[2] implies that at a point of observation the scattered wave field is specified by all rays passing through this point and, moreover, every such ray reflected from or scattered by the obstacle contributes to the scattered field (ray-tracing procedure). This contribution is formed as a result of the corresponding ray field interacting with some asymptotically small part of the scattering surface, say, near the point of reflection or close to the origin of the diffracted ray. The amplitudes and phases of the waves depend only on the local characteristics of the corresponding parts of the scatterer and are not affected by other remote parts of the scattering surface.

Relying upon such a powerful and physically intuitive principle, in the present monograph we study a set of canonical problems that enable us to develop mathematically justified procedures to compute the diffraction or excitation coefficients. These coefficients can easily be incorporated into any GTD-type ray-tracing procedure. The diffraction coefficients (scattering diagrams or amplitudes) specified from particular canonical problems, in view of the localization principle, can be applied in very diverse situations encountered in research or engineering.

The canonical problems studied in the present monograph are the problems of diffraction of acoustic and electromagnetic waves by wedges and cones with impedance boundary conditions. The latter conditions are considered more general and realistic than those ideal ones and hence are widely exploited in practice. On the other hand, the corresponding boundary value problems are more complex for analysis. In particular, we obtain exact but not explicit (i.e., in-quadrature)

---

[1]In this monograph, the term GTD embraces not only the original nonuniform version of Keller's but also its uniform versions like the uniform geometrical theory of diffraction (UTD) and the uniform asymptotic theory of diffraction (UAT); furthermore, it includes Ufimtsev's physical theory of diffraction (PTD), an extension of Kirchhoff's approach.

[2]V. A. Fock, "New methods in diffraction theory," *Philosph. Mag.*, vol. 39 (ser. 7), pp. 149–155, 1948. This article is included as Chapter 1 in V. A. Fock, *Electromagnetic Diffraction and Propagation Problems* (Int. Series of Monographs on Electromagnetic Waves 1). Oxford, UK: Pergamon Press, 1965.

solutions.[3] The proposed approaches usually require some numerical solution of a standard integral equation obtained after essential analytical work simplifying the problem at hand. The expressions for the diffraction coefficients follow from additional analytical work usually based on the asymptotic evaluation of integral representations of a solution.

To that end, we exploit different analytical tools of mathematical physics and of mathematical diffraction theory as a part of mathematical physics. Essential analytical efforts are usually rewarded by efficient (i.e., accurate and fast) solution algorithms, as demonstrated by numerical examples that accompany the analytical treatment in every chapter beside the first two (Introduction and Fundamentals).

Therefore, the present monograph should be useful, first of all, for specialists in diffraction theory, including scientific researchers or engineers studying high-frequency wave phenomena of different nature; in addition, it might be also of interest for graduate and postgraduate students with a university-level background in mathematics and physics. We hope that the monograph might serve as an advanced introduction to an important part of the modern theory of diffraction, that is, into scattering of waves by wedges and cones.

A few readers might doubt the value of such "anachronistic" approaches, a reaction that is at first sight understandable especially in view of today's powerful numerical techniques. To prove the contrary, let us mention a few real-world scattering and diffraction problems: planning cellular mobile networks, predicting mm-wave radar signatures of moving cars and other obstacles on a busy highway, and simulating echoes generated by concealed objects in THz range, all of which share in common their geometric scale in each of the three spatial dimensions exceeding a few hundred wavelengths. In this circumstance, it is necessary to use asymptotic methods or hybrid techniques that augment appropriately analytical methods with numerical approaches. The latter reflects, according to our observation, a new trend in the study of wave phenomena. As mentioned earlier, exact solutions and their asymptotic expressions for canonical bodies, two of which are to be studied in the present monograph, are building blocks of GTD. Furthermore, even when such a short-wavelength problem can be solved, enabled by a supercomputer and after a long computation time, in a purely numerical way, fast analytic and asymptotic solutions with an acceptable accuracy are always important for engineering purposes.

In addition, analytical approaches and numerical methods enrich and benefit from each other. As to be expounded in the monograph, the Fredholm integral equations are numerically solved by means of the quadrature method. Purely numerical techniques can be accelerated with the aid of new advances achieved in the analytic and asymptotic approaches.[4] In this way, both "old school" theorists and modern-day "number crunchers" can better achieve their common goal, namely, fast and accurate computation. Lastly, none of us, scientists and engineers alike, should ever forget that "the purpose of computing is insight, not numbers."[5] In this respect, analytical approaches, when reasonably combined at some stages with numerical

---

[3]Surely some important canonical problems have been solved explicitly like the famous Malyuzhinets problem.

[4]See for example B. Engquist, A. Fokas, E. Hairer and A. Iserles, Eds., *Highly Oscillatory Problems* (London Mathematical Society Lecture Note Series 366). Cambridge, UK: Cambridge Univ. Press, 2009.

[5]R. W. Hamming, *Numerical Methods for Scientists and Engineers* (Int. Series in Pure and Applied Mathematics). New York, NY: McGraw-Hill, 1962.

implementation of the solution procedure, happen to be more suitable than purely numerical approaches.

We are indebted to Professor P. L. E. Uslenghi, editor of the ISMB Series, for his kind invitation to write this monograph, and to the anonymous reviewers for their constructive comments. They are also grateful to Professors V. M. Babich (Mathematical V. A. Steklov Institute of the Russian Academy of Sciences, St. Petersburg branch), J. -M. L. Bernard (CEA, France), V. P. Smyshlyaev (University College of London), the late V. E. Grikurov (St. Petersburg State University), and A. V. Shanin (Moscow State University) for the numerous discussions on the subject of the book. The work could not be efficiently completed without permanent support of our colleagues from St. Petersburg and Stuttgart Universities. The authors are thankful to participants of the weekly Seminar on mathematical diffraction theory at the Mathematical V. A. Steklov Institute of the Russian Academy of Sciences (St. Petersburg branch) for the stimulating criticism and valuable advices. Lastly, we appreciate the support we were given from the staff at SciTech Publishing Inc., Raleigh, NC.

St. Petersburg, Russia; Stuttgart, Germany
March 2012

Mikhail A. Lyalinov
Ning Yan Zhu

Introduction

General and historical remarks

The problems of wave scattering by wedge-shaped or cone-shaped surfaces with impedance boundary conditions are traditionally the key problems considered in the framework of their applications in different versions of the GTD [0.1–0.12]. An efficient solution of such a problem, together with the general analysis of the wave phenomena, makes it possible for a researcher to exploit the corresponding diffraction coefficients in different practical situations. On the other hand, the methods of solving these problems are also really valuable and challenging from an analytical point of view. The corresponding problems of diffraction with impedance boundary conditions [0.13] on the surface of a wedge or of a cone are technically even more complicated than those with perfect boundary conditions. At the same time, they are more realistic and enable one to deal with new physical phenomena observed in experiments such as surface waves propagating from the points of irregularity of the boundary. Certainly, the impedance-type boundary conditions have their natural limitations of applicability; in particular, they are not formally valid in some vicinities of edges or conical points. However, the practical experience confirms a remarkable observation that the corresponding models, nevertheless, can be used giving reliable and accurate results even for scattering by impedance surfaces with edges or conical vertices.

In our work the procedure to solve a diffraction problem consists of three basic steps: (1) developing an efficient analytic representation; (2) discussing the solvability and uniqueness; and (3) deriving efficient formulas for the far-field asymptotics including as well their numerical computation.

The very first canonical body studied using a wave-based approach is probably a half-plane. Applying an approximate method introduced by him in 1882 [0.14], Kirchhoff obtained accurate results for diffracted waves in the transition region of incidence (see his Lectures on Mathematical Optics appeared posthumously in 1891, that is, four years after his death, under the editorship of K. Hensel, a mathematician [0.15]). In 1892 Poincaré published his results of asymptotic analysis of diffraction of a focused beam by a half-plane, which are valid outside the transition regions of incidence and reflection [0.16].

It is, however, Arnold Sommerfeld who succeeded as the very first in exactly solving wave diffraction by a half-plane [0.17]. For this reason, the year 1896, in which the seminal paper by Sommerfeld appeared, is usually regarded as the very beginning of the history of diffraction by obstacles with canonical shapes like wedges or cones. This date can also be considered as the year of birth of the Mathematical Diffraction Theory as a part of Mathematical Physics.

Six years later, H. M. Macdonald reported his investigation on diffraction by a perfectly conducting right-circular cone in [0.18].

The most widely used mathematical method in investigating wave diffraction in wedge-shaped regions is probably the Sommerfeld–Malyuzhinets technique named after its inventors ([0.17,0.19] and [0.20]; see also [0.21–0.23]). Its chief steps are (1) expressing the unknown solution of the Helmholtz equation in the Sommerfeld integrals, that is, as a linear superposition of plane waves, the simplest solutions to the wave equation, but with unknown amplitudes, the spectra[1]; (2) inserting the Sommerfeld integrals into the boundary conditions at the wedge's faces, inverting the resultant integral equations for the spectra and obtaining in this way a matrix difference equation for the spectra; (3) solving the matrix (or scalar) functional difference (FD) equation; and (4) evaluating the Sommerfeld integrals with the saddle-point method and deducing a first-order uniform asymptotic expression for the far field.

It is worth mentioning that in 1959 both Senior [0.24] and Williams [0.25] published their exact solutions to plane-wave diffraction by an imperfectly conducting wedge, a special case of the Malyuzhinets problem, for the electric properties of both faces of the wedge are characterized by one and the same surface impedance. Especially the approach employed by Williams resembles closely that of Malyuzhinets, namely, beginning with a Sommerfeld integral, converting the boundary-value problem to that of a difference equation. The latter, however, is solved by Williams with the aid of Barnes's double gamma function [0.26].

For a thorough coverage of this topic until the first half of the 1960s, appreciating contributions made by scientists from all around the world, the readers are referred to a fine book by Rubinowicz, a Polish disciple of Sommerfeld [0.27]; the development of the Sommerfeld–Malyuzhinets technique since then can be found, for instance, in several monographs [0.28–0.32] and review papers [0.33,0.34]. Tuzhilin made some interesting extensions of the technique and developed the so-called $\mathcal{S}$-integrals, a modified Fourier transform for tackling generalized Malyuzhinets functional difference equations [0.35]. Also, Bobrovnikov and Fisanov [0.36] introduced a generalized Malyuzhinets function (see also [0.37]).

Applied to diffraction of an electromagnetic plane wave by an impedance wedge, this technique leads to exact explicit (i.e., in-quadrature) solutions completely for normal incidence. However, for skew incidence such an explicit solution is available only in a few special cases, where either the opening angles of the wedge or the face impedances take particular values. In these cases, the matrix functional equation can be converted to an either fully or partially decoupled functional equation of the first order whose exact solution can be found in closed form [0.38–0.40]. A higher-order functional equation for one of the spectral functions, however, arises in the most general case. The problem for such an equation is equivalent in some sense to a boundary-value problem for analytic vectors. There are also connections between FD equations and singular integral equations [0.29].

It is worth noting that the functional equations[2] treated in this monograph have also been encountered in different branches of mathematical and theoretical physics, such as quantum three-body problems [0.42,0.43], in simple models of the stripping reaction in quantum

[1]We use the terminology that is widely accepted in diffraction theory, which differs from that in the traditional spectral theory of operators.

[2]A thorough historical review of difference equations until the beginning of the 1920s can be found in [0.41].

mechanics [0.44–0.48], in some models of quantum theory of a solid body known as Harper's equation (see [0.49]), in statistical mechanics as well as in the quantum analogue of the inverse scattering method in form of Baxter's functional equation (for instance [0.50]).

To solve this kind of functional difference equations, a procedure was proposed more than 50 years ago by Jost [0.44,0.45]. The basic idea lies in converting a functional difference equation with periodic coefficients in a simpler one defined on a Riemann surface and solving it there. This idea was further developed by Albeverio [0.46]. Since several years, this method has been resorted to and further advanced for studying electromagnetic wave scattering in wedged-shaped regions [0.51–0.60].

The problems of diffraction in a conical domain form the second class of canonical problems to be considered in this monograph. The study of this class of problems started from the analysis of the exact solutions as, for example, in [0.18] (see also [0.61–0.63]) for the ideal boundary conditions. Important advances for these problems have been achieved by Borovikov [0.64] and Jones [0.65] (see also [0.66–0.69]). Adapting and advancing some ideas of the work by Cheeger and Taylor [0.70,0.71], Smyshlyaev [0.72,0.73] developed an approach for diffraction by perfect cones in acoustics and electrodynamics. In particular, he has obtained some important integral representations for the diffraction coefficients of the spherical wave propagating from the vertex of the cone. The numerical implementation of these results and further analytical advances dealing with the Abel-Poisson summation for the diffraction coefficients have been made by Babich, Smyshlyaev, Samokish, and Dementiev [0.74–0.76].

The problems of diffraction by cones with impedance boundary conditions have been intensively studied in just the last 16 years. In 1997 Bernard exploited in his important paper [0.77] a natural and efficient approach: in studying wave diffraction by a circular impedance cone using incomplete separation of variables, he reduced the problem at hand to a second-order difference equation and then to an integral one. Since then, this approach has been generalized and applied to acoustic wave diffraction by an arbitrarily shaped impedance cone [0.78] and to the corresponding electromagnetic problem [0.79] (see also [0.80]). Some additional results for the problems at hand and references can be found in [0.81,0.82]. In this monograph we also study carefully the far-field asymptotics of the wave fields, as these have not been exhaustively considered in the aforementioned works.

Description of the content

Chapter 1 is devoted to some known basic equations, results, and techniques used extensively in the following chapters. We formulate basic equations, boundary, and other conditions. Then we briefly discuss different integral transforms and representations: Sommerfeld–Malyuzhinets, Kontorovich–Lebedev and Fourier transforms, Watson–Bessel integrals. We carefully consider a solution to the problem of diffraction by an impedance wedge at normal incidence of a plane wave (the Malyuzhinets problem). The theory of functional equations for one unknown function is discussed in the last section of the chapter.

Chapter 2 deals with the problem of diffraction of a skew-incident plane electromagnetic wave by a wedge with axially anisotropic impedance faces. By using the Sommerfeld integral, a matrix system of Malyuzhinets functional equations can be found that is then reduced to a second-order functional equation for one spectral function. The latter equation is studied by means of reduction to the corresponding integral equation. Far-field asymptotics is developed, and some numerical results are presented.

In Chapter 3 the wave field from an electric dipole over an impedance wedge is studied. Actually, the results of the second chapter together with the integral expansion with respect to plane waves enable us to give an integral representation of the solution. The asymptotic analysis of the solution leads to an asymptotic expression of the far field. In particular, the wave from the edge, the surface, and reflected waves are derived from the integral representation.

Chapter 4 is devoted to scattering of a transverse magnetic (TM) surface wave by an angular break of an impedance sheet. The technique developed in Chapter 2 is successfully applied to the problem at hand. Some numerical results are also presented. Uniqueness of the solution is briefly discussed and is based on a modified radiation condition taking into account the reflected and transmitted surface waves.

In Chapters 5 and 6, diffraction of a plane wave by a circular cone with impedance boundary conditions is studied for acoustic and electromagnetic cases, respectively. Incomplete separation of variables enables us to reduce the problems at hand to functional equations for the corresponding Fourier coefficients. The latter equations are then thoroughly studied using reduction to the Fredholm-type integral equations. By means of the appropriate integral representations the far-field analysis is then given. In particular, the diffraction coefficients of the spherical wave and the surface waves from the vertex are discussed. Some numerical results are also included.

Chapter 7 concludes this monograph with a brief summary and some comments.

CHAPTER 1

Fundamentals

The present chapter discusses several basic analytical tools to be repeatedly exploited in the main part of the monograph for dealing with diffraction of acoustic and electromagnetic waves by wedges and cones with impedance boundary conditions. We formulate basic equations, boundary, edge, and radiation conditions in the first three sections. Then Section 1.4 outlines different integral transforms and representations, including the Sommerfeld–Malyuzhinets and Kontorovich–Lebedev transforms. Section 1.5 expounds the solution to the so-called Malyuzhinets problem. The theory of Malyuzhinets functional difference equations for one unknown function is detailed in the last section of the chapter.

1.1 Equations for acoustic and electromagnetic waves

1.1.1 Acoustic waves

At a point $\boldsymbol{x}$ in a fluid like air, the density ϱ and the particle velocity $\boldsymbol{v}$ are related to each other via the equation of conservation of mass:

$$\frac{\partial \varrho}{\partial t} + \operatorname{div}(\varrho \boldsymbol{v}) = 0, \tag{1.1}$$

if the fluid is ideal, that is, free of viscosity, and in addition the body force is negligible, then Euler's equation of motion holds:

$$\varrho \left(\frac{\partial}{\partial t} + \boldsymbol{v} \cdot \operatorname{grad} \right) \boldsymbol{v} = -\operatorname{grad} p, \tag{1.2}$$

where p stands for the pressure and depends upon the density ϱ and the specific entropy s according to the equation of state:

$$p = p(\varrho, s). \tag{1.3}$$

In case of a disturbance to a fluid at rest, the pressure p can be expressed as a sum of the pressure at rest p_0 and the excess pressure p', namely, $p = p_0 + p'$. For a small disturbance, that is, $|p'| \ll p_0$ in case of p, a first-order linear approximation (1.1)–(1.3) leads to

$$\frac{\partial \varrho'}{\partial t} + \varrho_0 \operatorname{div} \boldsymbol{v}' = 0, \tag{1.4}$$

$$\varrho_0 \frac{\partial \boldsymbol{v}'}{\partial t} = -\operatorname{grad} p', \tag{1.5}$$

$$p' = c^2 \varrho', \quad c^2 = \left(\frac{\partial p}{\partial \varrho}\right)\bigg|_{\varrho=\varrho_0, s=s_0}, \tag{1.6}$$

with the speed of sound c. Furthermore, it is assumed in the previous relations that the fluid at rest is homogeneous and the entropy in the fluid is constant.

Eliminating ϱ' and $\boldsymbol{v}'$ from (1.4)–(1.6) we get

$$\operatorname{div} \operatorname{grad} p - \frac{1}{c^2} \frac{\partial^2 p}{\partial t^2} = 0, \tag{1.7}$$

the wave equation for the excess pressure p. Beginning from this equation, the accent on a symbol has been dropped whenever the meaning of the symbol is obvious.

It follows from (1.4)–(1.6) that the energy transported by a linear sound wave is conserved; that is,

$$\frac{\partial w}{\partial t} + \operatorname{div} \boldsymbol{I} = 0, \tag{1.8}$$

with the energy density w and the power-flux density $\boldsymbol{I}$ given by

$$w = \frac{1}{2} \varrho_0 \boldsymbol{v} \cdot \boldsymbol{v} + \frac{1}{2} \frac{p^2}{\varrho_0 c^2}, \quad \boldsymbol{I} = p\boldsymbol{v}. \tag{1.9}$$

At an interface between two fluids, the normal component of the velocity and the pressure are continuous:

$$\boldsymbol{v}_1 \cdot \boldsymbol{n} = \boldsymbol{v}_2 \cdot \boldsymbol{n}, \quad p_1 = p_2. \tag{1.10}$$

Here $\boldsymbol{n}$ denotes the unit normal to the interface, whereas the indices 1 and 2 refer to the fluids on both sides of the interface.

For a time-harmonic sound wave, where the time-dependence is e^{-ikct}, $k = 2\pi/\lambda$ is the wavenumber and λ is the wavelength, the linearized version of Euler's equation of motion (1.5) reduces to

$$-ikc\varrho_0 \boldsymbol{v} = -\operatorname{grad} p, \tag{1.11}$$

and the wave equation (1.7) reduces to the Helmholtz equation

$$\operatorname{div} \operatorname{grad} p + k^2 p = 0. \tag{1.12}$$

On use of (1.11) in (1.10), the boundary conditions for the pressure at an interface read

$$\frac{1}{\varrho_{01}} \frac{\partial p_1}{\partial n} = \frac{1}{\varrho_{02}} \frac{\partial p_2}{\partial n}, \quad p_1 = p_2. \tag{1.13}$$

For more on acoustics see [1.1,1.2].

1.1.2 Electromagnetic waves

Field equations

The behavior of an electromagnetic field at a point $\boldsymbol{x}$ is governed by Maxwell's equations:

$$\operatorname{curl} \boldsymbol{H} = \boldsymbol{J} + \frac{\partial \boldsymbol{D}}{\partial t}, \tag{1.14}$$

$$\operatorname{curl} \boldsymbol{E} = -\frac{\partial \boldsymbol{B}}{\partial t}, \tag{1.15}$$

$$\operatorname{div} \boldsymbol{D} = \varrho, \tag{1.16}$$

$$\operatorname{div} \boldsymbol{B} = 0. \tag{1.17}$$

In a homogeneous and isotropic medium, the displacement $\boldsymbol{D}$ and the induction $\boldsymbol{B}$ are related to the electric field strength $\boldsymbol{E}$ and the magnetic field strength $\boldsymbol{H}$ according to

$$\boldsymbol{D} = \varepsilon \boldsymbol{E}, \quad \boldsymbol{B} = \mu \boldsymbol{H}, \tag{1.18}$$

where ε is the dielectric constant (permittivity) and μ is the permeability.

Eliminating $\boldsymbol{B}$ from Faraday's law of induction (1.15) by means of the Ampère-Maxwell law of the magnetic field (1.14) and the material properties (1.18), one arrives at the wave equation for $\boldsymbol{E}$ in a source-free region

$$\Delta \boldsymbol{E} - \frac{1}{c^2}\frac{\partial^2 \boldsymbol{E}}{\partial t^2} = 0, \tag{1.19}$$

where Δ is $\Delta = \operatorname{grad}\operatorname{div} - \operatorname{curl}\operatorname{curl}$, and the velocity c is $c = 1/\sqrt{\mu\varepsilon}$. A similar equation can be derived for $\boldsymbol{H}$.

The relation (1.8) holds also for electromagnetic fields, as can be verified by (1.14) and (1.15), with w and $\boldsymbol{I}$ now given by

$$w = \frac{1}{2}\varepsilon \boldsymbol{E} \cdot \boldsymbol{E} + \frac{1}{2}\mu \boldsymbol{H} \cdot \boldsymbol{H}, \quad \boldsymbol{I} = \boldsymbol{E} \times \boldsymbol{H}. \tag{1.20}$$

At an interface between two media, none of which is perfectly conducting, both the tangential components of the electric and magnetic fields and the normal components of the displacement and induction are continuous:

$$\boldsymbol{E}_{\mathrm{t1}} = \boldsymbol{E}_{\mathrm{t2}}, \quad \boldsymbol{H}_{\mathrm{t1}} = \boldsymbol{H}_{\mathrm{t2}}, \quad D_{\mathrm{n1}} = D_{\mathrm{n2}}, \quad B_{\mathrm{n1}} = B_{\mathrm{n2}}. \tag{1.21}$$

For a time-harmonic electromagnetic field, where the time-dependence is again given by e^{-ikct}, the Ampère-Maxwell law (1.14) and Faraday's law of induction (1.15) go over to[1]

$$\operatorname{curl} \boldsymbol{H} = \boldsymbol{J} - \mathrm{i}kc\boldsymbol{D}, \tag{1.22}$$

$$\operatorname{curl} \boldsymbol{E} = \mathrm{i}kc\boldsymbol{B}, \tag{1.23}$$

---

[1] Remark that the application of div-operator to both sides of (1.22) and (1.23) leads to equations (1.16) and (1.17), implying that $\operatorname{div} \boldsymbol{J} = ikc\varrho$.

and the wave equation (1.19) goes over to the Helmholtz equation

$$\Delta \boldsymbol{E} + k^2 \boldsymbol{E} = 0. \tag{1.24}$$

Potentials

The absence of magnetic monopoles in nature (1.17) implies that the induction $\boldsymbol{B}$ can be expressed in terms of the so-called vector potential $\boldsymbol{A}$ according to

$$\boldsymbol{B} = \operatorname{curl} \boldsymbol{A}. \tag{1.25}$$

The divergence of the vector potential is, however, undetermined.

On use of (1.25) in Faraday's law of induction (1.15), it turns out that the electric field strength $\boldsymbol{E}$ can be represented by means of the vector potential $\boldsymbol{A}$ and a scalar potential ϕ:

$$\boldsymbol{E} = -\frac{\partial \boldsymbol{A}}{\partial t} - \operatorname{grad} \phi. \tag{1.26}$$

For $\boldsymbol{A}$ to be a solution to the wave equation (1.19), the Lorenz condition must be enforced:

$$\operatorname{div} \boldsymbol{A} + \mu\varepsilon \frac{\partial \phi}{\partial t} = 0. \tag{1.27}$$

This condition can be satisfied by relating $\boldsymbol{A}$ and ϕ to the eponymous Hertz vector $\boldsymbol{\Pi}$ in the following way:

$$\boldsymbol{A} = \mu\varepsilon \frac{\partial \boldsymbol{\Pi}}{\partial t}, \quad \phi = -\operatorname{div} \boldsymbol{\Pi}. \tag{1.28}$$

The Lorenz condition can also be met by using a magnetic Hertz vector $\boldsymbol{\Pi}_{\mathrm{m}}$:

$$\boldsymbol{A} = \operatorname{curl} \boldsymbol{\Pi}_{\mathrm{m}}, \quad \phi = 0. \tag{1.29}$$

Hence, an electromagnetic field in a source-free region is given in terms of $\boldsymbol{\Pi}$ and $\boldsymbol{\Pi}_{\mathrm{m}}$:

$$\boldsymbol{B} = \mu\varepsilon \operatorname{curl} \frac{\partial \boldsymbol{\Pi}}{\partial t} + \operatorname{curl} \operatorname{curl} \boldsymbol{\Pi}_{\mathrm{m}}, \tag{1.30}$$

$$\boldsymbol{E} = \operatorname{curl} \operatorname{curl} \boldsymbol{\Pi} - \operatorname{curl} \frac{\partial \boldsymbol{\Pi}_{\mathrm{m}}}{\partial t}. \tag{1.31}$$

In a spherical coordinate system, a special choice for the Hertz vectors, namely,

$$\boldsymbol{\Pi} = \boldsymbol{e}_r r u, \quad \boldsymbol{\Pi}_{\mathrm{m}} = \boldsymbol{e}_r r v, \tag{1.32}$$

proves to be particularly convenient, with the Debye potentials u and v obeying the scalar wave equation (1.7). In (1.32), $\boldsymbol{e}_r$ stands for the radial unit vector, and r for the distance of a point to the origin of the spherical coordinate system.

Therefore, a time-harmonic source-free field in a spherical coordinate system can be written with the aid of the Debye potentials u and v

$$\boldsymbol{B} = -\mathrm{i}\frac{k}{c}\,\mathrm{curl}\,(\boldsymbol{e}_r r u) + \mathrm{curl}\,\mathrm{curl}\,(\boldsymbol{e}_r r v)\,, \tag{1.33}$$

$$\boldsymbol{E} = \mathrm{curl}\,\mathrm{curl}\,(\boldsymbol{e}_r r u) + \mathrm{i}kc\,\mathrm{curl}\,(\boldsymbol{e}_r r v)\,. \tag{1.34}$$

For details on electromagnetics see [1.3,1.4].

1.2 Boundary conditions

In this book, diffraction by wedges and cones is studied. To be brief, this chapter addresses problems dealing with only wedges. The scattering of waves in a cone-shaped domain can be considered analogously and will be addressed in Section 1.4.5, Chapter 5, and Chapter 6.

Let the domain

$$\{(r,\varphi) : r > 0,\ -\Phi < \varphi < \Phi\} \tag{1.35}$$

be taken out from the coordinate plane. The remaining angular region $\{(r,\varphi) : r \geq 0,\ \Phi \leq \varphi \leq \pi \bigcup -\pi \leq \varphi \leq -\Phi\}$ will be called wedge;[2] in a diffraction problem this wedge is illuminated by a plane incident wave (Fig. 1.1).

The wave process is examined inside the domain (1.35). Its boundaries (wedge's sides or faces) are the half-lines (half-planes)

$$\varphi = -\Phi \quad \text{and} \quad \varphi = \Phi\,, \qquad r > 0\,. \tag{1.36}$$

Note that for $\Phi \leq \pi/2$ the angle $\Phi \leq \varphi \leq 2\pi - \Phi$ is not less than π; such an angle can be called a "wedge" only in the conventional sense.

The solution $u(x, y)$ of the scalar wave equation (1.12) (the Helmholtz equation) is assumed to have continuous second derivatives inside the domain (1.35).

Now let us turn to the boundary conditions. If, in addition to the aforementioned, the solution u is continuous up to the boundaries (1.36), the Dirichlet boundary conditions

$$u|_{\varphi=-\Phi} = u|_{\varphi=\Phi} = 0 \quad \text{for} \quad r > 0 \tag{1.37}$$

can be set. (In acoustics, a surface on which the Dirichlet boundary condition holds is called acoustically soft.) In turn, to postulate the Neumann condition (in acoustics, such a surface is termed acoustically hard)

$$\left.\frac{\partial u}{\partial n}\right|_{\varphi=-\Phi} = \left.\frac{\partial u}{\partial n}\right|_{\varphi=\Phi} = 0\,, \quad r > 0 \tag{1.38}$$

or the impedance condition

$$\left.\frac{\partial u}{\partial n} - \mathrm{i}k\sin\theta_{\pm} u\right|_{\varphi=\pm\Phi} = 0\,, \quad r > 0\,, \quad \theta_{\pm} = \mathrm{const}, \tag{1.39}$$

[2] The same term we shall also use for the real wedge in the 3D case, which does not lead to confusion.

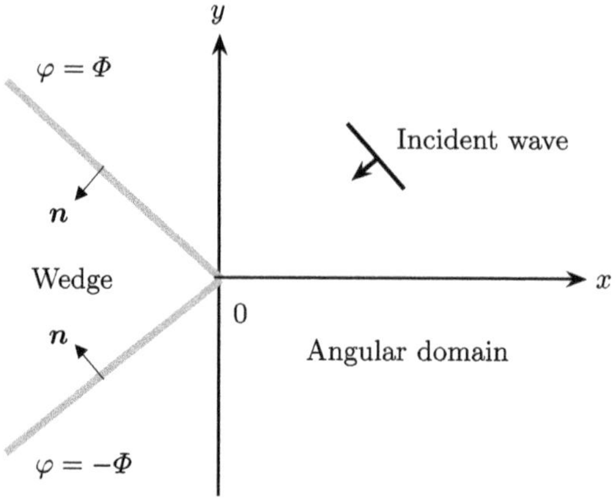

Figure 1.1 A plane wave illuminating a wedge.

for the boundaries one needs to assume the existence of continuous derivatives of the solution on the faces of the wedge. Note that in (1.38) and (1.39) the operator $\partial/\partial n = \pm(1/r)\,\partial/\partial\varphi$ means differentiation along the normal to the boundary outward with respect to the domain (1.35) (or inward with respect to the wedge); see Fig. 1.1. The constants $\theta_\pm$ in (1.39) are generally complex and sometimes called Brewster angles.[3]

In electromagnetic theory the impedance (Leontovich) boundary condition for the total field reads

$$\boldsymbol{E} - (\boldsymbol{E} \cdot \boldsymbol{n})\boldsymbol{n} = -\eta\, \boldsymbol{n} \times \boldsymbol{H}, \tag{1.40}$$

where $\boldsymbol{n}$ is the unit normal vector to a boundary surface S, and η is the (with respect to Z_0, the intrinsic impedance of the ambient medium) normalized surface impedance. Beginning from this section, $\boldsymbol{H}$ stands for the magnetic field vector multiplied by Z_0. (In the given form of the condition (1.40) we select $\boldsymbol{n}$ to point into the exterior of the domain, in which the wave processes are studied.) It is also convenient to introduce the Brewster angle θ by the equality $\sin\theta = \eta$. The Brewster angle θ is uniquely specified by η provided $0 \leq \operatorname{Re}\theta \leq \pi/2$. Remark that for an anisotropic impedance surface η is a tensor.

To simulate the effect of a thin material layer upon for instance the electromagnetic wave in the ambient medium, a so-called impedance sheet is usually used. The electric property of such a sheet of zero thickness is also given by a (with respect to Z_0) normalized impedance η. The conditions to be imposed on the electromagnetic fields read

$$[\boldsymbol{E} - (\boldsymbol{E} \cdot \boldsymbol{n})\boldsymbol{n}] = 0, \tag{1.41}$$

$$\boldsymbol{E} - (\boldsymbol{E} \cdot \boldsymbol{n}) = -\eta[\boldsymbol{n} \times \boldsymbol{H}]. \tag{1.42}$$

[3]The Dirichlet or Neumann boundary conditions are sometimes referred to as perfect (since they model the behavior of transverse-electric (TE) or transverse-magnetic (TM) electromagnetic waves on the surface of a perfect conductor); impedance boundary conditions (1.39) can be applied to simulate the surface of an imperfect conductor and are referred to as imperfect.

In (1.41) and (1.42), $\boldsymbol{n}$ denotes the unit normal to the wedge's faces, and the square brackets [.] signify the jump of the quantities inside the brackets, which may be discontinuous across the sheet.

Impedance-type boundary conditions like (1.39)–(1.42) allow us to account for, although approximately, the influence of a large class of more general and more realistic scattering bodies upon high-frequency wave fields outside them, which is of prime importance in practice. More on impedance boundary conditions can be found for instance in [1.32,1.18].

It is well-known from the general theory of the elliptic equations [1.5] that the solution to the scalar wave equation (1.12) satisfying one of the boundary conditions (1.37), (1.38), or (1.39) has derivatives of any order continuous up to the boundaries (1.36). The next section is devoted to a discussion of the solution's behavior in the vicinity of the vertex $r = 0$ as well as that at infinity $r \to \infty$.

1.3 Edge and radiation conditions

1.3.1 Vicinity of the edge and Meixner's condition

In a neighborhood of the wedge's edge (or vertex) $r = 0$ a solution to the Helmholtz equation with the classical boundary conditions cannot be smooth (except for some special cases); see [1.6]. The analysis of the solution behavior as $r \to 0$ is based on the assumption of finiteness of the wave-field energy in a bounded vicinity of the vertex.[4] Some additional details can be found in [1.7].

We define the density of the wave energy as

$$E := u_x\overline{u_x} + u_y\overline{u_y} + k^2 u\overline{u} \equiv \frac{\partial u}{\partial r}\overline{\frac{\partial u}{\partial r}} + \frac{1}{r^2}\frac{\partial u}{\partial \varphi}\overline{\frac{\partial u}{\partial \varphi}} + k^2 u\overline{u} \tag{1.43}$$

(here and later in Chapters 4–6 the bar means the complex conjugate). The motivation for the definition (1.43) is as follows.

Recall that a real function w satisfying the wave equation (1.7) obeys the relation

$$\frac{\partial}{\partial t}\mathcal{E} + \operatorname{div} \boldsymbol{S} = 0, \tag{1.44}$$

where

$$\mathcal{E} = \frac{1}{2}\left(\frac{\partial w}{\partial t}\right)^2 + \frac{c^2}{2}\left|\operatorname{grad} w\right|^2, \quad \boldsymbol{S} = -c^2\left(\frac{\partial w}{\partial t}\right)\operatorname{grad} w. \tag{1.45}$$

The formula (1.44) is known as the Umov-Poynting theorem in electrodynamics or acoustics. To obtain (1.44) one has to multiply the scalar wave equation by w_t and make simple transformations. See also Section 1.1.

The quantity $\mathcal{E}$ is naturally interpreted as the density of energy for the wave equation, whereas $\boldsymbol{S}$ is the Poynting vector (the energy-flux vector) provided w is understood as the real part of the complex solution $e^{-ikct}u(x, y)$. The substitution of $w = \operatorname{Re}\{e^{-ikct}u\}$ into (1.45)

---

[4]The behavior of the wave field near a conical singularity of the boundary is studied in a similar fashion.

yields

$$\mathcal{E} = \frac{1}{2}\left[\frac{\partial}{\partial t}\mathrm{Re}\left(u\,\mathrm{e}^{-ikct}\right)\right]^2 + \frac{c^2}{2}\left|\mathrm{grad}\,\mathrm{Re}\left(u\,\mathrm{e}^{-ikct}\right)\right|^2. \tag{1.46}$$

The expression (1.46) is a periodic function in time with the period $T = 2\pi/(kc)$. By averaging $\mathcal{E}$ over the period, that is, by calculating the integral

$$E \propto \frac{1}{T}\int_0^T \mathcal{E}\,\mathrm{d}t$$

one arrives at an expression that coincides with (1.43) up to the constant factor $c^2/4$.

Thus, with the help of (1.43) the energy of the wave field concentrated in the domain $\{0 < r < r_0 = \text{const},\ -\Phi < \varphi < \Phi\}$ can be written as

$$\int_{-\Phi}^{\Phi}\int_0^{r_0} E\,r\,\mathrm{d}r\,\mathrm{d}\varphi \equiv \int_{-\Phi}^{\Phi}\int_0^{r_0}\left(\left|\frac{\partial u}{\partial r}\right|^2 + \frac{1}{r^2}\left|\frac{\partial u}{\partial \varphi}\right|^2 + k^2\,|u|^2\right) r\,\mathrm{d}r\,\mathrm{d}\varphi. \tag{1.47}$$

We will say that the solution $u(r, \varphi)$ to the Helmholtz equation, which is subjected to the (perfect or imperfect) boundary conditions on the wedge's faces for $r > 0$, satisfies the Meixner condition if the integral (1.47) is finite for some $r_0 > 0$:

$$\int_{-\Phi}^{\Phi}\int_0^{r_0}\left(\left|\frac{\partial u}{\partial r}\right|^2 + \frac{1}{r^2}\left|\frac{\partial u}{\partial \varphi}\right|^2 + k^2\,|u|^2\right) r\,\mathrm{d}r\,\mathrm{d}\varphi < \infty. \tag{1.48}$$

The Meixner condition also has other formulations. The second one is as follows. Let the function $u(r, \varphi)$ satisfy the Helmholtz equation and the boundary conditions on the faces $\varphi = \pm\Phi$; then $u(r, \varphi)$ meets the Meixner condition if

$$|u| \le \text{const} < \infty \tag{1.49}$$

for $\{0 < r < r_0,\ -\Phi < \varphi < \Phi\}$ with some $r_0 > 0$.

Formulations (1.48) and (1.49) are equivalent, even when this fact is not immediately seen; cf. [1.7]. At the same time, we arrive there at the third equivalent formulation of the Meixner condition given below.

1.3.2 On the behavior of solutions to the Helmholtz equation in the angular domain as $r \to 0$

Now we write the third equivalent (and most frequently used) form of the Meixner condition. It can be manifested that for some $\delta > 0$

$$u = \mathrm{O}\left(r^{\delta}\right),\ \frac{\partial u}{\partial r} = \mathrm{O}\left(r^{\delta-1}\right),\ \frac{\partial^2 u}{\partial r^2} = \mathrm{O}\left(r^{\delta-2}\right),\ \left|\frac{\partial u}{\partial \varphi}\right| + \left|\frac{\partial^2 u}{\partial \varphi^2}\right| = \mathrm{O}\left(r^{\delta}\right) \tag{1.50}$$

for the Dirichlet case or

$$u = \text{const} + \mathrm{O}\left(r^{\delta}\right), \ \frac{\partial u}{\partial r} = \mathrm{O}\left(r^{\delta-1}\right), \ \frac{\partial^2 u}{\partial r^2} = \mathrm{O}\left(r^{\delta-2}\right), \ \left|\frac{\partial u}{\partial \varphi}\right| + \left|\frac{\partial^2 u}{\partial \varphi^2}\right| = \mathrm{O}\left(r^{\delta}\right) \quad (1.51)$$

for the Neumann or impedance cases.[5] As usual, by $\mathrm{O}(r^{\delta})$ a function satisfying the estimate $\left|\mathrm{O}(r^{\delta})\right| \leq Cr^{\delta}$ is understood; the constant C in this estimate depends on neither r nor φ, $0 \leq r \leq r_0$, $-\Phi \leq \varphi \leq \Phi$. Note that $\delta = \pi/(2\Phi)$ can be chosen if $2\Phi > \pi$.

The Meixner conditions also admit the following physical consequence:

$$\int_{-\Phi}^{\Phi} S_r|_{r=r_0}\, r_0 \mathrm{d}\varphi = \frac{c^2}{2} kc \int_{-\Phi}^{\Phi} \mathrm{Im}\left(\frac{\partial u}{\partial r}\overline{u}\right)\Bigg|_{r=r_0} r_0 \mathrm{d}\varphi \underset{r_0 \to 0}{\to} 0,$$

where S_r is the radial component of the Umov-Poynting vector (after averaging over the period). In other words, the previous formula means that the energy flux over the arc $\{-\Phi \leq \varphi \leq \Phi,\ r = r_0 > 0\}$ vanishes together with $r_0 \to 0$; the vertex is neither radiating nor absorbing.

1.3.3 Radiation conditions: Formulation of the problem

It is natural to assume that the diffracted field

$$u_{\mathrm{d}} = u - u_{\mathrm{g}}, \quad (1.52)$$

where u_{g} is the geometrical-optics (GO) field including incident and (from the wedge's faces) reflected waves, should satisfy the radiation conditions. The traditional form of these conditions is

$$u_{\mathrm{d}} \underset{r\to\infty}{=} \mathrm{O}\left(\frac{1}{\sqrt{r}}\right), \quad (1.53)$$

$$\frac{\partial u_{\mathrm{d}}}{\partial r} - iku_{\mathrm{d}} \underset{r\to\infty}{=} \mathrm{o}\left(\frac{1}{\sqrt{r}}\right). \quad (1.54)$$

We suggest using another (integral) form of the radiation condition:

$$\int_{-\Phi}^{\Phi} \left|\frac{\partial u_{\mathrm{d}}}{\partial r} - iku_{\mathrm{d}}\right|^2_{r=R} R\, \mathrm{d}\varphi \underset{R\to\infty}{\to} 0. \quad (1.55)$$

[6]

Clearly (1.55) follows from (1.54). However, the condition (1.55) is more general since the estimates (1.53)–(1.54) (contrary to (1.55)) are not valid in the penumbra, or half-shadow domains. Moreover, the condition (1.53) (the "boundedness condition," following the terminology by A. Sommerfeld) turns out to be unnecessary for all known problems. The electromagnetic analog of (1.55) will be briefly discussed in Chapter 6.

[5]Normally only the first of these relations is mentioned.

[6]The book [1.8, Chapter IV, §2] refers to W. Magnus concerning the condition (1.55). F. Rellich also exploited this form of the radiation condition extensively in his studies.

Now we are able to formulate the classical statement of the plane-wave diffraction problem by a wedge.

Problem 1.1. *The function (wave field) $u(r, \varphi)$ is a solution to the diffraction problem if it meets the following requirements:*

(i) *outside the wedge, $u \in \mathrm{C}^2\{r > 0,\ |\varphi| < \Phi\}$ satisfies the Helmholtz equation;*
(ii) *$u \in \mathrm{C}^1\{r > 0,\ |\varphi| \le \Phi\}$ satisfies on the wedge faces $\varphi = \pm\Phi$ one of the boundary conditions (1.37), (1.38), or (1.39);*
(iii) *the Meixner conditions in one of the equivalent forms hold as $r \to 0$;*
(iv) *the diffracted field (1.52) satisfies a radiation condition in the form (1.55).*

Remark: Radiation conditions and surface waves. In fact, the above statement of the diffraction problem does not work for all possible situations: the existence of surface waves violates the condition (1.55). Such waves appear for wedge diffraction problems with impedance (imperfect) boundary conditions (1.39) for some specific values of $\sin\theta_\pm$.[7]

Indeed, assume that the diffracted field (1.52) contains terms of the form

$$\text{const}\ \exp\left[ikr\cos(\varphi \mp \Phi \mp \theta_\pm)\right] \tag{1.56}$$

(such expressions obviously satisfy the Helmholtz equation and the impedance boundary conditions). Suppose that $\sin\theta_\pm$ are pure negatively imaginary numbers. In this case the expressions (1.56) are surface waves, that is, they are concentrated in a small vicinity of the faces $\varphi = \pm\Phi$ and propagate without attenuation along them. At the same time the contribution of these terms to integral (1.55) does not vanish as $R \to \infty$.

The difficulties arisen due to the surface waves can be overcome by some modification of the radiation condition (1.55). When it is necessary or convenient, however, we forbid the appearance of the surface waves by the restriction

$$\operatorname{Re}\sin\theta_\pm > 0 \quad (\text{or equivalently} \quad 0 < \operatorname{Re}\theta_\pm \le \pi/2\ ). \tag{1.57}$$

This additional condition allows us to keep working within the framework of Problem 1.1. Note that in Chapter 4 we shall indeed encounter the mentioned modified form of the radiation condition (see also [1.10] for the diffraction of a surface wave by an impedance cone).

1.3.4 The limiting-absorption principle

The *limiting-absorption principle* is known (see also [1.7]) as an alternative to the radiation conditions (1.53)–(1.55). In a sense it is more general than the radiation conditions. Briefly, its description is as follows. The wavenumber k is considered to be complex, $k \rightsquigarrow k + i\varepsilon$, where $\varepsilon = \text{const} > 0$ is sufficiently small. Items (i), (ii), and (iii) in Problem 1.1 remain the same for complex k, while instead of item (iv) the exponential decaying as $r \to \infty$ is required for both

---

[7] A similar difficulty arises in the problem of diffraction by an elastic wedge because of the appearance of Rayleigh waves (see [1.9]) and also for diffraction by an impedance cone [1.10].

the diffracted field u_d and its derivatives:

$$u_\mathrm{d},\ |\nabla u_\mathrm{d}| \underset{r\to\infty}{=} \mathrm{O}(\exp(-\varepsilon_1 r)), \tag{1.58}$$

where $\varepsilon_1 > 0$ is some (depending on ε) constant. Finally, Problem 1.1 is completed by the following

(v) *the solution $u(r, \varphi; k)$ for real $k > 0$ is understood as the limit*

$$u(r, \varphi; k) := \lim_{\varepsilon\to+0} u(r, \varphi; k + \mathrm{i}\varepsilon) \tag{1.59}$$

of the solution satisfying items (i) through (iv) in Problem 1.1. The limit in (1.59) is uniform in any finite sector $-\Phi \le \varphi \le \Phi$, $0 \le r \le r_0$ $(0 < r_0 < \infty)$.

It is remarkable that the corresponding limiting solution in general satisfies the aforementioned radiation conditions as well. For example, in the Malyuzhinets problem of diffraction by an impedance wedge to be discussed in Section 1.5, this fact can be directly verified from the explicit representation of the solution. In some more complex problems the existence of the limit (1.59) requires special study.

It can be shown that surface waves meet the limiting-absorption principle provided that the corresponding velocity vector is directed from the origin to infinity. Clearly, the surface waves propagating in the opposite direction violate the limiting-absorption principle.

The uniqueness theorem is valid for the Dirichlet, Neumann, or impedance absorbing diffraction problems based on the limiting-absorption principle (instead of the radiation conditions):

Theorem 1.1. (The uniqueness theorem.) *Let $u(r, \varphi; \varepsilon)$ satisfy the Helmholtz equation*

$$\Delta u + (k + \mathrm{i}\varepsilon)^2 u = 0 \quad (k > 0,\ \varepsilon > 0) \tag{1.60}$$

in the angular domain $\{(r, \varphi) :\ r > 0,\ -\Phi < \varphi < \Phi\}$ and have continuous first derivative up to the boundaries $\{r > 0,\ \varphi = \pm\Phi\}$; let on these boundaries either impedance

$$\left.\frac{\partial u}{\partial n} - \mathrm{i}(k + \mathrm{i}\varepsilon)\sin\theta_\pm u\right|_{\varphi=\pm\Phi} = 0 \tag{1.61}$$

with $\mathrm{Re}\,\sin\theta_\pm > 0$ *or perfect (Dirichlet or Neumann) boundary conditions be imposed on u; let $u(r, \varphi)$ obey the Meixner conditions as $r \to 0$ and for $r \ge$ const > 0 the estimates*

$$u,\ \frac{\partial u}{\partial r},\ \frac{\partial u}{\partial \varphi} = \mathrm{O}\left(\mathrm{e}^{-\varepsilon_1 r}\right) \tag{1.62}$$

hold with some (depending on ε) $\varepsilon_1 > 0$ uniformly with respect to φ, $-\Phi \le \varphi \le \Phi$. Then there exists such $\varepsilon_0 > 0$ that $u(r, \varphi; \varepsilon) \equiv 0$ for any $0 < \varepsilon < \varepsilon_0$.

The proof starts from the consideration of the integral

$$\int_{\mathcal{D}_{r_0 r_1}} \left[\Delta u + (k + \mathrm{i}\varepsilon)^2 u\right] \overline{u}\, \mathrm{d}x\mathrm{d}y$$

over the domain $\mathcal{D}_{r_0 r_1} = \{(r, \varphi) : \ 0 < r_0 < r < r_1, \ \ -\Phi < \varphi < \Phi\}$. In view of (1.60) this integral is equal to zero. By means of Green's formula one has

$$\int\limits_{\mathcal{D}_{r_0 r_1}} \left[-|\nabla u|^2 + (k + \mathrm{i}\varepsilon)^2 |u|^2\right] \mathrm{d}x\mathrm{d}y + \int\limits_{\partial\mathcal{D}_{r_0 r_1}} \frac{\partial u}{\partial n} \overline{u}\, \mathrm{d}s = 0. \tag{1.63}$$

Consider the imaginary part of (1.63)

$$2\varepsilon k \int\limits_{\mathcal{D}_{r_0 r_1}} |u|^2 \mathrm{d}x\mathrm{d}y + \mathrm{Im} \int\limits_{\partial\mathcal{D}_{r_0 r_1}} \frac{\partial u}{\partial n} \overline{u}\, \mathrm{d}s = 0 \tag{1.64}$$

and take the limits as $r_0 \to 0, r_1 \to \infty$ there. Note that the integrals over the arcs $\{(r, \varphi) : \ r = r_0, -\Phi < \varphi < \Phi\}$ and $\{(r, \varphi) : \ r = r_1, -\Phi < \varphi < \Phi\}$ vanish (by virtue of Meixner's conditions and the assumption (1.62), respectively). Thus

$$\begin{aligned} &2\varepsilon k \int_0^\infty \int_{-\Phi}^{\Phi} |u|^2 r\, \mathrm{d}r\mathrm{d}\varphi \\ &+ \mathrm{Im} \left[\mathrm{i}(k + \mathrm{i}\varepsilon) \left(\sin\theta_+ \int_0^\infty |u|^2\Big|_{\varphi=\Phi} \mathrm{d}r + \sin\theta_- \int_0^\infty |u|^2\Big|_{\varphi=-\Phi} \mathrm{d}r\right)\right] = 0 \end{aligned} \tag{1.65}$$

(for the Dirichlet or Neumann boundary conditions the second term is zero). The quantity

$$\mathrm{Im}\, (\mathrm{i}(k + \mathrm{i}\varepsilon) \sin\theta_\pm) = k\, \mathrm{Re}\, \sin\theta_\pm - \varepsilon\, \mathrm{Im}\, \sin\theta_\pm \tag{1.66}$$

is positive for a sufficiently small ε provided $\mathrm{Re}\, \sin\theta_\pm > 0$. Thus the second term in (1.65) for such ε is nonnegative and the equality (1.65) is possible only if $u \equiv 0$, which completes the proof. ■

The key point of this proof is the positiveness of the quantity (1.66) for a sufficiently small ε. It is obvious that this positiveness is achieved not only for the examined cases but also if

$$\mathrm{Re}\, \sin\theta_\pm = 0, \quad \mathrm{Im}\, \sin\theta_\pm < 0$$

(that is, for lossless boundaries of a special kind). Moreover, Theorem 1.1 remains valid if one replaces $k \sin\theta_\pm \rightsquigarrow g_\pm$ in the boundary conditions (1.61) (the values $g_\pm$ are independent of k but subjected to the inequalities $\mathrm{Re}\, g_\pm > 0$).

1.4 Integral transformations

This section describes integral transformations that are extensively exploited in this monograph. Some general formulas and properties are provided; for details the reader is referred to the corresponding literature.

1.4.1 Fourier transform and the convolution theorem

Throughout this book we employ a slightly modified form of the Fourier transform with the integration along the imaginary axis. Such a form is more convenient for the problems at hand

and reads

$$F(\alpha) = \frac{1}{\mathrm{i}} \int_{\mathrm{i}\mathbb{R}} f(\nu)\, \mathrm{e}^{\mathrm{i}\nu\alpha}\, \mathrm{d}\nu, \qquad f(\nu) = \frac{1}{2\pi \mathrm{i}} \int_{\mathrm{i}\mathbb{R}} F(\tau)\, \mathrm{e}^{-\mathrm{i}\nu\tau}\, \mathrm{d}\tau. \tag{1.67}$$

The classes of functions, to which the Fourier transform can be applied, are carefully discussed in [1.11]. By use of the modified Fourier transform the convolution theorem can be written as

$$\frac{1}{\mathrm{i}} \int_{\mathrm{i}\mathbb{R}} f(\nu) g(\nu)\, \mathrm{e}^{\mathrm{i}\nu\alpha} \mathrm{d}\nu \;=\; \frac{1}{2\pi \mathrm{i}} \int_{\mathrm{i}\mathbb{R}} F(\alpha - \tau) G(\tau)\, \mathrm{d}\tau. \tag{1.68}$$

It is important to note that the asymptotic behavior of the original $f(\nu)$ as $\nu \to \pm\mathrm{i}\infty$ may lead to specific singularities of $F(\alpha)$ being analytically continued onto a strip of the complex α-plane containing the imaginary axis. The study of such singularities and of the procedure of the analytic continuation plays a crucial role in derivation of the far-field asymptotics in Chapters 5 and 6.

1.4.2 The Sommerfeld integral

Solutions to a number of diffraction problems in angular domains can be found explicitly in the form of the so-called Sommerfeld integral [1.12]. Under this name the integrals of the following kind are traditionally understood:

$$\begin{aligned} u(kr, \varphi) &= \frac{1}{2\pi \mathrm{i}} \int_{\gamma_+} \mathrm{e}^{-ikr\cos z} \left[s(z+\varphi) - s(\varphi - z) \right] \mathrm{d}z \\ &= \frac{1}{2\pi \mathrm{i}} \int_{\gamma_+ \bigcup \gamma_-} \mathrm{e}^{-ikr\cos z} s(z+\varphi)\, \mathrm{d}z. \end{aligned} \tag{1.69}$$

(hereafter we assume $kr > 0$ unless otherwise noted.)

The integration contours $\gamma_\pm$ (the Sommerfeld contours) are shown in Fig. 1.2. To explain their construction consider the behavior of $\mathrm{e}^{-ikr\cos z}$. It is an entire 2π-periodic function of z that is rapidly decaying,

$$\text{const} \exp\left[-kr \sin(\operatorname{Re} z) \sinh(\operatorname{Im} z)\right], \quad |\operatorname{Im} z| \to \infty,$$

inside the half-strips $\{z : 2m\pi < \operatorname{Re} z < (2m+1)\pi,\ \operatorname{Im} z > 0\}$, $\{z : (2m-1)\pi < \operatorname{Re} z < 2m\pi,$ $\operatorname{Im} z < 0\}$, $m = 0, \pm 1, \ldots$ (these half-strips are shown in Fig. 1.2 in gray); $\exp(-ikr\cos z)$ grows rapidly in other half-strips.

The contour γ_+ lies in the upper half-plane and encompasses the half-strip $\{z : -\pi \leq \operatorname{Re} z \leq 0,\ \operatorname{Im} z \geq \text{const} > 0\}$; thus, the contour goes to infinity in the domains of decreasing of $|\exp(-ikr\cos z)|$. The contour γ_- is symmetric with γ_+ with respect to the origin.

The function $s(z)$ in (1.69) is assumed to be meromorphic with the following behavior at infinity:

$$|s(z)| \leq \text{const} \exp(\beta |\operatorname{Im} z|), \quad |\operatorname{Im} z| \to \infty. \tag{1.70}$$

For most of the known cases, the constant β is nonpositive; however, the convergence of the integral (1.69) is ensured for positive βs as well.

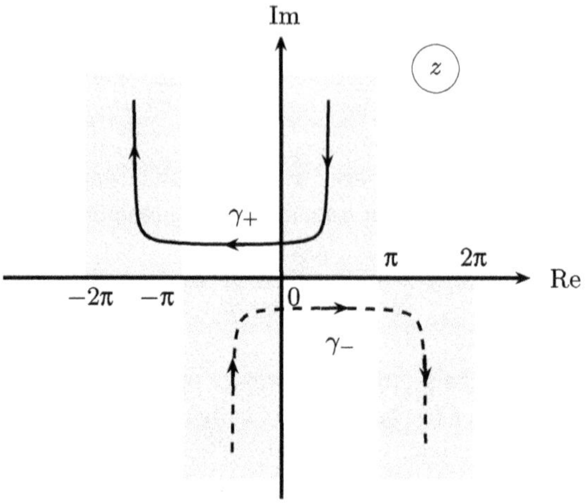

Figure 1.2 The Sommerfeld contours γ_+ and γ_-. The half-strips, where $|\exp(-ikr\cos z)|$ decreases (provided $kr > 0$), are shown in gray.

It turns out that the representation (1.69) for the solutions of diffraction problems is very natural and convenient. Its first remarkable application was given by A. Sommerfeld [1.12]. We demonstrate the efficiency of (1.69) by applying it to the Malyuzhinets problem of impedance wedge diffraction in Section 1.5.

It is possible to show that (1.69) is, in fact, a quite natural form for the solution: it satisfies the Helmholtz equation, the radiation conditions, and (for $\beta \geq 0$ in (1.70)) the Meixner conditions. Substituting (1.69) into the boundary conditions leads to similar integrals in which the expression $s(z+\varphi) - s(\varphi - z)$ is replaced by some operator acting on this expression computed at $\varphi = \pm\Phi$.[8]

A detailed discussion about the properties of the Sommerfeld integral can be found in [1.7].

1.4.3 Malyuzhinets's theorem: Sommerfeld–Malyuzhinets (SM) transform

This section deals with the Malyuzhinets theorem [1.7,1.13,1.14] (see also [1.15–1.18]), which plays the principal role for the approach.

Theorem 1.2. 1° *Let the function* $\Upsilon(z)$ *be regular inside the strip* $\{z : -\pi - \varepsilon_1 \leq \operatorname{Re} z \leq \varepsilon_1,\ \operatorname{Im} z \geq \varepsilon_2 > 0\}$, *where* ε_1 *and* ε_2 *are some positive constants;*[9]
2° *for some* $D = \text{const}$ *and for sufficiently large* $\operatorname{Im} z$ *the following estimate holds in this strip:*

$$|\Upsilon(z)| \leq \text{const}\,\exp\left(D\operatorname{Im} z\right); \tag{1.71}$$

[8]By means of this observation it turns out to be possible to reformulate the boundary conditions as functional equations on $s(z)$. The technique for solving such equations for various situations is developed in the next chapters.

[9]By the regularity in a closed domain we imply the regularity in some neighborhood of this domain.

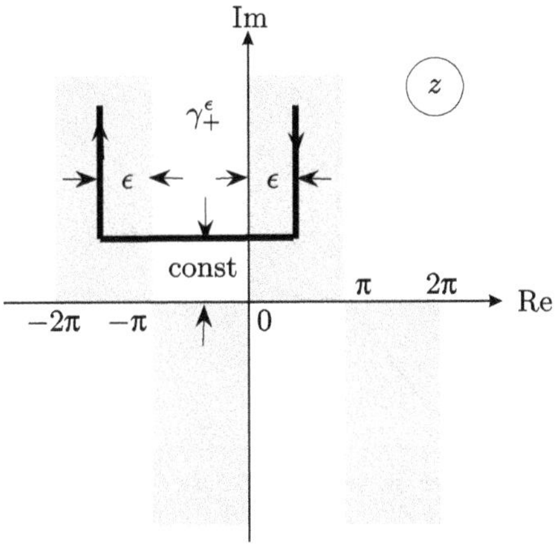

Figure 1.3 The contour γ_+^ε.

3° *for any* $R > 0$

$$\int_{\gamma_+^\varepsilon} e^{-iR\cos z}\Upsilon(z)\,dz = 0, \quad \varepsilon < \varepsilon_1. \tag{1.72}$$

Then

$$\Upsilon(z) \equiv 0 \qquad \text{if} \quad D < 1 \tag{1.73}$$

$$\Upsilon(z) = \sin z\left[c_0 + c_1\cos z + \cdots + c_{\ell-1}(\cos z)^{\ell-1}\right] \quad \text{if} \quad D \geq 1, \tag{1.74}$$

where $\ell = [D]$ *is the integral part of* D, $c_0, \ldots, c_\ell$ *are some constants.*

Let us first study the case $D < 1$.

Apply the change of variable $-i\cos z = \tau$ to the integral (1.72). We arrive at the identity

$$\frac{1}{2\pi i}\int_\Gamma e^{\tau R}\widetilde{\Upsilon}(\tau)\,d\tau = 0 \quad (\forall R > 0), \tag{1.75}$$

where the image of the narrowed contour γ_+, which is denoted by γ_+^ε and shown in Fig. 1.3, is the contour Γ in Fig. 1.4 [10] (the orientation is preserved so the domain inside the loop γ_+^ε turns out to be on the right from Γ) and

$$\widetilde{\Upsilon}(\tau) = -i\frac{\Upsilon(\arccos i\tau)}{\sqrt{\tau^2+1}}$$

[10] It can easily be shown that the images of the vertical infinite lines of γ_+^ε convert to the infinite arcs of hyperbole, whereas the horizontal line of Fig. 1.3 becomes the arc of an ellipse.

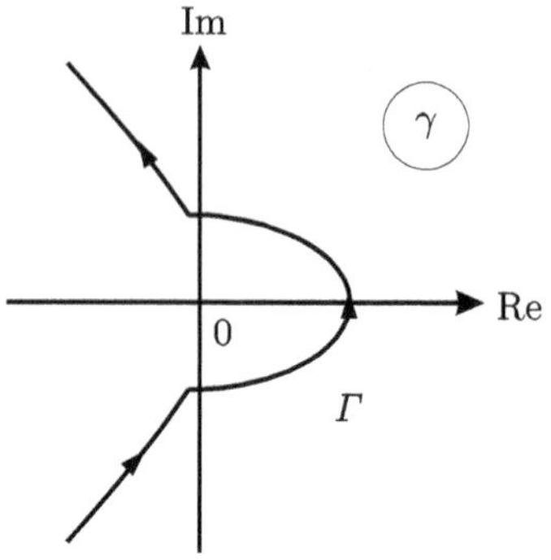

Figure 1.4 Contour Γ — the image of contour γ_+^ϵ under the mapping $-\mathrm{i}\cos z = \tau$.

is a regular function in the domain including some vicinity of Γ and all the points on the right from Γ, besides

$$\left|\widetilde{\Upsilon}(\tau)\right| \underset{|\tau|\to\infty}{=} \mathrm{O}\left(|\tau|^{D-1}\right). \tag{1.76}$$

Let ξ be an arbitrary point on the right from the contour Γ, that is, $\mathrm{Re}\,(\tau-\xi) < 0$, $\tau \in \Gamma$. Multiply (1.75) by $\exp(-R\xi)$ and integrate over R from 0 to ∞. By changing the order of integration one has

$$\frac{1}{2\pi\mathrm{i}}\int_\Gamma \widetilde{\Upsilon}(\tau)\int_0^\infty \mathrm{e}^{(\tau-\xi)R}\,\mathrm{d}R\,\mathrm{d}\tau = \frac{-1}{2\pi\mathrm{i}}\int_\Gamma \frac{\widetilde{\Upsilon}(\tau)}{\tau-\xi}\,\mathrm{d}\tau = 0. \tag{1.77}$$

Since

$$\frac{1}{2\pi\mathrm{i}}\int_\Gamma \frac{\widetilde{\Upsilon}(\tau)}{\tau-\xi}\,\mathrm{d}\tau = \underset{\tau=\xi}{\mathrm{res}}\,\frac{\widetilde{\Upsilon}(\tau)}{\tau-\xi} = \widetilde{\Upsilon}(\xi),$$

from (1.77) it can be argued that

$$\widetilde{\Upsilon}(\xi) = 0.$$

Thus, $\Upsilon(z) = 0$ inside the loop of γ_+^ε and consequently $\Upsilon(z) \equiv 0$ in the whole domain of regularity of $\Upsilon(z)$. That concludes the proof for the case $D < 1$. ■

Let us use the following simple but important lemma.

Lemma 1.1. *Let the function $\Upsilon(z)$ satisfy the conditions of Theorem 1.2. Then in any narrowed half-strip*

$$\{z: \; -\pi - \varepsilon_1' \leq \mathrm{Re}\,z \leq \varepsilon_1',\; \mathrm{Im}\,z \geq \varepsilon_2' > 0\}, \tag{1.78}$$

where $0 < \varepsilon_1' < \varepsilon_1$, $\varepsilon_2' > \varepsilon_2$, and for any integer $m > 0$

$$\left|\frac{\mathrm{d}^m\Upsilon(z)}{\mathrm{d}z^m}\right| \leq \mathrm{const}\,\mathrm{e}^{D|\mathrm{Im}\,z|}. \tag{1.79}$$

The proof of the lemma can be found in [1.7], Section 3.4.

Now we are in a position to complete the proof of the theorem (the case $D \geq 1$). Let us temporarily assume $1 \leq D < 2$. By integrating (1.72) by parts one obtains

$$\begin{aligned}\int_{\gamma_+} \mathrm{e}^{-\mathrm{i}R\cos z}\Upsilon(z)\,\mathrm{d}z &= \int_{\gamma_+} \frac{\Upsilon(z)}{\mathrm{i}R\sin z}\mathrm{d}\,\mathrm{e}^{-\mathrm{i}R\cos z} \\ &= \int_{\gamma_+} \mathrm{e}^{-\mathrm{i}R\cos z}\left[\frac{\mathrm{i}}{R}\frac{\mathrm{d}}{\mathrm{d}z}\frac{\Upsilon(z)}{\sin z}\right]\mathrm{d}z = 0.\end{aligned} \tag{1.80}$$

The expression $\Upsilon(z)/\sin z$ grows as $|\mathrm{Im}\, z| \to \infty$ not faster than const $\exp[(D-1)|\mathrm{Im}\, z|]$. By virtue of Lemma 1.1 the derivative $(\mathrm{d}/\mathrm{d}z)\,[\Upsilon(z)/\sin z]$ has the same order of growth. The theorem is already proved for $D-1<1$; thus,

$$\frac{\mathrm{d}}{\mathrm{d}z}\left[\frac{\Upsilon(z)}{\sin z}\right] = 0.$$

By induction, it can be argued that

$$\left[\frac{\mathrm{d}}{\mathrm{d}z}\frac{1}{\sin z}\right]^{\ell}\Upsilon(z) = 0, \quad \ell = [D]$$

which is followed by (1.74).

It is worth commenting on the so-called inversion formula to the Sommerfeld integral given by G. D. Malyuzhinets [1.14]. The Sommerfeld integral can be represented as

$$F(r) = \frac{1}{\pi\mathrm{i}}\int_{\gamma_+} f(\alpha)\,\mathrm{e}^{-ikr\cos\alpha}\mathrm{d}\alpha, \quad |\arg(-ik)| \leq \pi/2, \tag{1.81}$$

where $f(\alpha) = -f(-\alpha)$. The fundamental result in the theory of the Sommerfeld–Malyuzhinets transform is the following theorem,

Theorem 1.3. *Let the function $F(r)$ be regular for $0 < |r| < \infty$ in the sector $|\arg \mathrm{r}| < \delta$ and satisfy the inequality*

$$|F(r)| < M\,|r|^{-1+a}\,\mathrm{e}^{b|r|}, \tag{1.82}$$

where M, a, and b are positive constants. Let the end points of the contour γ_+ in the integral (1.81) coincide with $\alpha = \arg k + \epsilon + \mathrm{i}\infty$ and $\alpha = \arg k - \pi - \epsilon + \mathrm{i}\infty$, where $0 < \epsilon < \pi$, $\delta \leq \pi/2$. Then, among the functions that grow not faster than $\exp[(1-a_1)\,|\mathrm{Im}\,\alpha|]$, $a_1 > 0$ and are regular inside and on the contour γ_+ with possible exception of infinity, a unique odd solution $f(\alpha)$ of the integral equation (1.81) exists. For $\mathrm{Im}\,(k\cos\alpha) > b$, this solution is represented by the integral

$$f(\alpha) = \frac{ik}{2}\sin\alpha\int_0^{\infty} F(r)\,\mathrm{e}^{ikr\cos\alpha}\,\mathrm{d}r. \tag{1.83}$$

Inside the contour γ_+, the function $f(\alpha)$ satisfies the inequality

$$|f(\alpha)| < M_1\,\exp[(1-a)\,\mathrm{Im}\,\alpha], \quad \mathrm{Im}\,\alpha \to +\infty. \tag{1.84}$$

The formula (1.83) represents the inversion of the Sommerfeld–Malyuzhinets transform (1.81). The proof can be found in [1.14,1.18] and is similar to that given in this section for Theorem 1.2.

1.4.4 Kontorovich–Lebedev (KL) transform and its connection with the Sommerfeld integral

In this section we consider a complex form of the Kontorovich–Lebedev (KL) transform, whose original form was published by the eponyms of this transform in [1.19], and its connection with the SM transform for $|\arg(-ik)| < \pi/2$ [1.20]; see also the appendices in [1.21,1.22]. Let a, b, c, and d be positive numbers; let $0 < b < \pi/2$ and $|\arg(-ik)| < \pi/2$. We consider a function of r, $p(r) = \mathrm{O}(r^{a-1/2}\exp(ikr\cos b))$, analytic for positive value of r and also in the entire region $c < |kr| < \infty$, $|\arg(r)| < \epsilon_1$. Let

$$\widetilde{\Psi}(\alpha) = \frac{k}{-2i\pi}\int_0^{+\infty}\frac{p(r)}{\sqrt{-ikr}}\mathrm{e}^{ikr\cos\alpha}\,\mathrm{d}r. \tag{1.85}$$

From the theory of the SM transform, $\widetilde{\Psi}(\alpha)$ is an analytic function in the domain $\mathrm{Re}(-ik_0(\cos\alpha + \cos b)) > 0$, as $|\arg(-ik_0)| < |\arg(-ik)| + \epsilon_1$, and then in the strip $|\mathrm{Re}(\alpha)| < \pi/2 - |\arg(-ik)| + \epsilon_1$. Its behavior in these domains is $\mathrm{O}(\exp(-a|\mathrm{Im}\,\alpha|))$. The previous expression enables us to write

$$\frac{p(r)}{\sqrt{-ikr}} = \frac{1}{2i}\int_\gamma \mathrm{e}^{-ikr\cos\alpha}\sin\alpha\,\widetilde{\Psi}(\alpha)\,\mathrm{d}\alpha, \tag{1.86}$$

where γ is the Sommerfeld double-loop contour. This contour consists of two loops γ_+ and γ_-, which are symmetric with respect to the origin, $\gamma_+ = (i\infty + h + \epsilon, id + h + \epsilon] \cup [id + h + \epsilon, id + h - \pi - \epsilon] \cup [id + h - \pi - \epsilon, i\infty + h - \pi - \epsilon)$ with $d > 0$, $h = \arg(-ik) + \pi/2$, $0 < \epsilon < \epsilon_1 < \pi$.

Using the Fourier transform, we introduce the function

$$\Psi(\nu) = \frac{\nu\sin\pi\nu}{i}\int_{i\mathbb{R}}\widetilde{\Psi}(\alpha)\,\mathrm{e}^{i\nu\alpha}\,\mathrm{d}\alpha.$$

From the properties of $\widetilde{\Psi}(\alpha)$ given by (1.85), and the expression of the modified Bessel function K_ν, following:

$$\mathrm{K}_\nu(-ikr) = \frac{1}{2i}\int_{i\mathbb{R}}\mathrm{e}^{ikr\cos\alpha}\mathrm{e}^{i\nu\alpha}\,\mathrm{d}\alpha,$$

we obtain, for $|\mathrm{Re}\,\nu| < a$,

$$\Psi(\nu) = -\frac{k}{i\pi}\int_0^{+\infty}\frac{p(r)}{\sqrt{-ikr}}\nu\sin(\pi\nu)\,\mathrm{K}_\nu(-ikr)\,\mathrm{d}r, \tag{1.87}$$

Inversely, we can let

$$\widetilde{\Psi}(\alpha) = \frac{1}{2\mathrm{i}\pi} \int\limits_{\mathrm{i}\mathbb{R}} \frac{\Psi(\nu)}{\nu \sin \pi \nu} \mathrm{e}^{-\mathrm{i}\nu\alpha} \mathrm{d}\nu$$

and use it for (1.86), as α is imaginary, in connection with the expression [1.23, p. 16]

$$\frac{\mathrm{K}_\nu(-\mathrm{i}kr)}{-\mathrm{i}kr} = \frac{1}{4\nu \sin \pi \nu} \int\limits_{\gamma} \mathrm{e}^{-\mathrm{i}kr\cos\alpha} \sin \alpha \frac{\mathrm{e}^{\mathrm{i}\nu\alpha} + \mathrm{e}^{-\mathrm{i}\nu\alpha}}{2\mathrm{i}} \mathrm{d}\alpha.$$

By doing so, we obtain

$$p(r) = \frac{1}{\mathrm{i}\pi} \int\limits_{\mathrm{i}\mathbb{R}} \Psi(\nu) \frac{\mathrm{K}_\nu(-\mathrm{i}kr)}{\sqrt{-\mathrm{i}kr}} \mathrm{d}\nu. \tag{1.88}$$

The latter formula can be analytically continued for complex k, by studying the regularity of $\widetilde{\Psi}(\alpha)$. This task becomes particularly easy when $\widetilde{\Psi}(\alpha)$ is independent of k. In this case, we can let k vary as $|\arg(-\mathrm{i}k)| < \pi/2$, $\mathrm{Re}(-\mathrm{i}k(\cos\alpha + \cos b)) > 0$, to define the domain of regularity of this function, which contains the strip $|\mathrm{Re}\,\alpha| < \pi - b$. This then implies that $\Psi(\nu) = \mathrm{O}\left(\nu^d \sin(\pi\nu)/\cos\left[\nu(\pi - b)\right]\right)$ as $|\mathrm{Im}\,\nu| \to \infty$. By considering the behavior of $\Psi(\nu)$ and that of $\mathrm{K}_\nu(-\mathrm{i}kR)$, derived from [1.24],

$$\mathrm{K}_\nu(-\mathrm{i}kr) = \mathrm{O}(\nu^{a_0} \cos\left[\nu(\pi/2 + |\arg(-\mathrm{i}k)|)\right] / \sin(\nu\pi))$$

as $|\mathrm{Im}\,\nu| \to \infty$, $|\arg(\nu)| \to \pi/2$, $a_0 = |\mathrm{Re}\,\nu| - 1/2$, the expression (1.88) remains valid as $\pi/2 - b - |\arg(-\mathrm{i}k)| > 0$.

The formulas (1.87) and (1.88) give the form of the KL transform to which we refer in this book. By use of particular choice of k with $|\arg(-\mathrm{i}k)| \leq \pi/2$, we can then recover the form given in [1.3, p. 608], ($\arg(-\mathrm{i}k) = \pi/2$) or the one taken in [1.25] ($\arg(-\mathrm{i}k) = 0$).

Let us consider $\Psi_\mu(\nu) = [\mathrm{d}_{(w)}(\mu - \nu) + \mathrm{d}_{(w)}(\mu + \nu)]/2$ (μ is a parameter) with $\mathrm{d}_{(w)}(\nu) = -4w\mathrm{i}\pi/(\nu^2 - w^2)$, $w > 0$, $\mathrm{d}_{(w)}(\nu)$ tending to the Dirac distribution on the imaginary axis, $\delta(\nu)$, as $w \to 0^+$. This gives us from (1.88)

$$p(r) = \frac{1}{\mathrm{i}\pi} \frac{\mathrm{K}_\mu(-\mathrm{i}kr)}{\sqrt{-\mathrm{i}kr}},$$

as $w \to 0^+$. By use of the KL transform, we then have

$$\frac{\delta(\mu - \nu) + \delta(\mu + \nu)}{2} = \lim_{w \to 0+} \frac{k}{(\mathrm{i}\pi)^2} \int\limits_0^{+\infty} \nu \sin \pi \nu \frac{\mathrm{K}_\nu(-\mathrm{i}kr)}{\sqrt{-\mathrm{i}kr}} \frac{\mathrm{K}_\mu(-\mathrm{i}kr)}{\sqrt{-\mathrm{i}kr}} (-\mathrm{i}kr)^w \mathrm{d}r \tag{1.89}$$

in the sense of distribution. Other interesting developments on the KL transformation can be found in [1.26].

1.4.5 Watson–Bessel integral

In this section we study Green's function in a conical domain $\Omega \subset \mathrm{R}^3$ with the perfect boundary conditions on its boundary [1.27, App. A]. The Green function $G(\boldsymbol{x}, \boldsymbol{x}_0)$ fulfills the equation

$$(\operatorname{div}\operatorname{grad} + k^2)\, G(\boldsymbol{x}, \boldsymbol{x}_0) = -\boldsymbol{\delta}(\boldsymbol{x} - \boldsymbol{x}_0) \tag{1.90}$$

and one of the boundary conditions

$$G(\boldsymbol{x}, \boldsymbol{x}_0) = 0 \quad \text{or} \quad \partial_n G(\boldsymbol{x}, \boldsymbol{x}_0) = 0 \tag{1.91}$$

as $\boldsymbol{x} \in \partial\Omega$. The radiation condition reads

$$\left(\frac{\partial}{\partial r} - \mathrm{i}k\right) G(\boldsymbol{x}, \boldsymbol{x}_0) = \mathrm{o}(r^{-1}) \tag{1.92}$$

as $r = \|\boldsymbol{x}\| \to \infty$. Finally, Meixner's conditions are imposed

$$G(\boldsymbol{x}, \boldsymbol{x}_0) = \mathrm{O}(1), \qquad \operatorname{grad} G(\boldsymbol{x}, \boldsymbol{x}_0) = \mathrm{o}(r^{-1/2}) \tag{1.93}$$

when $r \to 0$.

The solution is constructed by means of separation of variables, implying that the spherical coordinates r, θ, φ are introduced, which leads to

$$\operatorname{div}\operatorname{grad} = r^{-2}\frac{\partial}{\partial r}r^2\frac{\partial}{\partial r} + r^{-2}\Delta_\omega,$$

where

$$\Delta_\omega = \frac{1}{\sin\theta}\frac{1}{\sin\theta}\sin\theta\frac{\partial}{\partial\theta} + \frac{1}{\sin^2\theta}\frac{\partial^2}{\partial\varphi^2}$$

is the Laplace–Beltrami operator, $\boldsymbol{\delta}(\boldsymbol{x} - \boldsymbol{x}_0) = r^{-2}\delta(r - r_0)\boldsymbol{\delta}(\boldsymbol{\omega} - \boldsymbol{\omega}_0)$ with $\boldsymbol{\omega} = (\theta, \varphi)$, $\boldsymbol{\omega}_0 = (\theta_0, \varphi_0)$. The delta function $\boldsymbol{\delta}(\boldsymbol{\omega} - \boldsymbol{\omega}_0)$ on the unit sphere S^2 can be written as an expansion

$$\boldsymbol{\delta}(\boldsymbol{\omega} - \boldsymbol{\omega}_0) = \sum_{j=1}^{\infty} \Phi_j(\boldsymbol{\omega})\Phi_j(\boldsymbol{\omega}_0) \tag{1.94}$$

with respect to the eigenfunction $\Phi_j(\boldsymbol{\omega})$ of the operator $\Delta_\omega - 1/4$ on the unit sphere with a hole cut out by the conical surface. The vertex of the cone coincides with the center of the sphere. The eigenfunctions satisfy the equation

$$(\Delta_\omega + \nu_j^2 - 1/4)\Phi_j(\boldsymbol{\omega}) = 0,$$

with the appropriate boundary condition (Dirichlet or Neumann) on the boundary σ of the hole. The series (1.94), understood in the distribution sense, is the direct consequence of the fact that $\{\Phi_j(\boldsymbol{\omega})\}$ form an orthonormal basis in $L_2(S^2)$.

Now we look for the solution of the problem (1.90)–(1.93) in the form

$$G(\boldsymbol{x}, \boldsymbol{x}_0) = G(r, \boldsymbol{\omega}; r_0, \boldsymbol{\omega}_0) = \sum_{j=1}^{\infty} g_j(r, r_0)\, \Phi_j(\boldsymbol{\omega})\Phi_j(\boldsymbol{\omega}_0). \tag{1.95}$$

Then we have from (1.90)

$$\left[\frac{\mathrm{d}}{\mathrm{d}r}\left(r^2\frac{\mathrm{d}}{\mathrm{d}r}\right) + k^2r^2 - \nu_j^2 + \frac{1}{4}\right] g_j(r, r_0) = -\delta(r - r_0).$$

Two linearly independent solutions of the latter equation with zero on the right-hand side are given by[11]

$$r^{-1/2}\mathrm{J}_{\nu_j}(kr) \quad \text{and} \quad r^{-1/2}\mathrm{e}^{-\mathrm{i}\pi\nu_j/2}\,\mathrm{K}_{\nu_j}(-\mathrm{i}kr),$$

where $\mathrm{J}_\nu(kr)$ and $\mathrm{K}_\nu(-\mathrm{i}kr)$ are the Bessel and modified Bessel (Macdonald) functions, respectively. From the radiation and Meixner's conditions we conclude that

$$g_j(r, r_0) = (rr_0)^{-1/2}\mathrm{e}^{-\mathrm{i}\pi\nu_j/2}\,\mathrm{J}_{\nu_j}(kr_<)\,\mathrm{K}_{\nu_j}(-\mathrm{i}kr_>),$$

where $r_< = \min\{r, r_0\}$, and $r_> = \max\{r, r_0\}$. Finally, the separation of variables leads to solution of the problem (1.90)–(1.93):

$$G(\boldsymbol{x}, \boldsymbol{x}_0) = \sum_{j=1}^{\infty} (rr_0)^{-1/2}\mathrm{e}^{-\mathrm{i}\pi\nu_j/2}\,\mathrm{J}_{\nu_j}(kr_<)\,\mathrm{K}_{\nu_j}(-\mathrm{i}kr_>)\,\Phi_j(\boldsymbol{\omega})\Phi_j(\boldsymbol{\omega}_0). \tag{1.96}$$

Now, by means of the Watson transformation, we can write

$$G(\boldsymbol{x}, \boldsymbol{x}_0) = -\frac{1}{\mathrm{i}\pi\sqrt{rr_0}}\int_C \nu\,\mathrm{e}^{-\mathrm{i}\pi\nu/2}\,\mathrm{J}_\nu(kr_<)\,\mathrm{K}_\nu(-\mathrm{i}kr_>)\,g(\boldsymbol{\omega}, \boldsymbol{\omega}_0, \nu)\,\mathrm{d}\nu, \tag{1.97}$$

where C is the integration contour, $C = (+\infty - \mathrm{i}h, -\mathrm{i}h] \cup [-\mathrm{i}h, \mathrm{i}h] \cup [\mathrm{i}h, +\infty + \mathrm{i}h)$, where h is positive. The contour C comprises the poles ν_j of the "spherical" Green function $g(\boldsymbol{\omega}, \boldsymbol{\omega}_0, \nu)$ satisfying

$$(\Delta_\omega + \nu^2 - 1/4)g(\boldsymbol{\omega}, \boldsymbol{\omega}_0, \nu) = \delta(\boldsymbol{\omega} - \boldsymbol{\omega}_0)$$

with the boundary condition $g|_{\omega\in\sigma} = 0$ or $\partial_m g|_{\omega\in\sigma} = 0$, where m is normal to σ on S^2.

A heuristic verification of (1.97) is based on the following formula, holding in the sense of distribution:

$$g(\boldsymbol{\omega}, \boldsymbol{\omega}_0, \nu) = \sum_{j=1}^{\infty} (\nu^2 - \nu_j^2)^{-1}\,\Phi_j(\boldsymbol{\omega})\Phi_j(\boldsymbol{\omega}_0)$$

and on the application of the theorem on residues. The function g is analytic with respect to ν having poles at ν_j located on the positive semi-axis.

---

[11]In this book we prefer to exploit $\mathrm{K}_\nu(-\mathrm{i}z) = (\mathrm{i}\pi/2)\exp(\mathrm{i}\pi\nu/2)\,\mathrm{H}_\nu^{(1)}(z)$, the modified Bessel (Macdonald) [1.24] function instead of the Hankel function $\mathrm{H}_\nu^{(1)}(z)$.

For the plane-wave incidence $u^{\mathrm{i}} = \exp(-\mathrm{i}\boldsymbol{\omega}_0 \cdot \boldsymbol{x})$, taking into account a relation for Green's function in free space,

$$G^0(\boldsymbol{x}, \boldsymbol{x}_0) = \frac{\mathrm{e}^{\mathrm{i}k\|\boldsymbol{x}-\boldsymbol{x}_0\|}}{4\pi\|\boldsymbol{x}-\boldsymbol{x}_0\|} = \frac{\mathrm{e}^{\mathrm{i}kr_0}}{4\pi r_0}\left[\mathrm{e}^{-\mathrm{i}\boldsymbol{\omega}_0\cdot\boldsymbol{x}} + \mathrm{o}(1)\right], \quad r_0 = \|\boldsymbol{x}_0\| \to \infty$$

and

$$\mathrm{K}_\nu(-\mathrm{i}kr_0) = \sqrt{\frac{\pi}{2}}\frac{\mathrm{e}^{\mathrm{i}kr_0}}{\sqrt{-\mathrm{i}kr_0}}\left[1 + \mathrm{O}\left(\frac{1}{kr}\right)\right],$$

we obtain for the total field $u(r, \boldsymbol{\omega}, \boldsymbol{\omega}_0)$

$$\begin{aligned} u(r, \boldsymbol{\omega}, \boldsymbol{\omega}_0) &= \lim_{r_0\to\infty} 4\pi r_0 \exp(-\mathrm{i}kr_0)\, G(\boldsymbol{x}, \boldsymbol{x}_0) \\ &= 2\mathrm{e}^{3\mathrm{i}\pi/4}\sqrt{\frac{2\pi}{kr}}\int_C \nu\,\mathrm{e}^{-\mathrm{i}\pi\nu/2}\,\mathrm{J}_\nu(kr)\, g(\boldsymbol{\omega}, \boldsymbol{\omega}_0, \nu)\,\mathrm{d}\nu \end{aligned} \tag{1.98}$$

by means of the standard limiting procedure ($r_0 \to \infty$) of reduction to the plane-wave incidence.

We call the integrals of the type (1.97) or (1.98) the Watson–Bessel integrals. In Chapters 5 and 6 the Watson–Bessel integrals are exploited in the problems of diffraction by impedance cones; see also [1.28].

1.5 Malyuzhinets's solution for the impedance wedge diffraction problem

The Sommerfeld integral is an efficient tool not only for the perfect boundary conditions [1.12] but also for impedance wedge diffraction; G. D. Malyuzhinets [1.13,1.14] was the first to apply this technique. Using Sommerfeld integral the impedance wedge diffraction problem is reduced to the eponymous Malyuzhinets functional equations for the Sommerfeld transformant. The Malyuzhinets equations turn out to be homogeneous with nonconstant coefficients. The key step for the solution of these equations is the invention of the special function (named after Malyuzhinets) that satisfies a specific functional equation. This function is discussed in detail. This section concludes with the far-field analysis of the solution.

1.5.1 Functional equations for the Malyuzhinets problem

Recall formulation of the problem of the plane-wave diffraction by an impedance wedge (the Malyuzhinets problem). The wave field in an angular domain $\{(r, \varphi) : r \geq 0, -\Phi < \varphi < \Phi\}$ is governed by the Helmholtz equation

$$\operatorname{div}\operatorname{grad} u + k^2 u = 0 \tag{1.99}$$

and is excited by the incident plane wave $u^{\mathrm{i}}(kr, \varphi) = \exp[-\mathrm{i}kr\cos(\varphi - \varphi_0)]$. The "wedge" is the domain $\{(r, \varphi) : r \geq 0,\ \Phi \leq \varphi \leq \pi \bigcup -\pi \leq \varphi \leq -\Phi\}$, and on the boundaries $\varphi = \Phi$

and $\varphi = -\Phi$ (which are the wedge's faces) the impedance (or third kind) boundary conditions

$$\left.\frac{\partial u}{\partial n} - ik\sin\theta_{\pm} u\right|_{\varphi=\pm\Phi} = 0, \quad r > 0, \quad \theta_{\pm} = \text{const} \tag{1.100}$$

are satisfied. The normalized impedances $\sin\theta_{\pm}$ of the wedge's faces are subject to the conditions $0 < \operatorname{Re}\theta_{\pm} \le \pi/2$. These conditions ensure dissipation of the energy by the wedge's faces, which, in accordance with the accepted terminology, are called absorbing. The solution to the problem with lossless ($\operatorname{Re}\theta_{\pm} = 0$) faces can be obtained by means of a natural limiting procedure when the absorption in the wedge's faces tends to zero.

Besides, the wave field should satisfy the Meixner condition at the edge

$$u(kr,\varphi) \underset{r\to 0}{=} \text{const} + \mathrm{O}(r^{\delta}), \quad \delta > 0 \tag{1.101}$$

(recall that the Meixner conditions (1.101) imply the boundedness of the energy near the edge, $\Phi \neq \pi/2$) and conditions at infinity. The latter can be taken in the form of either the integral Sommerfeld radiation conditions (1.55) or the limiting-absorption principle.

The existence of solution of the formulated problem follows from the construction of the present section.

The solution to the problem is sought in the form of the Sommerfeld integral

$$u(kr,\varphi) = \frac{1}{2\pi\mathrm{i}} \int_{\gamma_+} \mathrm{e}^{-\mathrm{i}kr\cos z}\,[s(\varphi+z) - s(\varphi-z)]\,\mathrm{d}z \tag{1.102}$$

where $s(z)$ is a meromorphic function and has finite limits as $\operatorname{Im} z \to \pm\infty$ uniformly with respect to the real part of z, $|\operatorname{Re} z| < 2\pi$. The latter requirement is in agreement with the Meixner condition. Substituting the integral into the boundary conditions (1.100) and applying the theorem on inversion of the Sommerfeld integral, we obtain the following Malyuzhinets functional equations:

$$\begin{cases} (\sin z + \sin\theta_+)s(z+\Phi) - (-\sin z + \sin\theta_+)s(-z+\Phi) = C_+\sin z, \\ (\sin z - \sin\theta_-)s(z-\Phi) - (-\sin z - \sin\theta_-)s(-z-\Phi) = C_-\sin z, \end{cases} \tag{1.103}$$

where $C_{\pm}$ are arbitrary constants.

Recall that (1) the existence of the finite limits $s(\pm\mathrm{i}\infty)$ follows from the Meixner conditions and from the relation between the behavior of the Sommerfeld integrals as $kr \to 0$ and $s(z)$ as $|\operatorname{Im} z| \to \infty$; and (2) the function $s(z)$ (the so-called Sommerfeld transformant or spectral function) can always be specified up to an arbitrary additive constant because the Sommerfeld integral of an even function is zero. In particular, by use of the shift $s(z) \rightsquigarrow s(z) - [s(\mathrm{i}\infty) + s(-\mathrm{i}\infty)]/2$ the equality $s(\mathrm{i}\infty) = -s(-\mathrm{i}\infty)$ can always be satisfied. Then, in (1.103) we have $C_{\pm} = 0$, which follows by comparing the left-hand and right-hand sides of equations (1.103) as $z \to \pm\mathrm{i}\infty$.

So we consider the homogeneous system (1.103):

$$\begin{cases} (\sin z + \sin\theta_+)s(z+\Phi) - (-\sin z + \sin\theta_+)s(-z+\Phi) = 0, \\ (\sin z - \sin\theta_-)s(z-\Phi) - (-\sin z - \sin\theta_-)s(-z-\Phi) = 0. \end{cases} \tag{1.104}$$

This system should be supplemented by the conditions specifying the analytical class of functions $s(z)$ in which the solution to (1.104) exists. These conditions are naturally inherited from the statement of the problem and from the properties of the Sommerfeld integral. Following the results of Section 1.4.2, we define the aforementioned class of functions as follows.

First, the solution to the homogeneous system (1.104) is sought among meromorphic functions,[12] which are regular in the strip

$$\Pi_\varepsilon = \{z : |\mathrm{Re}\, z| < \Phi + \varepsilon\}, \quad \varepsilon > 0 \quad (\varepsilon \text{ small}),$$

except the simple pole $z = \varphi_0$ with the unit residue. Second, as $\mathrm{Im}\, z \to \pm\infty$, the function $s(z)$ must satisfy the condition

$$|s(z) - s(\pm i\infty)| \leq \text{const} \exp(-\delta|\mathrm{Im}\, z|).^{13} \tag{1.105}$$

The aforementioned conditions for the class of functions on one hand ensure the required behavior of the solution at infinity and at the edge but on the other are not very restrictive so that the solution really exists.

1.5.2 The multiplication principle and the auxiliary solution $\Psi_0(z)$ to the functional equations (1.104)

Solution of the problem for functional equations (1.104) in the prescribed class of functions is generally not a simple task. However, this task can be split into a number of subproblems that are nontrivial but can be solved explicitly.

Let $\Psi_0(z)$ be a meromorphic function satisfying the system of equations (1.104). We do not require that $\Psi_0(z)$ has the behavior as in (1.105); however, it is regular in the strip $|\mathrm{Re}\, z| < \Phi + \varepsilon$. Provided one looks for the unknown function $s(z)$ in the form

$$s(z) = \Psi_0(z)\, \tilde{s}(z), \tag{1.106}$$

the function $\tilde{s}(z)$ fulfills the system of equations

$$\begin{cases} \tilde{s}(z + \Phi) - \tilde{s}(-z + \Phi) = 0, \\ \tilde{s}(z - \Phi) - \tilde{s}(-z - \Phi) = 0. \end{cases} \tag{1.107}$$

We have arrived at two subproblems: the choice of the fixed solution $\Psi_0(z)$ and derivation of the solution for the system (1.107). It is implied that the function $\tilde{s}(z)$ will correct the behavior of $\Psi_0(z)$ at infinity and such that the product (1.106) satisfies all the prescribed conditions. The determination of $\Psi_0(z)$ requires consideration of two additional subproblems, whereas the problem for $\tilde{s}(z)$ is simple.

It is remarkable that the solution $\Psi_0(z)$ to (1.104) can also be found in an explicit form. Its construction is based on the following simple but useful observation (which is valid not just for the functional equations in question).

---

[12]Recall that a function is meromorphic in a domain provided its only singularities are poles. A function is meromorphic if it is meromorphic on the whole complex plane.

[13]Recall that the inequality (1.105) implies the Meixner conditions (1.101).

Lemma 1.2. (**The multiplication principle.**) *Let $\sigma_1(z)$ and $\sigma_2(z)$ be solutions to the functional equations*

$$\sigma_1(z+\Phi) = a_1(z)\sigma_1(z), \quad \sigma_2(z+\Phi) = a_2(z)\sigma_2(z), \tag{1.108}$$

correspondingly. Then $\sigma(z) = \sigma_1(z)\sigma_2(z)$ solves the equation

$$\sigma(z+\Phi) = a(z)\sigma(z) \tag{1.109}$$

with $a(z) = a_1(z)a_2(z)$.

We turn to the construction of the auxiliary solution $\Psi_0(z)$. Consider the following auxiliary systems of equations:

$$\begin{cases} (\sin z + \sin\theta_+)s_1(z+\Phi) = (-\sin z + \sin\theta_+)s_1(-z+\Phi) \\ s_1(z-\Phi) = s_1(-z-\Phi), \end{cases} \tag{1.110}$$

$$\begin{cases} s_2(z+\Phi) = s_2(-z+\Phi) \\ (\sin z - \sin\theta_-)s_2(z-\Phi) = (-\sin z - \sin\theta_-)s_2(-z-\Phi). \end{cases} \tag{1.111}$$

In view of the multiplication principle, the product $\Psi_0(z) = s_1(z)s_2(z)$ solves (1.104) (note the important role of the equations $s_1(z-\Phi) = s_1(-z-\Phi)$ and $s_2(z+\Phi) = s_2(-z+\Phi)$ in (1.110)–(1.111)).

To motivate the steps to follow, we address the system (1.110); similar considerations can be applied to the system (1.111) as well. We rewrite (1.110) as

$$\begin{cases} \dfrac{s_1(z+\Phi)}{s_1(-z+\Phi)} = \dfrac{\sin\theta_+ - \sin z}{\sin\theta_+ + \sin z} = \dfrac{\cot\left[\frac{1}{2}(\theta_+ + z)\right]}{\cot\left[\frac{1}{2}(\theta_+ - z)\right]}, \\ \dfrac{s_1(z-\Phi)}{s_1(-z-\Phi)} = 1, \end{cases} \tag{1.112}$$

take the logarithmic derivatives of both sides of the previous equations, and introduce the notation

$$\chi_1(z) := \frac{\mathrm{d}}{\mathrm{d}z}\log s_1(z).$$

Thus, we obtain

$$\begin{cases} \chi_1(z+\Phi) + \chi_1(-z+\Phi) = -\dfrac{1}{\cos(z+\theta_+ - \pi/2)} - \dfrac{1}{\cos(z-\theta_+ + \pi/2)}, \\ \chi_1(z-\Phi) + \chi_1(-z-\Phi) = 0. \end{cases} \tag{1.113}$$

We then observe that it is sufficient to get the solution to the simpler system

$$\begin{cases} \eta_1(z+\Phi) + \eta_1(-z+\Phi) = -1/\cos z, \\ \eta_1(z-\Phi) + \eta_1(-z-\Phi) = 0,^{14} \end{cases} \tag{1.114}$$

[14]The system (1.114) has already been studied with regard to the wave field radiated by the vibrating faces of a wedge [1.29,1.30].

with the aid of which the solution $\chi_1(z)$ to (1.113) takes the form

$$\chi_1(z) = \eta_1(z + \theta_+ - \pi/2) + \eta_1(z - \theta_+ + \pi/2). \tag{1.115}$$

Finally, the solution to (1.110) is obtained by direct integration

$$s_1(z) = \text{const}\,\exp\left[\int_0^z \chi_1(z)\,\mathrm{d}z\right]. \tag{1.116}$$

To verify that equation (1.116) is indeed a solution of (1.110) we must check that $s_1(z)$ satisfies (1.110) at one point, say, at $z = 0$. The latter is obvious, and then $s_1(z)$ is the correct solution upon construction.

The solution to (1.114)–(1.116) is usually expressed in terms of the special function named after Malyuzhinets (see also [1.29–1.31]). In the next subsection, we introduce this function and study its main properties.

1.5.3 The Malyuzhinets function $\psi_\Phi(z)$ and its basic properties

Consider the meromorphic function $\psi_\Phi(z)$ satisfying the equation

$$\psi_\Phi(z + 2\Phi)\,/\,\psi_\Phi(z - 2\Phi) = \cot(z/2 + \pi/4). \tag{1.117}$$

Equation (1.117) does not define $\psi_\Phi(z)$ uniquely: in view of the multiplication principle (1.117) remains valid after multiplication of ψ_Φ by any meromorphic solution of the equation $f(z + 2\Phi) = f(z - 2\Phi)$, that is, by any meromorphic 4Φ-periodic function. It will be seen from the next subsections that the conditions

$$\psi_\Phi(z) \quad \text{is even}, \quad \psi_\Phi(0) = 1,^{15} \quad \psi_\Phi(z) \underset{\mathrm{Im}\,z\to\pm\infty}{\sim} \mathrm{e}^{c|\mathrm{Im}\,z|} \tag{1.118}$$

are sufficient to complete the definition of the Malyuzhinets function.

This function was introduced in [1.13]. Following tradition, we call it the Malyuzhinets function. First, we study its logarithmic derivative.

1.5.4 Examination of $(\mathrm{d}/\mathrm{d}z) \ln \psi_\Phi(z)$

Denote $\eta_\Phi(z) = (\mathrm{d}/\mathrm{d}z) \ln \psi_\Phi(z)$. Then from (1.117)–(1.118) it follows that $\eta_\Phi(z)$ satisfies the equations

$$\eta_\Phi(z + 2\Phi) - \eta_\Phi(z - 2\Phi) = -1/\cos z, \tag{1.119}$$

$$\eta_\Phi(z) + \eta_\Phi(-z) = 0 \tag{1.120}$$

and is bounded as $\mathrm{Im}\,z \to \pm\infty$. Clearly this system differs from (1.114) in the shift Φ of the argument of η_Φ.

---

[15] In fact, the normalization of the Malyuzhinets function can be any; this one is used to follow tradition.

Recall that the system (1.114) has arisen with regard to the problem for a wedge with a vibrating face. The odd solution to (1.119) can be written as $\eta_\Phi(z) = s(z - \Phi)$, where $s(z)$ satisfies this system. The solution to the system (1.119)–(1.120) is unique among functions bounded as $\operatorname{Im} z \to \pm\infty$ and is given by

$$\eta_\Phi(z) = -\frac{1}{4}\mathrm{VP}\int\limits_{\mathrm{i}\mathbb{R}} \frac{\mathrm{e}^{-\mathrm{i}z\zeta}}{\cos\frac{1}{2}\pi\zeta \sin 2\Phi\zeta}\mathrm{d}\zeta \tag{1.121}$$

$$= \frac{\mathrm{i}}{4}\int\limits_{\mathrm{i}\mathbb{R}} \frac{\sin z\zeta}{\cos\frac{1}{2}\pi\zeta \sin 2\Phi\zeta}\mathrm{d}\zeta = -\frac{1}{2}\int\limits_0^\infty \frac{\sinh z\zeta}{\cosh\frac{1}{2}\pi\zeta \sinh 2\Phi\zeta}\mathrm{d}\zeta.$$

The integral (1.121) converges absolutely if $|\operatorname{Re} z| < \pi/2 + 2\Phi$, and hence the function $\eta_\Phi(z)$ is regular in this strip. By repeated use of (1.119) the function can be continued to any values of z outside this strip. The points $\pm(\pi/2 + 2\Phi)$ are simple poles of $\eta_\Phi(z)$, as can be seen from (1.119). Therefore, $\eta_\Phi(z)$ is a meromorphic function with simple poles only. By virtue of (1.119) all these poles are translations of the points $\pm(\pi/2 + 2\Phi)$ either by π or by 4Φ; that is, the poles are

$$z_{mn} = \pm[(\pi/2)(2m-1) + 2\Phi(2n-1)] \qquad m, n = 1, 2, 3, \ldots. \tag{1.122}$$

It easily follows from (1.119) that

$$\operatorname*{res}_{z=z_{mn}} \eta_\Phi(z) = (-1)^{m-1}. \tag{1.123}$$

By means of straightforward calculations one obtains

$$\eta_\Phi\left(z + \frac{\pi}{2}\right) + \eta_\Phi\left(z - \frac{\pi}{2}\right) = -\int\limits_0^\infty \frac{\sinh z\zeta}{\sinh 2\Phi\zeta}\mathrm{d}\zeta = -\frac{\pi}{4\Phi}\tan\frac{z\pi}{4\Phi}, \tag{1.124}$$

$$\eta_\Phi(z + \Phi) + \eta_\Phi(z - \Phi) = -\int\limits_0^\infty \frac{\sinh z\zeta \cosh\Phi\zeta}{\sinh 2\Phi\zeta \cosh\frac{1}{2}\pi\zeta}\mathrm{d}\zeta$$

$$= -\frac{1}{2}\int\limits_0^\infty \frac{\sinh z\zeta}{\sinh\Phi\zeta \cosh\frac{1}{2}\pi\zeta}\mathrm{d}\zeta = \eta_{\frac{\Phi}{2}}(z). \tag{1.125}$$

By derivation, the relations (1.124)–(1.125) are valid in the strips $|\operatorname{Re} z| < 2\Phi$ and $|\operatorname{Re} z| < \pi/2 + \Phi$, respectively. However, since the left- and right-hand sides in (1.124)–(1.125) are meromorphic functions, these equalities are valid for any value of z. In turn, (1.124)–(1.125) can be used for the continuation of $\eta_\Phi(z)$ as well.

Since a translation by $-\Phi$ does not affect the asymptotic behavior as $\operatorname{Im} z \to \pm\infty$, the asymptotics of $\eta_\Phi(z)$ is

$$\eta_\Phi(z) \underset{z\to\pm i\infty}{=} \mp\frac{\pi i}{8\Phi} + \begin{cases} -i\dfrac{e^{\pm iz}}{\sin 2\Phi} + o\left(e^{\pm iz}\right) & , 2\Phi < \pi \\ \mp iz\dfrac{8}{3\pi^2}e^{\pm iz} + o\left(\mp iz\, e^{\pm iz}\right) & , 2\Phi = \pi \\ -\dfrac{i\pi}{4}\dfrac{e^{\pm\frac{\pi}{2\Phi}iz}}{\cos\frac{\pi^2}{4\Phi}} + o\left(e^{\pm\frac{\pi}{2\Phi}iz}\right) & , 2\Phi > \pi \end{cases} \tag{1.126}$$

1.5.5 The Malyuzhinets function $\psi_\Phi(z)$

From (1.121) one can directly obtain

$$\psi_\Phi(z) = \exp\left[\int_0^z \left(-\frac{1}{2}\int_0^\infty \frac{\sinh z\zeta}{\cosh\frac{1}{2}\pi\zeta \sinh 2\Phi\zeta}\mathrm{d}\zeta\right)\mathrm{d}z\right]. \tag{1.127}$$

Performing the integration on z, one arrives at

$$\psi_\Phi(z) = \exp\left[-\frac{1}{2}\int_0^\infty \frac{\cosh z\zeta - 1}{\zeta\cosh\frac{1}{2}\pi\zeta \sinh 2\Phi\zeta}\mathrm{d}\zeta\right], \tag{1.128}$$

The Malyuzhinets function is regular in the strip $|\operatorname{Re} z| < \pi/2 + 2\Phi$ (where the integral (1.128) converges). It can be continued to any value of z outside this strip by repeated use of (1.117). The points (1.122) are either simple poles (for m even) or zeros (for m odd) of $\psi_\Phi(z)$, depending on the sign of the residue of $\eta_\Phi(z)$ at these points. Therefore the Malyuzhinets function is meromorphic. Moreover, it exhibits the property $\psi_\Phi(\bar{z}) = \bar{\psi}_\Phi(z)$.

It is clearly seen from (1.124)–(1.125) that

$$\begin{aligned} \psi_\Phi\left(z+\frac{\pi}{2}\right)\psi_\Phi\left(z-\frac{\pi}{2}\right) &= C_1\exp\left\{\int_0^z\left[\eta_\Phi\left(z+\frac{\pi}{2}\right)+\eta_\Phi\left(z-\frac{\pi}{2}\right)\right]\mathrm{d}z\right\} \\ &= C_1\exp\left(-\int_0^z \frac{\pi}{4\Phi}\tan\frac{\pi}{4\Phi}z\,\mathrm{d}z\right) = C_1\cos\left(\frac{\pi}{4\Phi}z\right), \\ \psi_\Phi(z+\Phi)\psi_\Phi(z-\Phi) &= C_2\exp\left\{\int_0^z\left[\eta_\Phi(z+\Phi)+\eta_\Phi(z-\Phi)\right]\mathrm{d}z\right\} \\ &= C_2\exp\left[\int_0^z \eta_{\frac{\Phi}{2}}(z)\mathrm{d}z\right] = C_2\psi_{\frac{\Phi}{2}}(z). \end{aligned} \tag{1.129}$$

The constants C_1 and C_2 are found to be $C_1 = \psi_\Phi^2(\pi/2)$, $C_2 = \psi_\Phi^2(\Phi)$ by setting $z = 0$ in (1.129) and by use of the evenness of $\psi_\Phi(z)$. Thus,

$$\psi_\Phi\left(z+\frac{\pi}{2}\right)\psi_\Phi\left(z-\frac{\pi}{2}\right) = \psi_\Phi^2\left(\frac{\pi}{2}\right)\cos\left(\frac{\pi}{4\Phi}z\right), \tag{1.130}$$

$$\psi_\Phi(z+\Phi)\psi_\Phi(z-\Phi) = \psi_\Phi^2(\Phi)\psi_{\frac{\Phi}{2}}(z). \tag{1.131}$$

Now we discuss the asymptotic (as $|\mathrm{Im}\, z| \to \infty$) behavior of $\psi_\Phi(z)$. Let us rewrite (1.128) as

$$\begin{aligned}\psi_\Phi(z) &= \exp\left(-\frac{1}{4}\int_{\mathbb{R}} \frac{\cosh z\zeta - 1}{\zeta\cosh\frac{1}{2}\pi\zeta\sinh 2\Phi\zeta}\mathrm{d}\zeta\right)\\ &= \exp\left[-\frac{1}{4}\mathrm{VP}\int_{\mathbb{R}} \frac{\exp(z\zeta) - 1}{\zeta\cosh\frac{1}{2}\pi\zeta\sinh 2\Phi\zeta}\mathrm{d}\zeta\right].\end{aligned} \tag{1.132}$$

Consider the case $\mathrm{Im}\, z \to +\infty$ and shift the contour of integration up from the real axis:

$$-\frac{1}{4}\mathrm{VP}\int_{\mathbb{R}} \frac{\exp(z\zeta) - 1}{\zeta\cosh\frac{1}{2}\pi\zeta\sinh 2\Phi\zeta}\mathrm{d}\zeta = -\mathrm{i}\frac{\pi}{8\Phi}z - \frac{1}{4}\int_{\mathrm{i}\sigma+\mathbb{R}} \frac{\exp(z\zeta) - 1}{\zeta\cosh\frac{1}{2}\pi\zeta\sinh 2\Phi\zeta}\mathrm{d}\zeta, \tag{1.133}$$

where $0 < \sigma < \min\{1, \pi/2\Phi\}$.[16] The integral in the right-hand side of (1.133) can be, in turn, split as

$$\frac{1}{4}\int_{\mathrm{i}\sigma+\mathbb{R}} \frac{\exp(z\zeta)}{\zeta\cosh\frac{1}{2}\pi\zeta\sinh 2\Phi\zeta}\mathrm{d}\zeta - \frac{1}{4}\int_{\mathrm{i}\sigma+\mathbb{R}} \frac{1}{\zeta\cosh\frac{1}{2}\pi\zeta\sinh 2\Phi\zeta}\mathrm{d}\zeta. \tag{1.134}$$

The first term in (1.134) is of order $\mathrm{O}(\exp[\mathrm{i}z\min\{1, \pi/2\Phi\}])$ as $\mathrm{Im}\, z \to +\infty$ that can be argued by further shifting up the contour of integration. At the same time, the second term (1.134) does not depend on z. By collecting (1.132)–(1.134) one arrives at

$$\psi_\Phi(z) \underset{\mathrm{Im}\, z\to+\infty}{=} C(\Phi)\exp\{-\mathrm{i}\pi z/(8\Phi) + \mathrm{O}(\exp[\mathrm{i}z\min\{1, \pi/2\Phi\}])\}, \tag{1.135}$$

where

$$C(\Phi) = \exp\left(\frac{1}{4}\int_{\mathrm{i}\sigma+\mathbb{R}} \frac{1}{\zeta\cosh\frac{1}{2}\pi\zeta\sinh 2\Phi\zeta}\mathrm{d}\zeta\right).$$

This constant can be found independently, by taking the limit $\mathrm{Im}\, z \to +\infty$ in (1.130) with the use of (1.135). This yields $C(\Phi) = 2^{-1/2}\psi_\Phi(\pi/2)$.

The examination of the case $\mathrm{Im}\, z \to -\infty$ is quite analogous. Finally

$$\psi_\Phi(z) \underset{z\to\pm\mathrm{i}\infty}{=} \frac{1}{\sqrt{2}}\psi_\Phi\left(\frac{\pi}{2}\right)\exp\left(\mp\mathrm{i}\frac{\pi}{8\Phi}z\right)\left[1 + \mathrm{O}\left(\mathrm{e}^{\pm\mathrm{i}z\min\{1,\pi/2\Phi\}}\right)\right]. \tag{1.136}$$

---

[16] $\zeta = \mathrm{i}\min\{1, \pi/2\Phi\}$ is the pole of the integrand in the upper half-plane that is closest to the real axis.

1.5.6 Completion of the construction of $\Psi_0(z)$ and of $s(z)$

In view of (1.114)–(1.116) and (1.117) and (1.118) the solution $s_1(z)$ to equations (1.110) is expressed in terms of the Malyuzhinets function

$$s_1(z) = \Psi(z, \theta_+) \equiv \psi_\Phi(z + \Phi + \pi/2 - \theta_+)\psi_\Phi(z + \Phi - \pi/2 + \theta_+) \tag{1.137}$$

($\Psi(z, \theta_+)$ is the convenient notation of the product of the Malyuzhinets functions). One can solve equations (1.111) in a similar way and obtain

$$s_2(z) = \Psi(z - 2\Phi, \theta_-) = \psi_\Phi(z - \Phi + \pi/2 - \theta_-)\psi_\Phi(z - \Phi - \pi/2 + \theta_-). \tag{1.138}$$

Thus, in view of the multiplication principle (1.108)–(1.109) the function

$$\Psi_0(z) \equiv \Psi(z, \theta_+)\,\Psi(z - 2\Phi, \theta_-) \tag{1.139}$$

solves equations (1.104).[17]

Recall that we seek the solution $s(z)$ of (1.104), which behaves as

$$s(z) = s(\pm \mathrm{i}\infty) + \mathrm{O}\left(\exp\left(-\delta|\mathrm{Im}\, z|\right)\right), \quad \delta > 0 \tag{1.140}$$

as $\mathrm{Im}\, z \to \pm\infty$ (see (1.105)) and is regular inside the basic strip $|\mathrm{Re}\, z| < \Phi$ except the prescribed simple pole at the point $z = \varphi_0$ with the unit residue. The expression (1.139) does not meet these requirements. However, the expression (compare with (1.106))

$$s(z) = \frac{\Psi_0(z)}{\Psi_0(\varphi_0)}\tilde{s}(z), \tag{1.141}$$

does. Here

$$\tilde{s}(z) = \frac{\mu \cos \mu\varphi_0}{\sin \mu z - \sin \mu\varphi_0}, \quad \mu = \frac{\pi}{2\Phi} \tag{1.142}$$

is the solution to the simpler system (1.107). In view of (1.136) the growth of $\Psi_0(z)$ is exactly compensated by the decreasing of $\tilde{s}(z)$ and $\tilde{s}(z)$ has the simple pole $z = \varphi_0$ with the unit residue.[18]

This completes the construction of the sought-for function $s(z)$. Finally, the solution to the Malyuzhinets problem takes the form

$$u(kr, \varphi) = \frac{1}{2\pi \mathrm{i}} \int_\gamma \frac{\Psi_0(z + \varphi)}{\Psi_0(\varphi_0)} \frac{\mu \cos \mu\varphi_0}{\sin \mu(z + \varphi) - \sin \mu\varphi_0} \mathrm{e}^{-ikr\cos z} \mathrm{d}z. \tag{1.143}$$

---

[17]For the particular case $\theta_+ = \theta_-$ the expression for $\Psi_0(z)$ can be simplified by means of (1.131):

$$\Psi_0(z)|_{\theta_\pm = \theta} = \psi_\Phi^4(\Phi)\psi_{\Phi/2}\left(z + \pi/2 - \theta\right)\psi_{\Phi/2}\left(z - \pi/2 + \theta\right).$$

[18]Recall that we assumed the inequalities $0 < \mathrm{Re}\,\theta_\pm \le \pi/2$. The singularities $z = -\Phi - \theta_+$ and $z = \Phi - \theta_-$ of the function $\Psi_0(z)$ lie on the boundary of the basic strip if $\mathrm{Re}\,\theta_\pm = 0$. This can also be seen by shifting the argument by $\pm\Phi$ in equations (1.104).

Formula (1.143) is mainly used to compute numerically the near field, that is, the solution for $kr \lesssim 1$. Note that in the case of impedance boundary conditions, the solution behaves as

$$u(kr, \varphi) \underset{kr \to 0}{=} \text{const} + \mathrm{O}\left((kr)^{\delta}\right), \quad \delta > 0$$

with the only exception $\Phi = \pi/2$, $\sin\theta_+ \neq \sin\theta_-$ when the correction term is of the order $\mathrm{O}(kr \ln kr)$.

1.5.7 Far-field analysis of the exact solution

Of primary importance in the practice is the far-field behavior of diffracted waves. To carry out a far-field analysis of the exact solution means to evaluate the Sommerfeld integral (1.143) for $kr \to \infty$. To this end, we deform the Sommerfeld contour into the steepest-descent paths $\mathrm{S}(-\pi)$ and $\mathrm{S}(\pi)$. Thereby several poles of the spectrum $s(z)$ may be crossed (see Fig. 1.5). Herein we describe briefly the contributions of these poles and give their physical interpretations. To keep the discussion concise, we confine ourselves to the case $\pi/2 \leq \Phi \leq \pi$.

We start with the (with respect to φ) nonuniform asymptotic formula for the integral (1.143):

$$u(kr, \varphi) \underset{kr \to \infty}{\sim} \sum_n u_n + u_{\mathrm{e}}(kr, \varphi) + u_+ + u_-. \tag{1.144}$$

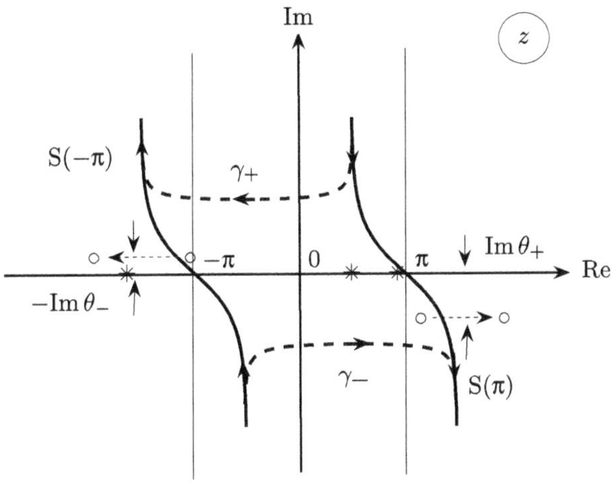

Figure 1.5 The Sommerfeld contour $\gamma_+ + \gamma_-$ (dashed), the steepest-descent contours $\mathrm{S}(\pm\pi)$ (solid), and the possible positions of poles of the integrand. The asterisks show (from left to right) the poles $(-2\Phi - \varphi_0) - \varphi$, $\varphi_0 - \varphi$, $(2\Phi - \varphi_0) - \varphi$ of $s(z + \varphi)$ that correspond to the zeros of the denominator $(\sin\mu(z + \varphi) - \sin\mu\varphi_0)$ (such a position of poles occurs if only the face $\varphi = \Phi$ is illuminated by the incident wave). The circles show the poles $\pm(\pi + \Phi + \theta_\pm) - \varphi$ of the integrand due to the Malyuzhinets function; the dotted arrows point the directions of movement of these poles when $\Phi - \varphi$ increases (right arrow) or $\Phi + \varphi$ increases (left arrow).

In the latter formula

$$u_e(kr,\varphi) = \widetilde{\mathbb{S}}(\varphi,\varphi_0)\frac{e^{ikr+i\pi/4}}{\sqrt{2\pi kr}}\left[1+O\left(\frac{1}{kr}\right)\right], \tag{1.145}$$

$$\widetilde{\mathbb{S}}(\varphi,\varphi_0) := s(\varphi-\pi) - s(\varphi+\pi), \quad s(z) = \frac{\Psi_0(z)}{\Psi_0(\varphi_0)}\frac{\mu\cos\mu\varphi_0}{\sin\mu z - \sin\mu\varphi_0}$$

describes a cylindrical wave (edge wave) outgoing from the edge of the wedge.[19]

The summands u_n describe the incident ($n = 0$) and reflected ($n \neq 0$) waves. The number of such waves is finite. It contains two or three terms provided $\pi/2 \leq \Phi \leq \pi$; the sum may also contain other terms corresponding to the waves multiply reflected from the faces if $0 \leq \Phi < \pi/2$. These terms have the form

$$\begin{gathered} u_n = a_r^n \exp\{-ikr\cos[\varphi - (-1)^n\varphi_0 - 2\Phi n]\}, \\ \text{if } |\varphi - 2\Phi n - (-1)^n\varphi_0| \lessapprox \pi, \quad n = 0, \pm 1, \ldots, \\ u_n = 0 \text{ otherwise,} \end{gathered} \tag{1.146}$$

where

$$a_r^n = (-1)^n \frac{\Psi_0\left((-1)^n\varphi_0 + 2\Phi n\right)}{\Psi_0(\varphi_0)}$$

are the reflection coefficients (except $n = 0$). Recall that $\Psi_0(z)$ solves the Malyuzhinets equations (1.104). Applying these equations for $z = \varphi_0 \mp \Phi$ one obtains

$$a_r^{\pm 1} = \frac{\sin(\Phi \mp \varphi_0) - \sin\theta_\pm}{\sin(\Phi \mp \varphi_0) + \sin\theta_\pm}. \tag{1.147}$$

The terms $u_\pm$ represent plane waves with complex phase functions that, as $\operatorname{Re}\theta_\pm > 0$, vanish exponentially with increasing of kr (the "surface waves"). These terms are computed by means of

$$u_\pm = \begin{cases} c_\pm \exp[ikr\cos(\Phi + \theta_\pm \mp \varphi)], & 0 \leq \Phi \mp \varphi < -\operatorname{gd}(\operatorname{Im}\theta_\pm) - \operatorname{Re}\theta_\pm, \\ 0 & \text{otherwise.} \end{cases} \tag{1.148}$$

where

$$\operatorname{gd}(x) = \operatorname{sgn}(x)\arccos(1/\cosh x) = 2\arctan(\exp x) - \pi/2 \tag{1.149}$$

[19]Note that the nonuniform diffraction coefficient $\widetilde{\mathbb{S}}(\varphi,\varphi_0)$ vanishes at the wedge's faces, $\widetilde{\mathbb{S}}(\pm\Phi) = 0$ (except the case of the Neumann boundary conditions, which is an elementary consequence of (1.104)). Thus, at the wedge's faces the correction term in (1.145) becomes the leading one.

Table 1.1 Poles and residues of $s(z+\varphi)$

n	Pole z_n	Residue R_n	Argument A_n
-1	$-2\Phi-\varphi_0-\varphi$	a_r^{-1}	$\pi-2\Phi-\varphi_0-\varphi$
0	$\varphi_0-\varphi$	a_r^0	$\pi-\lvert\varphi_0-\varphi\rvert$
$+1$	$2\Phi-\varphi_0-\varphi$	a_r^{+1}	$\pi-2\Phi+\varphi_0+\varphi$
2	$\pi+\Phi+\theta_+-\varphi$	c_+	$-\Phi+\varphi-\operatorname{Re}\theta_+-\operatorname{gd}(\operatorname{Im}\theta_+)$
3	$-\pi-\Phi-\theta_--\varphi$	c_-	$-\Phi-\varphi-\operatorname{Re}\theta_--\operatorname{gd}(\operatorname{Im}\theta_-)$

is the Gudermann function. Here the excitation coefficients of surface waves $c_\pm$ are equal to

$$c_\pm=\operatorname{res}s(z)|_{z=\pm(\pi+\Phi+\theta_\pm)}=\frac{\tilde{s}\left[\pm(\pi+\Phi+\theta_\pm)\right]}{\Psi_0(\varphi_0)}\Psi\left[-\Phi\pm(\pi+\theta_\pm),\theta_\mp\right] \tag{1.150}$$

$$\times\,\psi_\Phi\left(\pi/2+2\Phi+2\theta_\pm\right)\psi_\Phi\left(3\pi/2-2\Phi\right),$$ [20]

where $\tilde{s}(z)$ and $\Psi(z,\theta)$ are defined by (1.142) and (1.137), respectively.

Remark: Note that geometrical-optics (GO) and edge waves contribute to the far-field asymptotics (1.144) in power orders as $kr\to\infty$, whereas contributions of surface waves are exponentially small. However, retaining of these exponentially small terms has a physical sense since surface waves differ in phase from the GO parts and hence can be observed separately.

The notation $A \lesseqgtr B$ used in (1.146) implies that the nonuniform formula (1.144) fails in the vicinities of the directions φ for which $A\approx B$. A uniform asymptotic formula can be written with the aid of the uniform geometrical theory of diffraction (UTD) transition function $\mathrm{F_{KP}}(x)$, a variant of the Fresnel integral. Such a uniform expression reads (cf. [1.32])

$$u(kr,\varphi)\sim\sum_n u_n+\mathbb{S}(\varphi,\varphi_0)\frac{e^{ikr+i\pi/4}}{\sqrt{2\pi kr}}+u_++u_-. \tag{1.151}$$

The nonuniform diffraction coefficient $\widetilde{\mathbb{S}}$ in (1.145) is replaced with the following uniform one:

$$\mathbb{S}(\varphi,\varphi_0)=s(\varphi-\pi)-s(\varphi+\pi)-\sqrt{\frac{i}{2}}\sum_{n=-1}^{3}R_n\frac{1-\mathrm{F_{KP}}\left(ikrs_n^2\right)}{s_n} \tag{1.152}$$

with $s_n=-\sqrt{2i}\cos(\tilde{z}_n/2)$ where

$$\tilde{z}_n=\begin{cases} z_n & |\operatorname{Re}z_n|\le 2\pi \\ 2\pi\operatorname{sgn}(\operatorname{Re}z_n)+i\operatorname{Im}z_n, & \text{otherwise}\end{cases}$$

The poles of the spectral function $s(z+\varphi)$ and their residues are given in Table 1.1.

[20] Here we take into account the evenness of $\psi_\Phi(z)$ and the fact that, in view of (1.117),

$$\operatorname{res}|_{z=\pm(3\pi/2+2\Phi)}\psi_\Phi(z)=\psi_\Phi\left(3\pi/2-2\Phi\right).$$

The UTD transition function $\mathrm{F_{KP}}(z^2)$, introduced in [1.33], is defined as

$$\mathrm{F_{KP}}(z^2) = \pm 2\mathrm{i}z\mathrm{e}^{\mathrm{i}z^2}\int_{\pm z}^{\infty}\mathrm{e}^{-\mathrm{i}t^2}\mathrm{d}t, \quad \text{the lower sign for } \frac{\pi}{4} < \arg z < \frac{5\pi}{4}, \tag{1.153}$$

with the asymptotic expressions

$$\mathrm{F_{KP}}(z) \sim \begin{cases} \left(\sqrt{\pi z} - 2z\,\mathrm{e}^{\mathrm{i}\pi/4} - \dfrac{2}{3}z^2\mathrm{e}^{-\mathrm{i}\pi/4}\right)\mathrm{e}^{\mathrm{i}(\pi/4+z)}, & |z| \ll 1 \\ 1 + \mathrm{i}\dfrac{1}{2z} - \dfrac{3}{4}\dfrac{1}{z^2} - \mathrm{i}\dfrac{15}{8}\dfrac{1}{z^3} + \dfrac{75}{16}\dfrac{1}{z^4}, & |z| \gg 1 \end{cases} \tag{1.154}$$

Remark: Generally, the poles $\pm(\pi + \Phi + \theta_\pm) - \varphi$ never coincide with the saddle points $\pm\pi$ since $|\varphi| < \Phi$, and the inequalities

$$0 < \operatorname{Re}\theta_\pm < \pi/2, \quad \operatorname{Im}\theta_\pm < 0$$

are assumed. At the same time, the surface waves can noticeably contribute to the total far field only for moderate values of kr. For such kr the numerical values of $\Phi \mp \varphi + \theta_\pm$ become relevant. If these values are not sufficiently small, the corresponding UTD transition function (1.153) is used for accurate computations of the surface waves. It is obvious that, replacing all UTD transition functions by their large-argument expression (1.154), (1.151) and (1.152) reduce to (1.144) and (1.145).

1.6 Theory of Malyuzhinets functional equations for one unknown function

In this section we consider the more general Malyuzhinets-type equations. In this way, we follow basically the paper by A. A. Tuzhilin [1.34] in which Malyuzhinets's ideas were developed. The applied technique is based on the modified Fourier transform (the so-called $\mathcal{S}$-integrals). This technique is extensively exploited in subsequent chapters, where we derive and study functional equations for two unknown functions or functional equations of higher order.

1.6.1 General Malyuzhinets equations

We discuss here the theory of Malyuzhinets equations of more general form than (1.104). Such equations arise for diffraction problems in angular domains with more complicated boundary conditions.

Consider the following system of the functional (Malyuzhinets-type) equations:

$$\begin{cases} \displaystyle\prod_{q=1}^{n}\left[\ \sin z - (-1)^q \sin\theta_q^1\right] s(\ z + \Phi) \\ \displaystyle\quad - \prod_{q=1}^{n}\left[-\sin z - (-1)^q \sin\theta_q^1\right] s(-z + \Phi) = \sin z \sum_{p=0}^{n-1} a_p \cos^p z, \\ \displaystyle\prod_{q=1}^{m}\left[-\sin z - (-1)^q \sin\theta_q^2\right] s(\ z - \Phi) \\ \displaystyle\quad - \prod_{q=1}^{m}\left[\ \sin z - (-1)^q \sin\theta_q^2\right] s(-z - \Phi) = \sin z \sum_{p=0}^{m-1} b_p \cos^p z, \end{cases} \tag{1.155}$$

where $0 < \Phi \le \pi$, $a_p, b_p, \theta_q^{1,2}$ ($0 < \operatorname{Re}\theta_q^{1,2} \le \pi/2$) are complex constants. (Equations (1.104) are obviously included as a particular case.)

We construct solutions to the general equations (1.155) in several steps. First, we turn to the construction of the solution $s(z)$ of the homogeneous equations (1.155). We call the meromorphic function $s(z)$ the basic solution of the homogeneous equations provided it has neither zeros nor poles in the basic strip

$$\Pi(-\Phi, \Phi) = \{z : |\operatorname{Re} z| \le \Phi\}.$$

It follows from the functional equations that all the singularities of the basic solution are located in some strip parallel to the real axis. Generally, the existence of a finite limit of the basic solution as $\operatorname{Im} z \to \pm\infty$ is not assumed. Obviously the basic solution is not unique.

We apply the multiplication principle (1.108)–(1.109) to construct the basic solution to equations (1.155). Consider the following auxiliary systems of equations:

$$\begin{cases} (\sin z + \sin\theta)s_1(z+\Phi) = (-\sin z + \sin\theta)s_1(-z+\Phi) \\ s_1(z-\Phi) = s_1(-z-\Phi), \end{cases}$$
$$\begin{cases} (\sin z - \sin\theta)s_2(z+\Phi) = (-\sin z - \sin\theta)s_2(-z+\Phi) \\ s_2(z-\Phi) = s_2(-z-\Phi), \end{cases}$$
$$\begin{cases} s_3(z+\Phi) = s_3(-z+\Phi) \\ (\sin z - \sin\theta)s_3(z-\Phi) = (-\sin z - \sin\theta)s_3(-z-\Phi), \end{cases} \qquad (1.156)$$
$$\begin{cases} s_4(z+\Phi) = s_4(-z+\Phi) \\ (\sin z + \sin\theta)s_4(z-\Phi) = (-\sin z + \sin\theta)s_4(-z-\Phi). \end{cases}$$

Provided the solutions to these four systems are known, the solution to the homogeneous system (1.155) can be obtained with the aid of the multiplication principle.

1.6.2 Solution to the homogeneous Malyuzhinets equations

It has already been shown in Section 1.5.7 that the basic solution to the first among the simplest homogeneous systems of (1.156) can be expressed in terms of the Malyuzhinets function and is given by the formula

$$s_1(z) \equiv \Psi(z,\theta) = \psi_\Phi(z+\Phi+\pi/2-\theta)\psi_\Phi(z+\Phi-\pi/2+\theta). \qquad (1.157)$$

Analogously the solutions to the other systems from (1.156) are correspondingly the functions

$$s_2(z) \equiv 1/\Psi(z,\theta), \quad s_3(z) \equiv \Psi(z-2\Phi,\theta), \quad s_4(z) \equiv 1/\Psi(z-2\Phi,\theta). \qquad (1.158)$$

By the multiplication principle, the obtained basic solutions (1.157) and (1.158) of the systems (1.156) enable us to determine a basic solution to the system (1.155) in the form

$$\Psi_0(z) = \frac{\prod\limits_{\nu=0}^{[(n-1)/2]} \Psi\left(z, \theta_{2\nu+1}^1\right) \prod\limits_{\nu=0}^{[(m-1)/2]} \Psi\left(z-2\Phi, \theta_{2\nu+1}^2\right)}{\prod\limits_{\nu=1}^{[n/2]} \Psi\left(z, \theta_{2\nu}^1\right) \prod\limits_{\nu=1}^{[m/2]} \Psi\left(z-2\Phi, \theta_{2\nu}^2\right)}, \qquad (1.159)$$

where $[p]$ is the integral part of p, and by definition we assume that $\prod_{\nu=q}^{p}(\cdots) \equiv 1$ for $p < q$.

1.6.3 Solution to the inhomogeneous Malyuzhinets equations

Using the solution (1.159), one can simplify the system (1.155), reducing it to a system with constant coefficients. To this end, we substitute

$$s(z) = \Psi_0(z)\,\sigma(z), \tag{1.160}$$

into (1.155), which leads to the following system with constant coefficients:

$$\begin{cases} \sigma(z+\Phi) - \sigma(-z+\Phi) = h_1(z), \\ \sigma(z-\Phi) - \sigma(-z-\Phi) = h_2(z). \end{cases} \tag{1.161}$$

The right-hand sides in (1.161) are given by the formulas

$$h_1(z) = \frac{\sin z \sum\limits_{p=0}^{n-1} a_p \cos^p z}{\prod\limits_{\nu=1}^{n} \left[\sin z \;-\; (-1)^\nu \,\sin\theta_\nu^1\right] \Psi_0(z+\Phi)}, \tag{1.162}$$

$$h_2(z) = \frac{\sin z \sum\limits_{p=0}^{m-1} b_p \cos^p z}{\prod\limits_{\nu=1}^{m} \left[-\sin z \;-\; (-1)^\nu \,\sin\theta_\nu^2\right] \Psi_0(z-\Phi)}. \tag{1.163}$$

Clearly the solution to the system (1.161) can be represented in the form

$$\sigma(z) = s_0(z) + s_*(z), \tag{1.164}$$

where $s_0(z)$ is an arbitrary solution to the homogeneous and $s_*(z)$ is the particular one to the inhomogeneous system (1.161).

It is not difficult to obtain the solution $s_0(z)$ of the homogeneous system.[21] Hence, a further problem is constructing the solution $s_*(z)$ to the inhomogeneous system (1.161) in a closed form, satisfying the condition that the sum (1.164) belongs to the desired class of functions.

Solving the inhomogeneous equations is possible in different ways. The most natural approach is based on the modified Fourier transform; see Section 1.4.1. Being formally simple, this approach needs justification that is connected with the appropriate understanding of the convergence of the Fourier type integrals. On the other hand, the Fourier transform leads naturally to the so-called $\mathcal{S}$-integrals,[22] the direct use of which also enables us to justify the solution procedure. In the following subsections we discuss the rationale behind the introduction of $\mathcal{S}$-integrals and their use for solution to the inhomogeneous Malyuzhinets equations.

---

[21] In particular, any rational function of $\sin\mu z$, $\cos\mu z$ can be taken.

[22] Tuzhilin introduced the terminology [1.34].

1.6.4 Modified Fourier transform and S-integrals

The calculations in the present subsection are formal and serve mainly as the motivation to the $\mathcal{S}$-integrals. Let us first examine a special case of the system (1.161), namely,

$$\begin{cases} s_1(z+\Phi) - s_1(-z+\Phi) = h_1(z), \\ s_1(z-\Phi) - s_1(-z-\Phi) = 0, \end{cases} \tag{1.165}$$

with an odd meromorphic right-hand side $h_1(z)$. Extension to the general case of (1.161) is discussed later on.

The modified Fourier transform of the function $s(z)$ is described by the formulas (see also Section 1.4.1)

$$\tilde{s}(t) = \int_{\mathrm{i}\mathbb{R}} s(z)\,\mathrm{e}^{\mathrm{i}zt}\,\mathrm{d}z, \quad s(z) = -\frac{1}{2\pi}\mathrm{VP}\int_{\mathrm{i}\mathbb{R}} \tilde{s}(t)\,\mathrm{e}^{-\mathrm{i}zt}\,\mathrm{d}t. \tag{1.166}$$

Applying the transform (1.166) to the system (1.165) gives

$$\begin{cases} \tilde{s}_1(t)\,\mathrm{e}^{-\mathrm{i}\Phi t} - \tilde{s}_1(-t)\,\mathrm{e}^{\,\mathrm{i}\Phi t} = \tilde{h}_1(t), \\ \tilde{s}_1(t)\,\mathrm{e}^{\,\mathrm{i}\Phi t} - \tilde{s}_1(-t)\,\mathrm{e}^{-\mathrm{i}\Phi t} = 0, \end{cases}$$

and therefore

$$\tilde{s}_1(t) = \frac{\mathrm{i}}{2}\,\frac{\mathrm{e}^{-\mathrm{i}\Phi t}}{\sin 2\Phi t}\,\tilde{h}_1(t).$$

Substitution of this expression into the formula for the inverse Fourier transform yields

$$\begin{aligned} s_1(z) &= -\frac{1}{2\pi}\mathrm{VP}\int_{\mathrm{i}\mathbb{R}} \tilde{s}_1(t)\,\mathrm{e}^{-\mathrm{i}zt}\,\mathrm{d}t \\ &= -\frac{1}{2\pi}\mathrm{VP}\int_{\mathrm{i}\mathbb{R}} \mathrm{e}^{-izt}\left[\frac{\mathrm{i}\mathrm{e}^{-\mathrm{i}\Phi t}}{2\sin 2\Phi t}\int_{\mathrm{i}\mathbb{R}} h_1(\zeta)\,\mathrm{e}^{\mathrm{i}\zeta t}\,\mathrm{d}\zeta\right]\mathrm{d}t \\ &= -\frac{1}{2\pi}\int_{\mathrm{i}\mathbb{R}} h_1(\zeta)\left[\mathrm{VP}\int_{\mathrm{i}\mathbb{R}} \mathrm{e}^{-\mathrm{i}(z-\zeta)t}\,\frac{\mathrm{i}\mathrm{e}^{-\mathrm{i}\Phi t}}{2\sin 2\Phi t}\,\mathrm{d}t\right]\mathrm{d}\zeta \\ &= -\frac{\mathrm{i}\mu}{4\pi}\int_{\mathrm{i}\mathbb{R}} h_1(\zeta)\,\tan\frac{\mu(z-\zeta+\Phi)}{2}\,\mathrm{d}\zeta, \end{aligned}$$

where $\mu = \pi/(2\Phi)$. We have exploited the known equality

$$\int_{\mathrm{i}\mathbb{R}} \frac{\sin at}{2\sin(2\Phi t)}\,\mathrm{d}t = \mathrm{i}\frac{1}{2}\mu\tan\frac{1}{2}a\mu;$$

see [1.24, 3.511:2]. Finally, using the oddness of $h_1(z)$ and the identity

$$\tan\frac{\mu}{2}(a+b) - \tan\frac{\mu}{2}(a-b) = \frac{2\sin\mu b}{\cos\mu b + \cos\mu a}$$

we obtain the solution $s_1(z)$ that we seek in the form

$$s_1(z) = \frac{\mathrm{i}}{8\Phi} \int\limits_{\mathrm{i}\mathbb{R}} \frac{h_1(t)\sin\mu t}{\cos\mu t - \sin\mu z} \mathrm{d}t. \tag{1.167}$$

The integrals (1.167) are called, following Tuzhilin, $\mathcal{S}$-integrals.

1.6.5 The direct application of S-integrals

As has been expounded in the previous subsectoin the solution to the inhomogeneous Malyuzhinets equations can be expressed in the form of $\mathcal{S}$-integrals

$$s(z) = \frac{\mathrm{i}}{8\,\Phi} \int\limits_{\mathrm{i}\mathbb{R}} \mathrm{d}\tau \frac{F(\tau)}{\cos\mu\tau - \sin\mu z}, \qquad \mu = \frac{\pi}{2\Phi}, \tag{1.168}$$

where $F(\tau)$ is an even meromorphic function in the whole complex τ-plane, is regular in the vicinity of $\mathrm{i}\mathbb{R} := (-\mathrm{i}\infty, +\mathrm{i}\infty)$ and

$$F(\tau) \sim \mathrm{O}(\exp(\delta\,|\mathrm{Im}\,\tau|)), \quad \delta < \mu, \quad |\mathrm{Im}\,\tau| \to \infty. \tag{1.169}$$

The condition (1.169) ensures the uniform and absolute convergence of the integral (1.168) in any strip

$$(4n + 1)\,\Phi < \mathrm{Re}\,z < (4n + 5)\,\Phi, \quad n = 0, \pm 1, \ldots \tag{1.170}$$

and, hence, $s(z)$ is regular in any strip (1.170); that is, $s(z)$ is a piecewise regular, 4Φ-periodic function. Note that the regularity of the integrand in (1.168) is violated at the points where the denominator is zero and where they are exactly located on the boundaries of the strips (1.170).

In the particular case $F(\tau) \equiv 1$ in any strip (1.170) one has

$$\frac{\mathrm{i}}{8\Phi} \int\limits_{\mathrm{i}\mathbb{R}} \frac{\mathrm{d}\tau}{\cos\mu\tau - \sin\mu z} = \left(n + \frac{3}{4} - \frac{z}{4\Phi}\right) \frac{1}{\cos\mu z}. \tag{1.171}$$

(The latter equality is verified with the aid of change of the integration variable $\zeta = \exp(-\mathrm{i}\mu\tau)$, which leads to the integral of a rational function to be computed straightforward.)

We study the analytic continuation of $s(z)$ from any strip of regularity onto the whole complex plane. It is sufficient to consider the strip $-3\Phi < \mathrm{Re}\,z < \Phi$. Using (1.171), one can write

$$s(z) = \frac{\mathrm{i}}{8\,\Phi} \int\limits_{\mathrm{i}\mathbb{R}} \frac{F(\tau) - F(z - \Phi)}{\cos\mu\tau - \sin\mu z} \mathrm{d}\tau - \frac{z + \Phi}{4\Phi\,\cos\mu z} F(z - \Phi). \tag{1.172}$$

The second term in (1.172) is a meromorphic function on the whole z-plane. It is obvious that for any imaginary τ the integrand in the first summand in (1.172) can have singularities at the poles of $F(z - \Phi)$ but has no singularities on the line $\mathrm{Re}\,z = \Phi$, that is, the points $z = \Phi + \tau$ are regular points. This means that the integral in (1.172) specifies a meromorphic function in the strip

$$-3\,\Phi < \mathrm{Re}\,z < 5\,\Phi. \tag{1.173}$$

Therefore, the expression (1.172) provides the meromorphic continuation of $s(z)$ into the strip (1.173).

The expression (1.172) enables us to write

$$\begin{aligned} s(z+\Phi) - s(-z+\Phi) &= F(z)/\sin\mu z, \\ s(z-\Phi) - s(-z-\Phi) &= \end{aligned} \tag{1.174}$$

$$= -\frac{\mathrm{i}}{8\Phi}\int\limits_{\mathrm{i}\mathbb{R}} \frac{F(z-2\Phi)-F(-z-2\Phi)}{\cos\mu\tau+\cos\mu z}\mathrm{d}\tau - z\frac{F(z-2\Phi)-F(-z-2\Phi)}{4\Phi\,\sin\mu z} = 0.$$

Thus, the meromorphic continuation (1.172) fulfills equations (1.174).

Now we turn to solution to the inhomogeneous equation (1.161). Obviously, provided $s_1(z)$ and $s_2(z)$ are correspondingly the solutions to the systems

$$\begin{cases} s_1(z+\Phi) - s_1(-z+\Phi) = h_1(z), \\ s_1(z-\Phi) - s_1(-z-\Phi) = 0, \end{cases} \tag{1.175}$$

and

$$\begin{cases} s_2(z+\Phi) - s_2(-z+\Phi) = h_2(z), \\ s_2(z-\Phi) - s_2(-z-\Phi) = 0, \end{cases} \tag{1.176}$$

the function

$$s_*(z) = s_1(z) - s_2(z-2\Phi) \tag{1.177}$$

satisfies the system (1.161). Therefore, problem (1.161) is reduced to solution to the particular systems (1.175) and (1.176). As shown already, these solutions have the form of $\mathcal{S}$-integrals

$$s_{1,2}(z) = \frac{\mathrm{i}}{8\Phi}\int\limits_{\mathrm{i}\mathbb{R}} h_{1,2}(\tau)\frac{\sin\mu\tau}{\cos\mu\tau - \sin\mu z}\mathrm{d}\tau.$$

Substituting these integrals into (1.177), we obtain[23]

$$s_*(z) = \frac{\mathrm{i}}{8\Phi}\int\limits_{\mathrm{i}\mathbb{R}} h_1(\tau)\frac{\sin\mu\tau}{\cos\mu\tau - \sin\mu z}\mathrm{d}\tau - \frac{\mathrm{i}}{8\Phi}\int\limits_{\mathrm{i}\mathbb{R}} h_2(\tau)\frac{\sin\mu\tau}{\cos\mu\tau + \sin\mu z}\mathrm{d}\tau. \tag{1.178}$$

The formula (1.178) is obtained under the assumption on the behavior (1.169) of the function $F(\tau)$, which means that $h_{1,2}(z)$ vanish exponentially as $|\mathrm{Im}\, z| \to \infty$. This restriction,

---

[23]Note that to obtain the meromorphic continuation of the second term in (1.178) instead of (1.172) the following identity must be used:

$$\frac{\mathrm{i}}{8\Phi}\int\limits_{\mathrm{i}\mathbb{R}} \mathrm{d}\tau\,\frac{1}{\cos\mu\tau + \sin\mu z} = -\left(n + \frac{1}{4} - \frac{z}{4\Phi}\right)\frac{1}{\cos\mu z},$$

which is valid in the strips $(4n-1)\Phi < \mathrm{Re}\, z < (4n+3)\Phi$, $n = 0, \pm 1, \ldots,$

however, can be got rid of. Suppose $h_{1,2}(z)$ in (1.175) and (1.176) grow as

$$h_{1,2}(z) \underset{|\mathrm{Im}\, z|\to\infty}{=} \mathrm{O}\left(\exp\left(\delta|\mathrm{Im}\, z|\right)\right), \quad \delta < m_{1,2}.$$

Then the solutions to (1.175) and (1.176) are to be looked for in the form

$$s_{1,2}(z) = \sin^{m_{1,2}}(\mu z)\hat{s}_{1,2}(z). \tag{1.179}$$

The system (1.175), for example, then reads

$$\begin{cases} \hat{s}_1(z+\Phi) - \hat{s}_1(-z+\Phi) = h_1(z) / \cos^{m_1} \mu z, \\ \hat{s}_1(z-\Phi) - \hat{s}_1(-z-\Phi) = 0, \end{cases}$$

with a decaying right-hand side and

$$\hat{s}_1(z) = \sin^{m_1} \mu z \frac{\mathrm{i}}{8\Phi} \int_{\mathrm{i}\mathbb{R}} \frac{h_1(\tau)}{\cos^{m_1} \mu\tau} \frac{\sin \mu\tau}{\cos \mu\tau - \sin \mu z} \mathrm{d}\tau. \tag{1.180}$$

Therefore we come to the following statement.

Theorem 1.4. *General solution to the functional Malyuzhinets equations (1.155) has the form*

$$s(z) = \Psi_0(z)\left[s_0(z) + s_*(z)\right], \tag{1.181}$$

where the functions $\Psi_0(z)$ and $s_*(z)$ are defined by the formulas (1.159) and (1.178), (1.162)–(1.163) (see also (1.180)) correspondingly and $s_0(z)$ is the solution to the simplest homogeneous Malyuzhinets equations (1.161) (i.e., as $h_{1,2} = 0$).

CHAPTER 2

Diffraction of a skew-incident plane electromagnetic wave by a wedge with axially anisotropic impedance faces

In this chapter we present an exact solution to diffraction of a skew-incident plane electromagnetic wave by a wedge with axially anisotropic impedance faces. Applying the Sommerfeld–Malyuzhinets technique to the boundary-value problem under study yields a coupled system of difference equations for the spectra. On elimination, a difference equation of higher order for one spectrum arises. After simplification in terms of a generalized Malyuzhinets function and accounting for Meixner's edge condition as well as the poles and residues of the spectrum in the basic strip of the complex plane, the functional difference equation is converted, via the $\mathcal{S}$-integrals, to an integral equivalent. For points on the imaginary axis which is inside the basic strip the integral equivalent becomes a Fredholm equation of the second kind with a non-singular, wavenumber-free and exponentially decreasing kernel. Solving this integral equation by the quadrature method the spectrum can be determined by integral extrapolation and by analytical continuation. A first-order uniform asymptotic solution follows from evaluating the Sommerfeld integrals with the saddle-point method. Comparison with available exact solutions in several special cases shows that this approach leads to a fast and accurate solution of the problem under study.

2.1 Introduction

Ever since the seminal paper of Sommerfeld [2.1], diffraction of electromagnetic waves in wedge-shaped regions remains the subject matter of many works and is often analyzed with the Sommerfeld–Malyuzhinets technique; see Chapter 1 and [2.2–2.7].

The key step of the Sommerfeld–Malyuzhinets technique lies in solving the resulting functional difference equations of higher order. Recently, it has been shown that a generalized Malyuzhinets function χ_{Φ} [2.3, 2.8] and the $\mathcal{S}$-integral [2.9] can be combined to convert a functional difference equation of the second order to a Fredholm integral equation of the second kind with the integral equation solved numerically by means of the quadrature method [2.10–2.12]. Furthermore, this approach has been extended to solve a functional difference equation of higher order resulting from diffraction of a skew-incident plane electromagnetic wave by a wedge with scalar impedance faces [2.13]. See also Chapter 11 of [2.7].

In this chapter we apply this approach, with due modification, to a slightly more general diffraction problem, namely, diffraction of a skew-incident plane electromagnetic wave by a wedge with axially anisotropic impedance faces. The extensions are direct and within the framework of the general procedure for the isotropic impedance faces.

Recently, this problem has been tackled using a probabilistic random-walk method [2.14], a generalized Wiener-Hopf technique [2.15], and the Sommerfeld–Malyuzhinets technique, but solving the resulting functional equation in a different way [2.16].

These three approaches deal with more general anisotropic impedance faces than the approach used in the present chapter. This notwithstanding, we do believe that for problems solvable using all four approaches, the present one is at least as efficient as the other three in terms of accuracy and computational speed (see Section 2.7), thanks to the nonsingular, wavenumber-free, and rapidly decreasing kernel of the resulting Fredholm integral equation obtained on use of, among other things, the special function χ_{Φ}. In addition, a closed-form explicit solution to the problem under study is to be found yet (for such solutions in special cases see for instance [2.17–2.19]); however, we believe that it is not possible for the traditional meaning of the term explicit solution. Hence, this solution procedure deserves to be included in this book. Furthermore, this chapter serves as the point of departure for the scattering problem to be tackled in Chapter 3.

The material contained in this chapter is partly based on our paper [2.20].

2.2 Statement of the problem and uniqueness

2.2.1 Statement of the problem

A wedge is placed in a circular cylindrical coordinate system (r, φ, z) in such a way that its edge coincides with the z-axis and its faces are the half-planes $\varphi = \pm\Phi$ with $\pi > \Phi > 0$ (Fig. 2.1). On the wedge's faces the following conditions must be met by the tangential components of the electric (E_z, E_r) and the magnetic $(H_z/Z_0, H_r/Z_0)$ fields

$$\begin{aligned} E_r(r, \pm\Phi, z) &= \mp a_{12}^{\pm} H_z(r, \pm\Phi, z), \\ E_z(r, \pm\Phi, z) &= \pm a_{21}^{\pm} H_r(r, \pm\Phi, z). \end{aligned} \tag{2.1}$$

Here H_z and H_r are the respective magnetic field components multiplied with Z_0, and Z_0 denotes the intrinsic impedance of the surrounding medium, $a_{12}^{\pm}$ and $a_{21}^{\pm}$ are the (in terms of Z_0) normalized axially anisotropic surface impedances. The passivity requires that $\operatorname{Re} a_{12}^{\pm} \geq 0$ and $\operatorname{Re} a_{21}^{\pm} \geq 0$ hold good. Obviously, a scalar impedance wedge with $a_{12}^{\pm} = a_{21}^{\pm}$ studied in [2.13] represents a special case of the present problem.

Let a plane electromagnetic wave fall on this wedge. The z-components of the incident field are given by [1]

$$\begin{aligned} &\left[H_z^{\text{inc}}(r, \varphi; z)\ E_z^{\text{inc}}(r, \varphi; z)\right]^{\mathrm{T}} = \overline{U}_0 \exp\left[-\mathrm{i}k'r\cos(\varphi - \varphi_0) + \mathrm{i}k''z\right], \\ &\overline{U}_0 = [U_{10}\ U_{20}]^{\mathrm{T}}, \quad k' = k\sin\theta_0, \quad k'' = k\cos\theta_0. \end{aligned} \tag{2.2}$$

[1] Remember that a time-dependence $\exp(-\mathrm{i}kct)$ has been assumed, but it is dropped in this book.

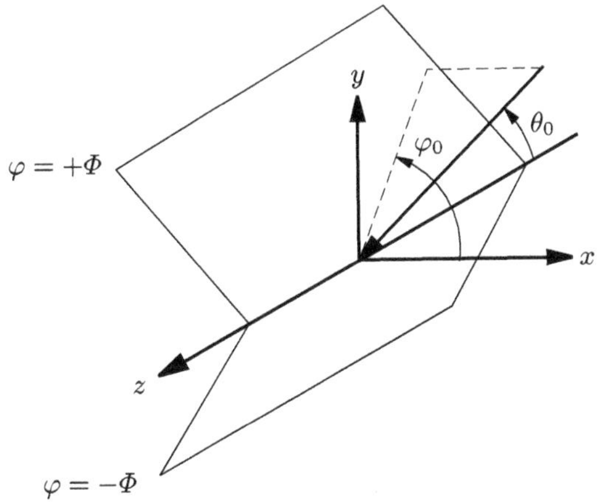

Figure 2.1 A wedge illuminated obliquely by a plane electromagnetic wave.

Here k stands for the wavenumber in the surrounding medium, in which a small loss has been assumed, that is, $k = k_0 + \mathrm{i}\kappa$, $\kappa > 0$. The angles θ_0 and φ_0 characterize the incident direction of the plane wave, and U_{10}/Z_0 and U_{20} denote the amplitude of the magnetic and electric components along the edge, respectively.

Therefore, the z-components of the total field take the form

$$\begin{aligned} &[H_z(r,\varphi;z)\ \ E_z(r,\varphi;z)]^{\mathrm{T}} = \overline{U}(r,\varphi)\exp(\mathrm{i}k''z), \\ &\overline{U}(r,\varphi) = [U_1(r,\varphi)\ \ U_2(r,\varphi)]^{\mathrm{T}}, \end{aligned} \tag{2.3}$$

where $\overline{U}(r,\varphi)$ meet outside the wedge of the two-dimensional Helmholtz equation

$$\left[\frac{1}{r}\frac{\partial}{\partial r}\left(r\frac{\partial}{\partial r}\right) + \frac{1}{r^2}\frac{\partial^2}{\partial\varphi^2} + (k')^2\right]\overline{U}(r,\varphi) = 0, \tag{2.4}$$

and on the wedge's faces following conditions, see (3) and (4) of [2.17, 2.18]:

$$\begin{aligned} \overline{\overline{I}}\cdot\frac{\mathrm{i}\partial\overline{U}(r,\pm\Phi)}{kr\partial\varphi} &= \mp\sin^2\theta_0\,\overline{\overline{\mathcal{A}}}^{\pm}\cdot\overline{U}(r,\pm\Phi) + \\ &\quad + \cos\theta_0\,\overline{\overline{\mathcal{B}}}\cdot\frac{\mathrm{i}\partial\overline{U}(r,\pm\Phi)}{k\partial r}, \end{aligned} \tag{2.5}$$

$$\overline{\overline{I}} = \begin{bmatrix} 1 & 0 \\ 0 & 1 \end{bmatrix}, \quad \overline{\overline{\mathcal{A}}}^{\pm} = \begin{bmatrix} a_{12}^{\pm} & 0 \\ 0 & 1/a_{21}^{\pm} \end{bmatrix}, \quad \overline{\overline{\mathcal{B}}} = \begin{bmatrix} 0 & -1 \\ 1 & 0 \end{bmatrix}. \tag{2.6}$$

As $r \to 0$, $\overline{U}(r,\varphi)$ remain finite (Meixner's edge condition; see Section 1.3.2):

$$\overline{U}(r,\varphi) = [C_1 + \mathrm{O}(r^{\delta})\ \ C_2 + \mathrm{O}(r^{\delta})]^{\mathrm{T}},\ \delta > 0. \tag{2.7}$$

Here C_1 and C_2 stand for two constants.

To complete the formulation of the problem, a radiation condition of the type

$$\left|\overline{U}(r,\varphi) - \overline{U}^{\mathrm{go}}(r,\varphi)\right| \le \exp(-a\kappa r), \quad a > 0 \tag{2.8}$$

has to be imposed on the unknown vector $\overline{U}(r,\varphi)$, where $\overline{U}^{\mathrm{go}}(r,\varphi)$ is the easily obtained ray-optical part of $\overline{U}(r,\varphi)$. See Section 2.6.

It is known that a classical solution to the present problem, if exists, is unique [2.17]; in addition, it can be shown that a limiting ($\kappa \to 0$) solution of the problem at hand for real-valued wavenumbers exists and is unique as well [2.13, 2.7].

2.2.2 On uniqueness of a solution

First, we assume that the ambient medium is slightly lossy. In that case the wavenumber has a small positive imaginary part, that is, $k = k_0 + \mathrm{i}\kappa$ with $\kappa > 0$. Let $\overline{W} = [W_1\ W_2]^{\mathrm{T}}$ be the difference of two solutions.

The following proposition is valid.

Proposition 2.1. (Uniqueness, $\kappa_0 > \kappa > 0$.) *Let $\overline{W}$ ($\overline{W}^{\mathrm{go}} \equiv 0$) be a classical solution of (2.4)–(2.8) and*

$$\operatorname{Re} a_{12}^{\pm} \geq 0, \quad \operatorname{Re} a_{21}^{\pm} \geq 0, \tag{2.9}$$

then it is zero for any sufficiently small κ.

We observe that $\overline{W}$ satisfies the Helmholtz equation (2.4), the boundary conditions (2.5), Meixner's conditions (2.7), and the condition at infinity $r \to \infty$ (see also (2.8))

$$|W_j| \leq \exp(-a\kappa r), \quad a > 0, \quad j = 1, 2. \tag{2.10}$$

We compute the scalar product of (2.4) and $\overline{W}$, integrate over the domain $\Omega_{\epsilon,R}$, which is a part of the angular domain $|\varphi| < \Phi$, $r > 0$, terminated by two circular arcs of small $r = \epsilon$ and large $r = R$ radii, correspondingly. Using the Green formula, we obtain

$$\int\limits_{\partial\Omega_{\epsilon,R}} \left\langle \frac{\partial \overline{W}}{\partial n}, \overline{W} \right\rangle \mathrm{d}S$$

$$- \int\limits_{\Omega_{\epsilon,R}} \left(|\operatorname{grad} W_1|^2 + |\operatorname{grad} W_2|^2\right) \mathrm{d}\Omega + (k_0 + \mathrm{i}\kappa)^2 (\sin\theta_0)^2 \int\limits_{\Omega_{\epsilon,R}} \langle \overline{W}, \overline{W} \rangle \, \mathrm{d}\Omega = 0.$$

Here $< \overline{a}, \overline{b} >$ is the scalar product of two vectors $\overline{a} = [a_1\ a_2]^{\mathrm{T}}$ and $\overline{b} = [b_1\ b_2]^{\mathrm{T}}$ with $< \overline{a}, \overline{b} > = a_1 b_1^* + a_2 b_2^*$. $\partial\Omega_{\epsilon,R}$ is the boundary of $\Omega_{\epsilon,R}$, consisting of two circular segments S_ϵ and S_R and of segments $S_{\pm}^{\epsilon,R}$ of the faces $S_\pm$ between these arcs.

Let ϵ tend to 0 and R to ∞; computing the imaginary part of the previous equality we thus have

$$\operatorname{Im} \int\limits_{S_+ \cup S_-} \left\langle \frac{\partial \overline{W}}{\partial n}, \overline{W} \right\rangle \mathrm{d}S + 2k_0\kappa \sin^2\theta_0 \int\limits_{\Omega_{0,\infty}} \langle \overline{W}, \overline{W} \rangle \, \mathrm{d}\Omega = 0. \tag{2.11}$$

Because the second term on the left-hand side is nonnegative, we turn to the first term. Using the boundary conditions (2.5) for this term leads to

$$\operatorname{Im} \int_{S_+ \cup S_-} \left\langle \frac{\partial \overline{W}}{\partial n}, \overline{W} \right\rangle \mathrm{d}S = \int_{S_+} \operatorname{Im} \left[\mathrm{i}(k_0 + \mathrm{i}\kappa) \sin^2 \theta_0 \left\langle \overline{\overline{\mathcal{A}}}^{+} \overline{W}, \overline{W} \right\rangle \right] \mathrm{d}S$$

$$+ \int_{S_-} \operatorname{Im} \left[\mathrm{i}(k_0 + \mathrm{i}\kappa) \sin^2 \theta_0 \left\langle \overline{\overline{\mathcal{A}}}^{-} \overline{W}, \overline{W} \right\rangle \right] \mathrm{d}S$$

$$+ \cos \theta_0 \operatorname{Im} \left\langle \overline{\overline{\mathcal{B}}} \overline{W}, \overline{W} \right\rangle_{\varphi=\Phi} \Bigg|_{\epsilon \to 0}^{R \to \infty}$$

$$- \cos \theta_0 \operatorname{Im} \left\langle \overline{\overline{\mathcal{B}}} \overline{W}, \overline{W} \right\rangle_{\varphi=-\Phi} \Bigg|_{\epsilon \to 0}^{R \to \infty}.$$

The last two terms on the right-hand side vanish in view of the conditions at the edge and at infinity (2.10). So we turn to the sign of the integral terms. For the integrand one has

$$\operatorname{Im} \left[\mathrm{i}(k_0 + \mathrm{i}\kappa) \sin^2 \theta_0 \left\langle \overline{\overline{\mathcal{A}}}^{\pm} \overline{W}, \overline{W} \right\rangle \right]_{\phi=\pm\Phi}$$

$$= k_0 \sin^2 \theta_0 \operatorname{Re} \left\langle \overline{\overline{\mathcal{A}}}^{\pm} \overline{W}, \overline{W} \right\rangle - \kappa \sin^2 \theta_0 \operatorname{Im} \left\langle \overline{\overline{\mathcal{A}}}^{\pm} \overline{W}, \overline{W} \right\rangle, \tag{2.12}$$

where κ is small. It is obvious that ($\pm$ is omitted)

$$\operatorname{Re} \left\langle \overline{\overline{\mathcal{A}}} \overline{W}, \overline{W} \right\rangle = \operatorname{Re} a_{12} \, |W_1|^2 + \operatorname{Re} a_{21} \, |W_2|^2 / |a_{21}|^2.$$

For nonactive wedge faces we have that $\operatorname{Re} \left\langle \overline{\overline{\mathcal{A}}} \overline{W}, \overline{W} \right\rangle \geq 0$, which follows from (2.9). Therefore, for small $\kappa > 0$ each term on the left-hand side of (2.11) is nonnegative, which is followed by $\overline{W} = 0$.

In following sections of this chapter we construct an exact solution for any small absorption, $\kappa > 0$. The limiting-absorption principle (see Section 1.3.4) enables us to specify a solution as the limit when $\kappa \to +0$. The limit exists indeed, which follows from analytic representations of the solution valid also for $k > 0$.

The uniqueness theorem for positive wavenumbers is of the utmost importance in scattering. It may also be ensured in the framework of formulation based on the radiation conditions.

Proposition 2.2. **(Uniqueness, $\kappa = 0$.)** *Let $k > 0$ and passivity conditions be valid*

$$\operatorname{Re} a_{12}^{\pm} > 0, \quad \operatorname{Re} a_{21}^{\pm} > 0.$$

If there exists a classical solution of the homogeneous diffraction probblem satisfying the radiation conditions

$$\int_{S_R} \left| \frac{\partial \overline{W}}{\partial r} - \mathrm{i}k \overline{W} \right|^2 \mathrm{d}S \to 0, \quad R \to \infty, \tag{2.13}$$

where S_R is the arc of radius R, then it is zero.

We consider some details of the proof. As in Proposition 2.1 we arrive at

$$\operatorname{Im} \int_{\partial\Omega_{\epsilon,R}} \left\langle \frac{\partial \overline{W}}{\partial n}, \overline{W} \right\rangle \mathrm{d}S = 0. \tag{2.14}$$

Making use of Meixner's conditions (2.7), one has

$$\begin{aligned} 0 &\le \operatorname{Im} \int_{S_+^{0,R} \cup S_-^{0,R}} \left\langle \frac{\partial \overline{W}}{\partial n}, \overline{W} \right\rangle \mathrm{d}S \\ &= -\operatorname{Im} \int_{S_R} \left\langle \frac{\partial \overline{W}}{\partial R}, \overline{W} \right\rangle \mathrm{d}S. \end{aligned} \tag{2.15}$$

The left-hand side is nonnegative, which is verified in the same manner as in the previous proposition.

The right-hand side of (2.15) is reduced to

$$\begin{aligned} &-\operatorname{Im} \int_{S_R} \left\langle \frac{\partial \overline{W}}{\partial R}, \overline{W} \right\rangle \mathrm{d}S \\ &= -k \int_{S_R} |\overline{W}|^2 \mathrm{d}S - \operatorname{Im} \int_{S_R} \left\langle \frac{\partial \overline{W}}{\partial R} - ik\overline{W}, \overline{W} \right\rangle \mathrm{d}S \end{aligned} \tag{2.16}$$

and then

$$\begin{aligned} k \int_{S_R} |\overline{W}|^2 \mathrm{d}S &\le -\operatorname{Im} \int_{S_R} \left\langle \frac{\partial \overline{W}}{\partial R} - ik\overline{W}, \overline{W} \right\rangle \mathrm{d}S \\ &\le \left(\int_{S_R} \left| \frac{\partial \overline{W}}{\partial R} - ik\overline{W} \right|^2 \mathrm{d}S \right)^{1/2} \left(\int_{S_R} |\overline{W}|^2 \mathrm{d}S \right)^{1/2}. \end{aligned} \tag{2.17}$$

By means of the radiation condition (2.13) we conclude that

$$\int_{S_R} |\overline{W}|^2 \mathrm{d}S \to 0,$$

as $R \to \infty$. Hence, from (2.15) we arrive at

$$\begin{aligned} 0 \le \operatorname{Im} \int_{S_+^{0,R} \cup S_-^{0,R}} \left\langle \frac{\partial \overline{W}}{\partial n}, \overline{W} \right\rangle \Bigg|_{\phi=\pm\Phi} \mathrm{d}S &\le -k \int_{S_R} |\overline{W}|^2 \mathrm{d}S \\ + \left(\int_{S_R} \left| \frac{\partial \overline{W}}{\partial R} - ik\overline{W} \right|^2 \mathrm{d}S \right)^{1/2} &\left(\int_{S_R} |\overline{W}|^2 \mathrm{d}S \right)^{1/2}. \end{aligned} \tag{2.18}$$

As a result, one has as $R \to \infty$

$$\operatorname{Im} \int_{S_+ \cup S_-} \left\langle \frac{\partial \overline{W}}{\partial n}, \overline{W} \right\rangle \mathrm{d}S = 0,$$

and then

$$\int_{S_+ \cup S_-} \operatorname{Re}\left[k \left\langle \overline{\overline{A}}^{\pm} \overline{W}, \overline{W} \right\rangle\right] \mathrm{d}S = 0$$

or equivalently

$$\overline{W}|_{\varphi=\pm\Phi} = 0.$$

Because the corresponding homogeneous Dirichlet problem has only a trivial solution (see, for instance [2.7]), we conclude that $\overline{W} \equiv 0$.

From the exact solution constructed as follows it is directly verified that the solution derived by use of the limiting-absorption principle also satisfies the radiation condition. We remark that the uniqueness of a solution for the point-source illumination to be considered in the following chapter can be treated analogously and hence is omitted in this book.

2.3 Sommerfeld integral and functional equations

To seek the solution to the boundary-value problem formulated in Section 2.2, we follow Sommerfeld and represent the unknown vector $\overline{U}(r, \varphi)$ by means of his eponymous integrals ([2.1–2.7] and Chapter 1):

$$\overline{U}(r, \varphi) = \frac{1}{2\pi \mathrm{i}} \int_{\gamma} \overline{\overline{f}}(\alpha + \varphi) \cdot \overline{U}_0 \, \mathrm{e}^{-\mathrm{i}k'r\cos\alpha} \mathrm{d}\alpha, \tag{2.19}$$

with

$$\overline{\overline{f}}(\alpha) = \begin{bmatrix} f_{11}(\alpha) & f_{12}(\alpha) \\ f_{21}(\alpha) & f_{22}(\alpha) \end{bmatrix}.$$

As in Chapter 1, γ is the Sommerfeld double loops, and $\overline{\overline{f}}(\alpha)$ is the spectra of the z-components (the Sommerfeld transformants), save for the known z-dependence, of the total field. In this way, the variables r and φ have been separated.

Inserting the previous expressions for $\overline{U}(r, \varphi)$ into the boundary condition (2.5) and using the inversion formula (Theorems 1.2 and 1.3) leads to a matrix equation for the spectra $\overline{\overline{f}}(\alpha)$:

$$\begin{aligned} &\overline{\overline{A}}^{\pm}(\alpha) \cdot \overline{\overline{f}}(\alpha \pm \Phi) = \overline{\overline{A}}^{\pm}(-\alpha) \cdot \overline{\overline{f}}(-\alpha \pm \Phi), \\ &\overline{\overline{A}}^{\pm}(\alpha) = \sin\alpha \, \overline{\overline{I}} \pm \sin\theta_0 \, \overline{\overline{\mathcal{A}}}^{\pm} - \cos\theta_0 \cos\alpha \, \overline{\overline{\mathcal{B}}}. \end{aligned} \tag{2.20}$$

The radiation condition (2.8) demands that

$$\overline{\overline{f}}(\alpha) - \overline{\overline{I}}/(\alpha - \varphi_0) \text{ be regular in } \Pi(-\Phi, \Phi), \tag{2.21}$$

that is, inside a strip in the complex α-plane with $-\Phi \leq \operatorname{Re}\alpha \leq \Phi$.

Equation (2.20) can be rewritten as

$$\overline{\overline{f}}(\alpha \pm 2\Phi) = \overline{\overline{B}}^{\pm}(\alpha) \cdot \overline{\overline{f}}(-\alpha),$$
$$\overline{\overline{B}}^{\pm}(\alpha) = \left[\overline{\overline{A}}^{\pm}(\alpha \pm \Phi)\right]^{-1} \cdot \overline{\overline{A}}^{\pm}(-\alpha \mp \Phi) = \begin{bmatrix} b_1^{\pm}(\alpha) & b_2^{\pm}(\alpha) \\ b_3^{\pm}(\alpha) & b_4^{\pm}(\alpha) \end{bmatrix}, \tag{2.22}$$

where

$$b_j^{\pm} = \frac{N_j^{\pm}(\alpha)}{D^{\pm}(\alpha)}, \quad j = 1, 2, 3, 4, \tag{2.23}$$

and

$$N_1^{\pm}(\alpha) = \cot^2\theta_0 \left[1 - \frac{a_{12}^{\pm}}{a_{21}^{\pm}} - 2\sin^2(\alpha \pm \Phi)\right] - \left[\sin(\alpha \pm \Phi) \mp \frac{a_{12}^{\pm}}{\sin\theta_0}\right]\left[\sin(\alpha \pm \Phi) \pm \frac{\csc\theta_0}{a_{21}^{\pm}}\right], \tag{2.24}$$

$$N_2^{\pm}(\alpha) = 2\cot\theta_0 \csc\theta_0 \sin(\alpha \pm \Phi)\cos(\alpha \pm \Phi), \tag{2.25}$$

$$N_3^{\pm}(\alpha) = -N_2^{\pm}(\alpha), \tag{2.26}$$

$$N_4^{\pm}(\alpha) = \cot^2\theta_0 \left[1 - \frac{a_{12}^{\pm}}{a_{21}^{\pm}} - 2\sin^2(\alpha \pm \Phi)\right] - \left[\sin(\alpha \pm \Phi) \pm \frac{a_{12}^{\pm}}{\sin\theta_0}\right]\left[\sin(\alpha \pm \Phi) \mp \frac{\csc\theta_0}{a_{21}^{\pm}}\right], \tag{2.27}$$

$$D^{\pm}(\alpha) = \left[\sin(\alpha \pm \Phi) \pm \sin\theta^{\pm}\right]\left[\sin(\alpha \pm \Phi) \pm \sin\chi^{\pm}\right]. \tag{2.28}$$

Auxiliary angles $\theta^{\pm}$ and $\chi^{\pm}$ were already introduced according to

$$\sin\theta^{\pm} = \frac{\csc\theta_0}{2}\left\{a_{12}^{\pm} + \frac{1}{a_{21}^{\pm}} + \left[\left(a_{12}^{\pm} - \frac{1}{a_{21}^{\pm}}\right)^2 + 4\cos^2\theta_0\left(\frac{a_{12}^{\pm}}{a_{21}^{\pm}} - 1\right)\right]^{1/2}\right\}, \tag{2.29}$$

$$\sin\chi^{\pm} = \frac{\csc\theta_0}{2}\left\{a_{12}^{\pm} + \frac{1}{a_{21}^{\pm}} - \left[\left(a_{12}^{\pm} - \frac{1}{a_{21}^{\pm}}\right)^2 + 4\cos^2\theta_0\left(\frac{a_{12}^{\pm}}{a_{21}^{\pm}} - 1\right)\right]^{1/2}\right\}, \tag{2.30}$$

with

$$0 \le \operatorname{Re}\theta^{\pm} < \pi/2 \quad \text{and} \quad 0 \le \operatorname{Re}\chi^{\pm} < \pi/2.$$

To characterize fully $\overline{\overline{f}}(\alpha)$ in the basic strip $\Pi(-2\Phi, 2\Phi)$, knowledge about all its poles and their corresponding residues is required. From the pole of the incident wave and its residue (2.21) and with the aid of (2.22), the principal parts of $\overline{\overline{f}}(\alpha)$ at possible poles in the basic strip are given by

$$\overline{\overline{f}}(\alpha) = \frac{\overline{\overline{I}}}{\alpha - \alpha_0^{go}} + \cdots, \quad \overline{\overline{f}}(\alpha) = \frac{\mathrm{H}(\pm\varphi_0)\overline{\overline{R}}^{\pm}}{\alpha - \alpha_{\pm 1}^{go}} + \cdots, \tag{2.31}$$

where $\alpha_j^{go} = 2j\Phi + (-1)^j\varphi_0,\ j = \cdots, -2, -1, 0, 1, 2, \ldots$ are the geometrical-optics poles, $\mathrm{H}(\cdot)$ is the Heaviside unit-step function, and $\overline{\overline{R}}^{\pm}$, the reflection coefficients, is defined by

$$\overline{\overline{R}}^{\pm} = -\overline{\overline{B}}^{\pm}(-\varphi_0). \tag{2.32}$$

To facilitate the solution procedure, let us derive in the next section a difference equation of higher order for one of the spectra, say, $f_{1\ell}(\alpha),\ \ell = 1, 2$.

2.4 A functional difference equation of higher order

2.4.1 A difference equation for one spectrum

To this end, it is expedient to rewrite (2.22) as a system of linear equations for $f_{1\ell}(\alpha)$ and $f_{2\ell}(\alpha)$:

$$\begin{aligned} f_{1\ell}(\alpha + 2\Phi) - b_1^+(\alpha) f_{1\ell}(-\alpha) - b_2^+(\alpha) f_{2\ell}(-\alpha) &= 0, \\ f_{2\ell}(\alpha + 2\Phi) - b_3^+(\alpha) f_{1\ell}(-\alpha) - b_4^+(\alpha) f_{2\ell}(-\alpha) &= 0, \\ f_{1\ell}(\alpha - 2\Phi) - b_1^-(\alpha) f_{1\ell}(-\alpha) - b_2^-(\alpha) f_{2\ell}(-\alpha) &= 0, \\ f_{2\ell}(\alpha - 2\Phi) - b_3^-(\alpha) f_{1\ell}(-\alpha) - b_4^-(\alpha) f_{2\ell}(-\alpha) &= 0. \end{aligned} \tag{2.33}$$

From the third of it, $f_{2\ell}(-\alpha)$ can be expressed in terms of $f_{1\ell}(\alpha)$, that is,

$$f_{2\ell}(-\alpha) = \left[f_{1\ell}(\alpha - 2\Phi) - b_1^-(\alpha) f_{1\ell}(-\alpha) \right] / b_2^-(\alpha).$$

Using this expression in the first of (2.33), we get a difference equation of higher order for $f_{1\ell}(\alpha)$ [2]

$$\begin{aligned} & f_{1\ell}(\alpha + 2\Phi) - b_2^+(\alpha) f_{1\ell}(\alpha - 2\Phi)/b_2^-(\alpha) = q_1(\alpha) f_{1\ell}(-\alpha), \\ & q_1(\alpha) = b_1^+(\alpha) - b_2^+(\alpha) b_1^-(\alpha)/b_2^-(\alpha). \end{aligned} \tag{2.34}$$

After having obtained a solution for $f_{1\ell}(\alpha)$ in the strip $\Pi(-2\Phi, 2\Phi)$, $f_{2\ell}(\alpha)$ in the same strip is given by, on use of the first and the third of (2.33)

$$\begin{aligned} f_{2\ell}(\alpha) &= \frac{f_{1\ell}(-\alpha \pm 2\Phi) - b_1^{\pm}(-\alpha) f_{1\ell}(\alpha)}{b_2^{\pm}(-\alpha)}, \\ & \alpha \in \Pi\left[(-1 \pm 1)\Phi, (1 \pm 1)\Phi\right] \end{aligned} \tag{2.35}$$

The spectra $\overline{\overline{f}}(\alpha)$ can be analytically extended to α outside the basic strip $\Pi(-2\Phi, 2\Phi)$ by virtue of (2.22).

---

[2] For the discussion on equivalence of such reduction we refer the reader to Section 2.5.2.

Only one nonconstant coefficient exists on the left-hand side of (2.34), namely,

$$\frac{b_2^+(\alpha)}{b_2^-(\alpha)} = \frac{\sin(\alpha+\Phi)\cos(\alpha+\Phi)}{\sin(\alpha-\Phi)\cos(\alpha-\Phi)} \times \frac{\sin(\alpha-\Phi)-\sin\theta^-}{\sin(\alpha+\Phi)+\sin\theta^+}\,\frac{\sin(\alpha-\Phi)-\sin\chi^-}{\sin(\alpha+\Phi)+\sin\chi^+}, \tag{2.36}$$

and if this very coefficient can be written in an equivalent form,

$$\frac{b_2^+(\alpha)}{b_2^-(\alpha)} = -\frac{\Psi_0(\alpha+2\Phi)}{\Psi_0(\alpha-2\Phi)}, \tag{2.37}$$

then the difference equation (2.34) can be further simplified. For this purpose a special function $\chi_\Phi(\alpha)$ proves to be particularly useful.

2.4.2 The generalized Malyuzhinets function $\chi_\Phi(\alpha)$

In their study of wave diffraction by either a perfectly conducting or a scalar impedance wedge surrounded by a gyroelectric medium, Bobrovnikov and Fisanov [2.3] introduced this special function $\chi_\Phi(\alpha)$, which is defined by the difference equation of the first order:

$$\chi_\Phi(\alpha+2\Phi) = \cos(\alpha/2)\,\chi_\Phi(\alpha-2\Phi). \tag{2.38}$$

As shown by Avdeev in [2.8], the meromorphic function $\chi_\Phi(\alpha)$ can be given in an explicit way for $|\mathrm{Re}\,\alpha| < \pi + 2\Phi$

$$\ln\chi_\Phi(\alpha) = \frac{1}{2}\int_0^\infty \frac{1}{s\,\sinh(\pi s)}\left[\frac{\alpha}{2\Phi} - \frac{\sinh(\alpha s)}{\sinh(2\Phi s)}\right] \mathrm{d}s. \tag{2.39}$$

With its help, the following relationships can be inferred [2.3,2.8]

$$\begin{aligned} &\chi_\Phi(\alpha)\,\chi_\Phi(-\alpha) = 1, \quad [\chi_\Phi(\alpha)]^* = \chi_\Phi(\alpha^*), \\ &\frac{\chi_\Phi(\alpha+\pi)}{\chi_\Phi(\alpha-\pi)} = 2^{1-\mu}\cos(\nu\alpha), \quad \mu = \frac{\pi}{2\Phi}, \quad \nu = \frac{\pi}{4\Phi} = \frac{\mu}{2}. \end{aligned} \tag{2.40}$$

Worthy of mentioning is its relation to the Malyuzhinets function $\psi_\Phi(\alpha)$ [2.3, 2.8]

$$\psi_\Phi(\alpha) = \frac{1}{[\chi_\Phi(\pi/2)]^2}\,\frac{\chi_\Phi(\alpha+\pi/2)}{\chi_\Phi(\alpha-\pi/2)}, \tag{2.41}$$

which can be used as an alternative way for evaluating the Malyuzhinets function. How to calculate the generalized Malyuzhinets function in an efficient manner will be discussed in detail in Section 2.8 (see also [2.17]).

Of relevance are the zeros α_{mn}^{z} and poles α_{mn}^{p} of this function

$$\alpha_{mn}^{\mathrm{z}} = -\alpha_{mn}^{\mathrm{p}} = \pi(2m+1) + 2\Phi(2n+1), \quad m, n = 0, 1, \ldots, \tag{2.42}$$

and its asymptotic behavior as $\operatorname{Im}\alpha \to \pm\infty$ [2.8]

$$\chi_{\Phi}(\alpha) = \exp\left[\mp\frac{\mathrm{i}\alpha^2}{16\Phi} - \frac{\ln 2}{4\Phi}\alpha \pm \frac{1}{2}\left(\Phi + \frac{\pi^2}{4\Phi}\right) + \mathrm{O}\left(\mathrm{e}^{\pm\mathrm{i}\mu\alpha}\right) + \mathrm{O}\left(\mathrm{e}^{\pm\mathrm{i}\alpha}\right)\right]. \tag{2.43}$$

Equipped with the basic facts about the generalized Malyuzhinets function we can continue the analysis.

2.4.3 Simplifying the functional difference equation of higher order

To use the generalized Malyuzhinets function $\chi_{\Phi}(\alpha)$ in obtaining the very equivalent expression (2.37), let us factor the numerator and denominator on the right-hand side of (2.36) into cosine functions with suitable arguments. Then each of the cosine function is replaced with the ratio of two generalized Malyuzhinets functions according to (2.38).

In this way it turns out that the auxiliary function $\Psi_0(\alpha)$ is given by

$$\begin{aligned}\Psi_0(\alpha) = &\frac{\chi_{\Phi}(\alpha+\Phi-\pi)\,\chi_{\Phi}(\alpha+\Phi)}{\chi_{\Phi}(\alpha-\Phi+\pi)\,\chi_{\Phi}(\alpha-\Phi)}\\ &\times\frac{\chi_{\Phi}(\alpha+\Phi+\pi/2)\,\chi_{\Phi}(\alpha+\Phi-\pi/2)}{\chi_{\Phi}(\alpha-\Phi-\pi/2)\,\chi_{\Phi}(\alpha-\Phi+\pi/2)}\\ &\times\frac{\chi_{\Phi}(\alpha-\Phi-\chi^-+\pi)\,\chi_{\Phi}(\alpha-\Phi+\chi^-)}{\chi_{\Phi}(\alpha+\Phi+\chi^+-\pi)\,\chi_{\Phi}(\alpha+\Phi-\chi^+)}\\ &\times\frac{\chi_{\Phi}(\alpha-\Phi-\theta^-+\pi)\,\chi_{\Phi}(\alpha-\Phi+\theta^-)}{\chi_{\Phi}(\alpha+\Phi+\theta^+-\pi)\,\chi_{\Phi}(\alpha+\Phi-\theta^+)}.\end{aligned} \tag{2.44}$$

In the case of $\Phi > \pi/2$, two zeros of $\Psi_0(\alpha)$ lie inside the basic strip $\Pi(-2\Phi, 2\Phi)$. They can be eliminated, however, in the following fashion:

$$F_0(\alpha) = \Psi_0(\alpha)\csc\left[\nu(\alpha-\Phi-\pi/2)\right]\csc\left[\nu(\alpha+\Phi+\pi/2)\right], \tag{2.45}$$

leading to a new function $F_0(\alpha)$ which is free of both zeros and poles in the basic strip.

Such a function $F_0(\alpha)$ allows us to introduce a new spectrum $\mathcal{F}_{1\ell}(\alpha)$ according to $f_{1\ell}(\alpha) = F_0(\alpha)\mathcal{F}_{1\ell}(\alpha)$. It is conspicuous that $\mathcal{F}_{1\ell}(\alpha)$ obeys a difference equation with constant coefficients on the left-hand side

$$\begin{aligned}&\mathcal{F}_{1\ell}(\alpha+2\Phi) + \mathcal{F}_{1\ell}(\alpha-2\Phi) = Q_1(\alpha)\,\mathcal{F}_{1\ell}(-\alpha),\\ &Q_1(\alpha) = q_1(\alpha)F_0(-\alpha)/F_0(\alpha+2\Phi).\end{aligned} \tag{2.46}$$

Before continuation, let us estimate the behavior of the right-hand side $H_{1\ell}(\alpha) = Q_1(\alpha)\,\mathcal{F}_{1\ell}(-\alpha)$ as $\operatorname{Im}\alpha \to \pm\infty$, by using, among other things, the property of the generalized Malyuzhinets function $\chi_{\Phi}(\alpha)$ (2.43), the definition of $F_0(\alpha)$ (2.45), and Meixner's edge condition (2.7), which implies $f_{1\ell}(\pm\mathrm{i}\infty)$ being finite. It reveals

$$H_{1\ell}(\alpha) = \mathrm{O}\left(\exp\left[\pm\mathrm{i}(1-\nu)\alpha\right]\right). \tag{2.47}$$

Hence, the right-hand side of the difference equation (2.46) declines with increasing $|\mathrm{Im}\,\alpha|$ in an exponential way as long as $\nu < 1$, that is, $\Phi > \pi/4$, which is assumed throughout this chapter.

The functional difference equation (2.46) can be transformed by using a technique originally proposed in [2.10] and further developed in [2.13] (see also Section 1.6).

2.5 Second-order functional difference equation and Fredholm integral equation of the second kind

2.5.1 An integral equivalent to the difference equation

To apply the technique outlined in [2.10,2.13] and Section 1.6, let us express at first both the spectrum $\mathcal{F}_{1\ell}(\alpha)$ and the right-hand side of (2.46) in their even and odd parts, that is, $\mathcal{F}_{1\ell}(\alpha) = \mathcal{F}_{1\ell}^{\mathrm{e}}(\alpha) + \mathcal{F}_{1\ell}^{\mathrm{o}}(\alpha)$, $H_{1\ell}(\alpha) = Q_1(\alpha)\mathcal{F}_{1\ell}(-\alpha) = H_{1\ell}^{\mathrm{e}}(\alpha) + H_{1\ell}^{\mathrm{o}}(\alpha)$. Equation (2.46) is then obviously identical to

$$\mathcal{F}_{1\ell}^{\mathrm{e}}(\alpha + 2\Phi) + \mathcal{F}_{1\ell}^{\mathrm{e}}(\alpha - 2\Phi) = H_{1\ell}^{\mathrm{e}}(\alpha), \tag{2.48}$$

$$\mathcal{F}_{1\ell}^{\mathrm{o}}(\alpha + 2\Phi) + \mathcal{F}_{1\ell}^{\mathrm{o}}(\alpha - 2\Phi) = H_{1\ell}^{\mathrm{o}}(\alpha). \tag{2.49}$$

To get the particular solution to the functional difference equation (2.49), which is equivalent to a system of equations

$$\mathcal{F}_{1\ell}^{\mathrm{o}}(\alpha + 2\Phi) - \mathcal{F}_{1\ell}^{\mathrm{o}}(-\alpha + 2\Phi) = H_{1\ell}^{\mathrm{o}}(\alpha), \tag{2.50}$$

$$\mathcal{F}_{1\ell}^{\mathrm{o}}(\alpha - 2\Phi) - \mathcal{F}_{1\ell}^{\mathrm{o}}(-\alpha - 2\Phi) = H_{1\ell}^{\mathrm{o}}(\alpha), \tag{2.51}$$

we resort to the $\mathcal{S}$-integral discussed in Chapter 1; see also [2.9, 2.6, 2.7, 2.10, 2.13]. The particular solution $\mathcal{F}_{1\ell,\mathrm{p}}^{\mathrm{o}}(\alpha)$ in the basic strip $\Pi\,(-2\Phi, 2\Phi)$ thus reads

$$\mathcal{F}_{1\ell,\mathrm{p}}^{\mathrm{o}}(\alpha) = \sin(\nu\alpha)\frac{\mathrm{i}}{4\Phi}\int_{\mathrm{i}\mathbb{R}} \frac{\sin(\nu t)H_{1\ell}^{\mathrm{o}}(t)}{\cos(\mu t) + \cos(\mu\alpha)}\,\mathrm{d}t, \tag{2.52}$$

because of a property of $H_{1\ell}(\alpha)$ (2.47) and hence of $H_{1\ell}^{\mathrm{o}}(\alpha)$.

To find the particular solution to (2.48) by using the same procedure, it is advantageous to introduce a new odd function $\mathcal{G}_{1\ell}^{\mathrm{o}}(\alpha)$ via $\mathcal{F}_{1\ell}^{\mathrm{e}}(\alpha) = \cot(\nu\alpha)\,\mathcal{G}_{1\ell}^{\mathrm{o}}(\alpha)$. The respective difference equation follows from (2.48),

$$\mathcal{G}_{1\ell}^{\mathrm{o}}(\alpha + 2\Phi) + \mathcal{G}_{1\ell}^{\mathrm{o}}(\alpha - 2\Phi) = -\cot(\nu\alpha)H_{1\ell}^{\mathrm{e}}(\alpha), \tag{2.53}$$

and can be dealt with in the same way as (2.49). Hence, the particular solution $\mathcal{F}_{1\ell,\mathrm{p}}^{\mathrm{e}}(\alpha)$ in the basic strip of the complex α-plane turns out to be

$$\mathcal{F}_{1\ell,\mathrm{p}}^{\mathrm{e}}(\alpha) = -\cos(\nu\alpha)\frac{\mathrm{i}}{4\Phi}\int_{\mathrm{i}\mathbb{R}} \frac{\cos(\nu t)H_{1\ell}^{\mathrm{e}}(t)}{\cos(\mu t) + \cos(\mu\alpha)}\,\mathrm{d}t. \tag{2.54}$$

Therefore, the particular solution to (2.46) in the same basic strip is given by

$$\mathcal{F}_{1\ell,\mathrm{p}}(\alpha) = \mathcal{F}^{\mathrm{e}}_{1\ell,\mathrm{p}}(\alpha) + \mathcal{F}^{\mathrm{o}}_{1\ell,\mathrm{p}}(\alpha) = -\frac{\mathrm{i}}{8\Phi}\int_{\mathrm{i}\mathbb{R}} \frac{Q_1(-t)\mathcal{F}_{1\ell}(t)}{\cos\left[\nu(\alpha+t)\right]}\mathrm{d}t. \tag{2.55}$$

The sought-for equivalent integral representation of (2.46) in the basic strip $\Pi\,(-2\Phi, 2\Phi)$ reads

$$\begin{aligned}\mathcal{F}_{1\ell}(\alpha) = {} & \frac{\nu\delta_{1\ell}/F_0(\alpha_0^{\mathrm{go}})}{\sin\left[\nu(\alpha-\alpha_0^{\mathrm{go}})\right]} + \frac{\nu\mathrm{H}(\varphi_0)R^{+}_{1\ell}/F_0(\alpha^{\mathrm{go}}_{+1})}{\sin\left[\nu(\alpha-\alpha^{\mathrm{go}}_{+1})\right]} + \frac{\nu\mathrm{H}(-\varphi_0)R^{-}_{1\ell}/F_0(\alpha^{\mathrm{go}}_{-1})}{\sin\left[\nu(\alpha-\alpha^{\mathrm{go}}_{-1})\right]} \\ & + C^{+}_{1\ell}\,\mathrm{e}^{-\mathrm{i}\nu\alpha} + C^{-}_{1\ell}\,\mathrm{e}^{\mathrm{i}\nu\alpha} - \frac{\mathrm{i}}{8\Phi}\int_{\mathrm{i}\mathbb{R}} \frac{Q_1(-t)\mathcal{F}_{1\ell}(t)}{\cos\left[\nu(\alpha+t)\right]}\mathrm{d}t.\end{aligned} \tag{2.56}$$

Here $\delta_{1\ell}$ denotes the Kronecker delta, and $R^{\pm}_{1\ell}$ are entries of the matrix reflection coefficient $\overline{\overline{R}}^{\pm}$ defined in (2.32). The nonintegral terms on the right-hand side are homogeneous solutions to the difference equation (2.46) that recover precisely the geometrical-optics poles and the asymptotic behavior of $\mathcal{F}_{1\ell}(\alpha)$ at infinity.

For (2.56) to be viewed as an integral equivalent to the difference equation (2.46), the constants $C^{\pm}_{1\ell}$ contained on its right-hand side have to be fixed.

2.5.2 Determining the constants $C^{\pm}_{1\ell}$

To this end we call upon (2.35), which represents the spectrum $f_{2\ell}(\alpha)$ in the basic strip by means of $f_{1\ell}(\alpha)$ there. For instance, for $\alpha \in \Pi(0, 2\Phi)$, $f_{2\ell}(\alpha)$ is proportional to the inverse of $b_2^{+}(-\alpha)$. According to its definition, there is

$$b_2^{+}(-\alpha) \sim \sin(-2\alpha + 2\Phi),$$

implying that $b_2^{+}(-\alpha)$ vanishes at $\alpha_m = \Phi + m\pi/2, \quad m = 0, \pm 1, \pm 2, \ldots$.

That $f_{2\ell}(\alpha)$ possesses no physical poles[3] at α_m leads to the following conditions to be imposed on $f_{1\ell}(\alpha)$:

$$\left[f_{1\ell}(-\alpha + 2\Phi) - b_1^{+}(-\alpha) f_{1\ell}(\alpha)\right]_{\alpha_m} = 0, \text{ for } \alpha_m \in \Pi(-\Phi, \Phi), \tag{2.57}$$

which eliminate the nonphysical poles.

For $\alpha_0 = \Phi$ condition (2.57) is always met thanks to $b_1^{+}(-\Phi) = 1$. The next two conditions for $m = \pm 1$ are actually one:

$$f_{1\ell}\,(\Phi - \pi/2) = b_1^{+}\,(-\Phi - \pi/2)\, f_{1\ell}\,(\Phi + \pi/2)\,, \tag{2.58}$$

because of the relation

$$b^{+}\left(-\Phi - \frac{\pi}{2}\right) = \frac{1}{b^{+}\,(-\Phi + \pi/2)} = -\frac{\left(\csc\theta_0 - a^{+}_{12}\right)\left(\csc\theta_0 + 1/a^{+}_{21}\right)}{(1 - \sin\theta^{+})(1 - \sin\chi^{+})}. \tag{2.59}$$

[3] By definition, physical poles are those that do not lead to violation of the conditions in the formulation of the problem, that is, of the radiation condition in our case.

For $|m| > 1, \alpha_m$ lies outside the basic strip, the respective conditions (2.57) need not be imposed.

A similar consideration applied to the expression for $f_{2\ell}(\alpha)$ (2.35) valid in the remaining half of the basic strip leads to a second condition to be satisfied by $f_{1\ell}(\alpha)$

$$f_{1\ell}(-\Phi - \pi/2) = b_1^-(\Phi - \pi/2)\, f_{1\ell}(-\Phi + \pi/2)\,, \tag{2.60}$$

where

$$b^-\left(\Phi - \frac{\pi}{2}\right) = -\frac{\left(\csc\theta_0 - a_{12}^-\right)\left(\csc\theta_0 + 1/a_{21}^-\right)}{(1+\sin\theta^-)(1+\sin\chi^-)}. \tag{2.61}$$

In this way, it has been demonstrated that (2.56), augmented with the two conditions (2.58) and (2.60), is an integral equivalent to the difference equation for the spectrum $f_{1\ell}(\alpha)$ (2.46).

2.5.3 Fredholm integral equation of the second kind

As indicated clearly by (2.56), especially by the last term on its right-hand side (the particular solution), $\mathcal{F}_{1\ell}(\alpha)$ in the basic strip depends upon its value along the imaginary axis of the complex plane. In line with (2.58) and (2.60), the same is true for the constants $C_{1\ell}^{\pm}$. Therefore, for points on the imaginary axis of the complex α-plane, (2.56), together with (2.58) and (2.60), amounts to a Fredholm integral equation of the second kind.

The kernel of the integral equation (2.56), $Q_1(-t)/\cos[\nu(\alpha+t)]$, is free from singularities and does not contain the wavenumber k, which may be a large parameter, as can be verified by recalling the definition of $Q_1(t)$ and the properties of $F_0(t)$ and $q_1(t)$. Furthermore, this kernel decreases exponentially with $|\mathrm{Im}\, t|$, or more precisely,

$$Q_1(-t)/\cos[\nu(\alpha+t)] = \mathrm{O}(\exp[\pm\mathrm{i}(1+\nu)t])\,, \tag{2.62}$$

as $t \to \pm\mathrm{i}\infty$. It is precisely these properties that render the above described solution procedure a very efficient, that is, fast and accurate one; see Section 2.7.

From the asymptotic behavior of $Q_1(\alpha)$ and $\mathcal{F}_{1\ell}(\alpha)$ it can be inferred that the kernel of (2.56) is square integrable. Hence, its solvability follows from the uniqueness of the solution to the present problem, which has been shown in Section 2.2.2; see also [2.10, 2.13].

In the next section, we derive a first-order uniform asymptotic expression for the solution to the boundary-value problem at hand, with a numerical solution of the integral equation (2.56) being postponed to Section 2.7.

2.6 Uniform asymptotic solution

2.6.1 Poles and residues

Here we confine ourselves to the case $\Phi > \pi/2$, because other cases can be studied in a similar way with double and multiple reflections having to be accounted for; see also [2.10, 2.13].

At first let us find out the domain of the complex angle α inside which the spectra $\overline{\overline{f}}(\alpha)$ must be available. As shown in Chapter 1, the Sommerfeld double loops γ lie within the strip $\Pi(-3\pi/2, 3\pi/2)$. As the azimuthal angle φ belongs to $(-\Phi, \Phi)$, the domain of the argument of the spectra in the Sommerfeld integrals (2.19), that is, $\Pi(-3\pi/2 - \Phi, 3\pi/2 + \Phi)$ is always wider than the basic strip $\Pi(-2\Phi, 2\Phi)$. As a consequence, the spectra $\overline{\overline{f}}(\alpha)$ obtained as a solution to the integral equation (2.56) have to be extended to the strip $\Pi(-4\Phi, 4\Phi)$ to cover the "worst" case with $\Phi = \pi/2$.

Such an extension can be carried out, repeatedly if necessary, by making use of the matrix equation for the spectra (2.22). Since this matrix equation is the transform of the boundary conditions at the wedge's faces, such a procedure of analytical continuation corresponds to the (multiple) reflection of waves at the wedge's faces. For example, using (2.22) once leads to the spectra in the strips $\Pi(\Phi, 3\Phi)$ and $\Pi(-3\Phi, -\Phi)$:

$$\overline{\overline{f}}(\alpha) = \overline{\overline{B}}^{\pm}(\alpha \mp 2\Phi) \cdot \overline{\overline{f}}(-\alpha \pm 2\Phi); \tag{2.63}$$

by using (2.63) in its own right-hand side, we get the expression for $\Pi(3\Phi, 5\Phi)$ and $\Pi(-5\Phi, -3\Phi)$:

$$\overline{\overline{f}}(\alpha) = \overline{\overline{B}}^{\pm}(\alpha \mp 2\Phi) \cdot \overline{\overline{B}}^{\mp}(-\alpha \pm 4\Phi) \cdot \overline{\overline{f}}(\alpha \mp 4\Phi). \tag{2.64}$$

In a second step we consider the poles and the residues of the spectra $\overline{\overline{f}}(\alpha)$. The principal parts for the first three geometrical-optics terms are already given in (2.31), Section 2.3, so we need to study only poles that describe an entirely different type of waves, namely, excited surface waves. These waves propagate along the wedge's faces and are confined to their vicinity. Obviously the surface-wave poles must lie outside the strip $\Pi(-\Phi, \Phi)$; otherwise the radiation condition would be broken (see (2.21)).

In the strips of width 2Φ next to $\Pi(-\Phi, \Phi)$, (2.63), the expression for $\overline{\overline{f}}(\alpha)$, suggests that the possible surface-wave poles must stem from the matrices $\overline{\overline{B}}^{\pm}(\alpha \mp 2\Phi)$, that is, the zeros of the common denominators of the matrix entries. It can be shown that four of these zeros, namely,

$$\alpha^{\mathrm{sw}}_{\theta^\pm} = \pm\left(\pi + \Phi + \theta^{\pm}\right), \quad \alpha^{\mathrm{sw}}_{\chi^\pm} = \pm\left(\pi + \Phi + \chi^{\pm}\right), \tag{2.65}$$

may be captured when deforming the Sommerfeld double loops γ and hence are important for our analysis.

Using a standard procedure, the principal parts of the spectra at these poles are found to be

$$\overline{\overline{f}}(\alpha) = \frac{\overline{\overline{R}}^{\mathrm{s}}_{\theta^\pm} \cdot \overline{\overline{f}}\left[\pm(\Phi - \pi - \theta^{\pm})\right]}{\alpha - \alpha^{\mathrm{sw}}_{\theta^\pm}} + \dots, \quad \overline{\overline{f}}(\alpha) = \frac{\overline{\overline{R}}^{\mathrm{s}}_{\chi^\pm} \cdot \overline{\overline{f}}\left[\pm(\Phi - \pi - \chi^{\pm})\right]}{\alpha - \alpha^{\mathrm{sw}}_{\chi^\pm}} + \dots. \tag{2.66}$$

Here, the excitation coefficients of the surface waves are given by

$$\overline{\overline{R}}^{\mathrm{s}}_{\theta^\pm} = \frac{\pm \sec\theta^{\pm}}{\sin\theta^{\pm} - \sin\chi^{\pm}} \begin{bmatrix} R_1^{\pm} & R_2^{\pm} \\ R_3^{\pm} & R_4^{\pm} \end{bmatrix} \tag{2.67}$$

with

$$R_1^{\pm} = \cot^2\theta_0\left(1 - a_{12}^{\pm}/a_{21}^{\pm} - 2\sin^2\theta^{\pm}\right) \\ - \left(\sin\theta^{\pm} + a_{12}^{\pm}/\sin\theta_0\right)\left(\sin\theta^{\pm} - \csc\theta_0/a_{21}^{\pm}\right), \tag{2.68}$$

$$R_2^{\pm} = \pm 2\cot\theta_0\csc\theta_0\sin\theta^{\pm}\cos\theta^{\pm}, \tag{2.69}$$

$$R_3^{\pm} = -R_2^{\pm}, \tag{2.70}$$

$$R_4^{\pm} = \cot^2\theta_0\left(1 - a_{12}^{\pm}/a_{21}^{\pm} - 2\sin^2\theta^{\pm}\right) \\ - \left(\sin\theta^{\pm} - a_{12}^{\pm}/\sin\theta_0\right)\left(\sin\theta^{\pm} + \csc\theta_0/a_{21}^{\pm}\right). \tag{2.71}$$

The explicit expression for $\overline{\overline{R}}^{\,s}_{\chi^{\pm}}$ results from interchanging $\theta^{\pm}$ and $\chi^{\pm}$ with each other in the previous formulas for $\overline{\overline{R}}^{\,s}_{\theta^{\pm}}$.

Now, we are capable of deriving a uniform asymptotic expression for the electromagnetic field $\overline{U}(r,\varphi)$ from the exact solution obtained already.

2.6.2 First-order uniform asymptotics

Deforming the Sommerfeld double loops γ to the steepest-descent paths $S(\pm\pi)$ (see Fig. 1.5) leads to an alternative exact expression for $\overline{U}(r,\varphi)$

$$\overline{U}(r,\varphi) = \frac{1}{2\pi i}\int_{S(-\pi)+S(\pi)} \overline{\overline{f}}(\alpha+\varphi)\cdot\overline{U}_0\, e^{-ik'r\cos\alpha}\,d\alpha + \sum_{n=1}^{7}\overline{r}_n, \tag{2.72}$$

where

$$\overline{r}_n = H(A_n)\overline{\overline{R}}_n\cdot\overline{U}_0\, e^{-ik'r\cos\alpha_n}.$$

The parameters appeared in the above equation are defined in Table 2.1.

Another elementary special function has been used in Table 2.1, namely, the Gudermann function gd(·), which was given in Chapter 1.

Table 2.1 Poles and residues of $\overline{\overline{f}}(\alpha+\varphi)$

n	Pole α_n	Residue $\overline{\overline{R}}_n$	Argument A_n
1	$\varphi_0 - \varphi$	$\overline{\overline{I}}$	$\pi - \lvert\varphi_0 - \varphi\rvert$
2	$2\Phi - \varphi_0 - \varphi$	$\overline{\overline{R}}^{+}$	$\pi - 2\Phi + \varphi_0 + \varphi$
3	$-2\Phi - \varphi_0 - \varphi$	$\overline{\overline{R}}^{-}$	$\pi - 2\Phi - \varphi_0 - \varphi$
4	$\pi + \Phi + \theta^+ - \varphi$	$\overline{\overline{R}}^{\,s}_{\theta^+}\cdot\overline{\overline{f}}(\Phi - \pi - \theta^+)$	$-\Phi + \varphi - \operatorname{Re}\theta^+ - \operatorname{gd}\left(\operatorname{Im}\theta^+\right)$
5	$-\pi - \Phi - \theta^- - \varphi$	$\overline{\overline{R}}^{\,s}_{\theta^-}\cdot\overline{\overline{f}}(-\Phi + \pi + \theta^-)$	$-\Phi - \varphi - \operatorname{Re}\theta^- - \operatorname{gd}\left(\operatorname{Im}\theta^-\right)$
6	$\pi + \Phi + \chi^+ - \varphi$	$\overline{\overline{R}}^{\,s}_{\chi^+}\cdot\overline{\overline{f}}(\Phi - \pi - \chi^+)$	$-\Phi + \varphi - \operatorname{Re}\chi^+ - \operatorname{gd}\left(\operatorname{Im}\chi^+\right)$
7	$-\pi - \Phi - \chi^- - \varphi$	$\overline{\overline{R}}^{\,s}_{\chi^-}\cdot\overline{\overline{f}}(-\Phi + \pi + \chi^-)$	$-\Phi - \varphi - \operatorname{Re}\chi^- - \operatorname{gd}\left(\operatorname{Im}\chi^-\right)$

Such a deformation of the integration contour has a clear meaning in physics, namely, splitting the total field into its different parts: the diffracted waves $\overline{U}^{\mathrm{d}}(r, \varphi)$, represented by the integral along the steepest-descent paths $\mathrm{S}(\pm\pi)$, the geometrical-optics parts $\overline{U}^{\mathrm{go}}(r, \varphi)$, and the surface waves $\overline{U}^{\mathrm{sw}}(r, \varphi)$, expressed in the sum on the right-hand side of (2.72). More precisely,

$$\overline{U}^{\mathrm{d}}(r, \varphi) = \frac{1}{2\pi \mathrm{i}} \int_{\mathrm{S}(-\pi)+\mathrm{S}(\pi)} \overline{\overline{f}}(\alpha + \varphi) \cdot \overline{U}_0 \, \mathrm{e}^{-\mathrm{i}k' r \cos\alpha} \, \mathrm{d}\alpha, \tag{2.73}$$

$$\overline{U}^{\mathrm{go}}(r, \varphi) = \sum_{n=1}^{3} \overline{r}_n, \tag{2.74}$$

$$\overline{U}^{\mathrm{sw}}(r, \varphi) = \sum_{n=4}^{7} \overline{r}_n. \tag{2.75}$$

In the following, we seek a first-order uniform expression for the diffracted field $\overline{U}^{\mathrm{d}}(r, \varphi)$. As usual, we make use of the saddle-point method (its uniform version) [2.5] and obtain

$$\overline{U}^{\mathrm{d}}(r, \varphi) = \frac{\mathrm{e}^{\mathrm{i}k' r}}{\sqrt{r}} \overline{\overline{Q}}(r, \varphi) \cdot \overline{U}_0 + \mathrm{O}\left[\left(k' r\right)^{-3/2}\right], \tag{2.76}$$

where $\overline{\overline{Q}}(r, \varphi)$, the matrix diffraction coefficient, is given by

$$\overline{\overline{Q}}(r, \varphi) = \sqrt{\frac{\mathrm{i}}{2\pi k'}} \left[\overline{\overline{f}}(\varphi - \pi) - \overline{\overline{f}}(\varphi + \pi)\right] - \frac{\mathrm{i}}{\sqrt{4\pi k'}} \sum_{n=1}^{7} \overline{\overline{R}}_n \frac{1 - \mathrm{F_{KP}}(\mathrm{i}k' r s_n^2)}{s_n}.$$

The UTD transition function $\mathrm{F_{KP}}(\cdot)$, a variant of the Fresnel integral, is defined in (1.153) [2.21], and

$$s_n = -\sqrt{2\mathrm{i}} \cos\frac{\widetilde{\alpha}_n}{2}, \quad \text{where } \widetilde{\alpha}_n = \begin{cases} \alpha_n & |\mathrm{Re}\,\alpha_n| \le 2\pi \\ 2\pi \,\mathrm{sgn}(\mathrm{Re}\,\alpha_n) + \mathrm{i}\,\mathrm{Im}\,\alpha_n & \text{otherwise} \end{cases}.$$

In this way, we get the first-order uniform asymptotic solution for the z-components of the total field

$$\begin{aligned} &[H_z(r, \varphi; z)\ E_z(r, \varphi; z)]^{\mathrm{T}} \sim \\ &\left[\overline{U}^{\mathrm{go}}(r, \varphi) + \overline{U}^{\mathrm{sw}}(r, \varphi) + \overline{U}^{\mathrm{d}}(r, \varphi)\right] \exp\left(\mathrm{i}k'' z\right), \end{aligned} \tag{2.77}$$

where the superscripts "go," "sw," and "d" stand for the geometrical-optics part, the surface waves, and diffracted waves, respectively. The formulas for these wave ingredients are explicitly given in this subsection.

The next section discusses how to solve the integral equation (2.56) and presents some numerical examples.

2.7 Numerical results

2.7.1 Numerical computation of the spectra

To facilitate the numerical solution of (2.56), subject to (2.58) and (2.60), it proves beneficial to explicitly take into account (2.56), the asymptotic behavior of $\mathcal{F}_{1\ell}(\alpha)$, in such a way

$$\mathcal{F}_{1\ell}(\alpha) = \cos(\nu\alpha)\mathcal{L}_{1\ell}(\alpha). \tag{2.78}$$

The asymptotic behavior of $\mathcal{F}_{1\ell}(\alpha)$ implies that $\mathcal{L}_{1\ell}(\pm\mathrm{i}\infty)$ remain finite.

Therefore, the integral equation for $\mathcal{L}_{1\ell}(\alpha)$ results from (2.56)

$$\begin{aligned}\mathcal{L}_{1\ell}(\alpha) = &\frac{\mathcal{F}^{\mathrm{go}}_{1\ell}(\alpha)}{\cos(\nu\alpha)} + \frac{C^-_{1\ell}\mathrm{e}^{\mathrm{i}\nu\alpha} + C^+_{1\ell}\mathrm{e}^{-\mathrm{i}\nu\alpha}}{\cos(\nu\alpha)}\\ &- \frac{1}{\cos(\nu\alpha)}\frac{\mathrm{i}}{8\Phi}\int_{\mathrm{i}\mathbb{R}} \frac{\cos(\nu t)Q_1(-t)\mathcal{L}_{1\ell}(t)}{\cos\left[\nu(\alpha+t)\right]}\mathrm{d}t.\end{aligned} \tag{2.79}$$

Here $\mathcal{F}^{\mathrm{go}}_{1\ell}$ stands for the first line on the right-hand side of the integral equation (2.56). Evidently there is $C^{\pm}_{1\ell} = \mathcal{L}_{1\ell}(\pm\mathrm{i}\infty)/2$.

In a second step we transform the imaginary axis into a finite domain:

$$t = \frac{1}{\nu}\arctan(\mathrm{i}\tau) = \frac{\mathrm{i}}{2\nu}\ln\frac{1+\tau}{1-\tau}, \quad -1 \le \tau \le 1.$$

This converts the last term on the right-hand side of (2.79) to

$$\frac{1}{\cos(\nu\alpha)}\frac{1}{2\pi}\int_{-1}^{+1} \frac{Q_1[-t(\tau)]\mathcal{L}_{1\ell}[t(\tau)]}{[\cos(\nu\alpha) - \mathrm{i}\tau\sin(\nu\alpha)](1-\tau^2)}\mathrm{d}\tau.$$

Using

$$\alpha = \frac{1}{\nu}\arctan(\mathrm{i}\beta) = \frac{\mathrm{i}}{2\nu}\ln\frac{1+\beta}{1-\beta}, \quad -1 \le \beta \le 1$$

and $l_{1\ell}(\beta) = \mathcal{L}_{1\ell}[\alpha(\beta)]$ in (2.79), we obtain

$$\begin{aligned} l_{1\ell}(\beta) &- \frac{1-\beta^2}{2\pi}\int_{-1}^{+1} \frac{Q_1[-t(\tau)]l_{1\ell}(\tau)}{(1+\beta\tau)(1-\tau^2)}\mathrm{d}\tau - \left[C^-_{1\ell}(1-\beta) + C^+_{1\ell}(1+\beta)\right]\\ &= (1-\beta^2)\left[\frac{\nu\delta_{1\ell}/F_0(\varphi_0)}{\mathrm{i}\beta\cos(\nu\varphi_0) - \sin(\nu\varphi_0)}\right.\\ &\left. + \frac{\nu\mathrm{H}(\varphi_0)R^+_{1\ell}/F_0(\alpha^{\mathrm{go}}_{+1}) - \nu\mathrm{H}(-\varphi_0)R^-_{1\ell}/F_0(\alpha^{\mathrm{go}}_{-1})}{\mathrm{i}\beta\sin(\nu\varphi_0) - \cos(\nu\varphi_0)}\right].\end{aligned} \tag{2.80}$$

To solve (2.80), we use the quadrature method (for example, [2.22, 2.23]). For the numerical integration, we follow Nyström [2.24] and choose an N-point Gauss-Legendre rule where x_n is the n-th abscissa, and w_n is its respective weight. For short we use l_n for $l_{1\ell}(x_n)$, $n = 1, 2, \ldots N$.

Enforcing equation (2.80) at each abscissa x_m, $m = 1, 2, \ldots, N$, (2.80) is converted to a system of linear algebraic equations

$$
\begin{aligned}
l_m &- \left(1 - x_m^2\right) \sum_{n=1}^{N} \frac{P_n l_n}{1 + x_m x_n} - \left[C_{1\ell}^{-}(1 - x_m) + C_{1\ell}^{+}(1 + x_m)\right] \\
&= (1 - x_m^2) \cdot \left[\frac{\nu \delta_{1\ell} / F_0(\varphi_0)}{\mathrm{i} x_m \cos(\nu\varphi_0) - \sin(\nu\varphi_0)} \right. \\
&\quad \left. + \frac{\nu \mathrm{H}(\varphi_0) R_{1\ell}^{+} / F_0(\alpha_{+1}^{\mathrm{go}}) - \nu \mathrm{H}(-\varphi_0) R_{1\ell}^{-} / F_0(\alpha_{-1}^{\mathrm{go}})}{\mathrm{i} x_m \sin(\nu\varphi_0) - \cos(\nu\varphi_0)} \right],
\end{aligned}
\tag{2.81}
$$

with

$$
P_n = \frac{w_n Q_1[-t(x_n)]}{2\pi(1 - x_n^2)}.
$$

Evaluating the integrals there with the same N-point Gauss-Legendre rule, we get the corresponding discretized forms of the two constraints (2.58) and (2.60). Together with (2.81), we arrive at a matrix equation of dimension $N + 2$ for the N discrete values of the auxiliary function $\mathcal{L}_{1\ell}\left[t(x_n)\right], n = 1, 2, \ldots, N$ on the imaginary axis and the two constants $C_{1\ell}^{\pm}$.

After having solved this matrix equation, $\mathcal{F}_{1\ell}(\alpha)$ in the basic strip $\Pi(-2\Phi, 2\Phi)$ can be calculated via

$$
\begin{aligned}
\mathcal{F}_{1\ell}(\alpha) = {} & \frac{\nu \delta_{1\ell} / F_0(\varphi_0)}{\sin\left[\nu(\alpha - \varphi_0)\right]} - \frac{\nu \mathrm{H}(\varphi_0) R_{1\ell}^{+} / F_0(\alpha_{+1}^{\mathrm{go}}) - \nu \mathrm{H}(-\varphi_0) R_{1\ell}^{-} / F_0(\alpha_{-1}^{\mathrm{go}})}{\cos\left[\nu(\alpha + \varphi_0)\right]} \\
& + C_{1\ell}^{-} \mathrm{e}^{\mathrm{i}\nu\alpha} + C_{1\ell}^{+} \mathrm{e}^{-\mathrm{i}\nu\alpha} + \sum_{n=1}^{N} \frac{p_n}{\cos(\nu\alpha) - \mathrm{i} x_n \sin(\nu\alpha)},
\end{aligned}
\tag{2.82}
$$

and $p_n = P_n l_n$. Multiplying it with the meromorphic function $F_0(\alpha)$, we get the spectral function $f_{1\ell}(\alpha)$ in the basic strip $\Pi(-2\Phi, 2\Phi)$.

To determine numerically the coefficients of the previous matrix equation, or more precisely the meromorphic function $F_0(\alpha)$, it is necessary to compute repeatedly the generalized Malyuzhinets function $\chi_\Phi(\alpha)$. It proves to be advantageous to use an efficient procedure described in detail in Section 2.8 and [2.17].

Up to this point, this chapter, to a large extent, is dedicated to solving the functional difference equation (2.34) obeyed by $f_{1\ell}(\alpha)$. The remaining spectral function $f_{2\ell}(\alpha)$ can be obtained for instance via (2.35). But we prefer to get $f_{2\ell}(\alpha)$ in an analogous way as in the case of $f_{1\ell}(\alpha)$. For details the readers are referred to Appendix B of [2.13]. After having found them on the imaginary axis, their values in the strip $\Pi(-\Phi, \Phi)$ are extrapolated in a similar way as (2.82), for the remaining part of the complex plane the boundary condition (2.22) will be called upon, if necessary also repeatedly.

2.7.2 Examples

To verify the previously described solution procedure as well as its efficiency, that is, the accuracy of the obtained results and required CPU time, we take recourse to the closed-form exact solution given in [2.17, 2.18]. According to a detailed convergence test reported in [2.20], the present solution procedure is indeed very efficient; for example, a relative error of less than one part out of one million has been achieved by using 40 abscissae in the quadrature method (see Table 1 of [2.20]). Furthermore, the efficiency depends only weakly upon the skewness of the incidence $\cos\theta_0$ and the surface impedances (see Tables 2 and 3 of [2.20]).

The achieved efficiency rewards the analytical efforts described in this chapter, namely, to decouple the original functional difference equations (2.20) or (2.22) and to simplify the resultant equation (2.34) to (2.46) in terms of the generalized Malyuzhinets function $\chi_\Phi(\alpha)$, to convert (2.46) into its integral equivalent (2.56) by virtue of the $\mathcal{S}$-integrals and lastly to obtain a Fredholm integral equation of the second kind for points on the imaginary axis of the α-plane.

Figures 2.2 and 2.3 display the total fields for three different cases, where the skewness is fixed with $\theta_0 = 30°$, but the anisotropic surface impedances vary, with the amplitude of the imaginary parts decreasing from Case 1 via Case 2 to Case 3 (see the legend of Fig. 2.2 for details). For copolarization, this change in the surface impedance is visible merely in regions where reflected waves exist (in Fig. 2.2 $\varphi \geq 63°$), it is discernible almost everywhere for cross-polarization (Fig. 2.3). Except in Case 1 with $a_{12}^{\pm} = 1/a_{21}^{\pm}$, the amplitudes of the cross-polarized fields are not polarization-reciprocal, that is,

$$\frac{|U_1(r,\varphi)|_{U_{10}=0\,\mathrm{V/m}}}{U_{20}} \neq \frac{|U_2(r,\varphi)|_{U_{20}=0\,\mathrm{V/m}}}{U_{10}}. \tag{2.83}$$

2.8 Appendix: Computation of the generalized Malyuzhinets function

2.8.1 Numerical integration

Because of the important part played by the generalized Malyuzhinets function $\chi_\Phi(\alpha)$ in the solution procedure outlined already (see also Chapter 3) we need an efficient procedure for its numerical calculation. Such a procedure has been proposed in [2.17] which is briefly described in this section.

Let us take the properties of the generalized Malyuzhinets function $\chi_\Phi(\alpha)$, summarized in Section 2.4.2, as the departing point for its numerical computation. In view of (2.38) and (2.40), it is permissible to confine our attention to the computation of $\chi_\Phi(\alpha)$ in the semi-infinite strip $0 \leq \operatorname{Re}\alpha \leq \min\{\pi, 2\Phi\}$ and $\operatorname{Im}\alpha \geq 0$. In addition, we assume that $\Phi = p\pi/(2q)$ holds with p and q being noncommensurable positive integers. This assumption, however, imposes no practical limitation on the numerical computation because of the finite precision of today's digital computers.

The experiences gathered in the calculation of the Malyuzhinets function ([2.5, 2.25–2.27] and the references cited there) suggest dividing the semi-infinite strip into two parts: $\operatorname{Im}\alpha \leq a$ where a is a yet unknown constant; and the remaining part of the strip. For the first part a numerical quadrature seems to be the most convenient choice, whereas for the second part an asymptotic expansion proves to be beneficial.

In [2.8], Avdeev demonstrated that $\chi_{p\pi/(2q)}(\alpha)$ can be represented by a sum of pq finite integrals in total and recommended its use for the numerical evaluation. Here, a single although infinite integral (2.39) is used instead. To this end, (2.39) is put into the following form:

$$\ln \chi_\Phi(\alpha) = \int_0^\infty \mathrm{e}^{-t} f(t;\alpha,\Phi)\,\mathrm{d}t, \tag{2.84}$$

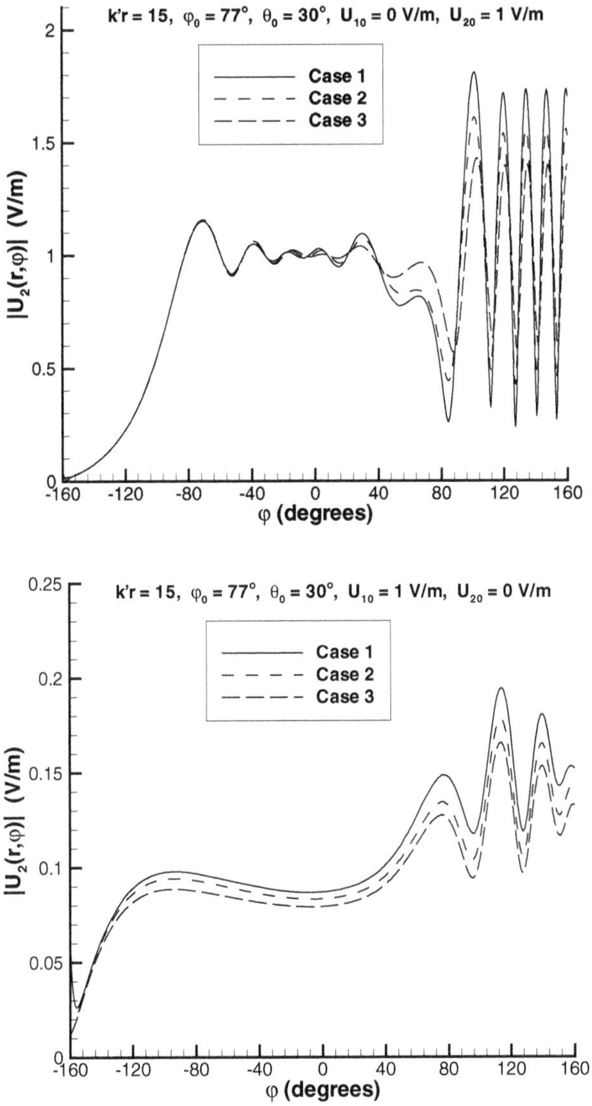

Figure 2.2 Total fields of copolarization as a function of the azimuthal angle φ and the anisotropic impedance faces (Case 1: $a_{12}^{+}=1/a_{21}^{+}=0.3+\mathrm{i}0.4$, $a_{12}^{-}=1/a_{21}^{-}=0.1-\mathrm{i}0.2$; Case 2: $a_{12}^{+}=0.3+\mathrm{i}0.2$, $a_{21}^{+}=1.2-\mathrm{i}0.8$, $a_{12}^{-}=0.1-\mathrm{i}0.1$, $a_{21}^{-}=2+\mathrm{i}2$; Case 3: $a_{12}^{+}=0.3$, $a_{21}^{+}=1.2$, $a_{12}^{-}=0.1$, $a_{21}^{-}=2$) for $\Phi=160^{\circ}$.

with

$$f(t;\alpha,\Phi)=\frac{1}{t\left(1-\mathrm{e}^{-2t}\right)}\left[\frac{\alpha}{2\Phi}-\frac{\mathrm{e}^{-(2\Phi-\alpha)t/\pi}-\mathrm{e}^{-(2\Phi+\alpha)t/\pi}}{1-\mathrm{e}^{-4\Phi t/\pi}}\right].$$

Hence, as indicated by the integrand of (2.84), the most natural quadrature scheme for its numerical evaluation is that due to Gauss and Laguerre (see [2.28]). It is worth mentioning

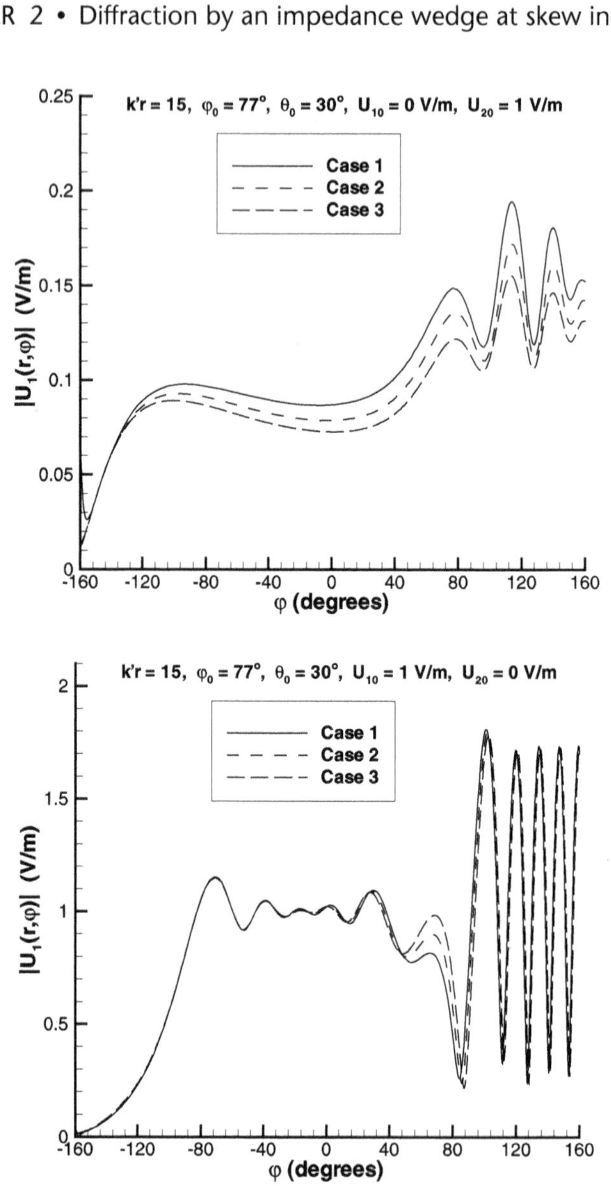

Figure 2.3 Total fields of cross-polarization as a function of the azimuthal angle φ and the anisotropic impedance faces (for details see the legend of Fig. 2.2).

that for a similar reason this scheme has been used for a highly accurate computation of the Malyuzhinets function in [2.29].

The accuracy of this procedure is affected by both the number of nodes and the imaginary part of α. To reconcile accuracy with computation time, we use the 20-point Gauss-Laguerre scheme for $\text{Im}\,\alpha \leq a = 4.2$. A comparison with a series expansion to be given in the next section reveals that in this way the relative error of the algorithm does not exceed one part out of 10^6 for $\pi/30 \leq \Phi \leq \pi$.

2.8.2 Series representation

According to [2.8], $\ln \chi_{p\pi/(2q)}(\alpha)$, the logarithm of the generalized Malyuzhinets function, can be expressed in the following way:

$$\ln \chi_{p\pi/(2q)}(\alpha) = \frac{\mathrm{i}\pi}{24}\left(\frac{p}{q}+\frac{q}{p}\right) - \frac{q\alpha}{2\pi p}\left(\ln 2 + \frac{\mathrm{i}\alpha}{4}\right) + \frac{\alpha}{2\pi p}\ln\left[1-(-1)^{p+q}\mathrm{e}^{\mathrm{i}q\alpha}\right]$$
$$-\frac{\mathrm{i}}{2\pi pq}\sum_{\ell=1}^{\ell_{\mathrm{M}}}\frac{(-1)^{(p+q)\ell}}{\ell^2}\mathrm{e}^{\mathrm{i}\ell q\alpha} + \mathrm{i}\sum_{\substack{n=1\\ n\neq \ell q}}^{n_{\mathrm{M}}}\frac{(-1)^n\mathrm{e}^{\mathrm{i}n\alpha}}{2n\sin(pn\pi/q)}$$
$$+\mathrm{i}\sum_{\substack{m=1\\ m\neq \ell p}}^{m_{\mathrm{M}}}\frac{(-1)^m\mathrm{e}^{\mathrm{i}qm\alpha/p}}{2m\sin(\pi qm/p)} + R_{\ell_{\mathrm{M}}}(\alpha) + R_{m_{\mathrm{M}}}(\alpha) + R_{n_{\mathrm{M}}}(\alpha), \tag{2.85}$$

where the last three terms on the right-hand side denote the remainders of the series, which converges for $\mathrm{Im}\,\alpha > a > 0$ when $\ell_{\mathrm{M}}, m_{\mathrm{M}}$, and n_{M} tend to infinity. Other than a purely numerical evaluation of (2.84), the expansion (2.85) allows us to easily estimate the achieved relative accuracy by using the remainders of the series that decrease rapidly with $\mathrm{Im}\,\alpha$. To obtain the same relative accuracy as mentioned at the end of Section 2.8.1 also for the second part of the strip with $\mathrm{Im}\,\alpha > a = 4.2$, it suffices to take $\ell_{\mathrm{M}} = 3$, $m_{\mathrm{M}} = 7$, and $n_{\mathrm{M}} = 4$.

CHAPTER 3

Scattering of waves from an electric dipole over an impedance wedge

This chapter deals with scattering of the electromagnetic wave emitted by a Hertzian dipole by an impedance wedge. Using a plane-wave expansion of the dipole field and our newly obtained exact solution of diffraction of a skew-incident plane electromagnetic wave by the same canonical body (see Chapter 2) enables us to deduce an integral representation for the total field. Then by evaluating the multiple integral the far-field expressions are derived and studied. The formulation and the basic steps of analysis are presented. In particular, the far-field expressions for the geometrical-optics and the edge-diffracted waves, including a uniform asymptotic version of the far field are given. The numerical calculation and physical interpretation of the analysis are also carried out.

3.1 Formulation of the problem and plane-wave expansion of the incident field

3.1.1 Statement of the problem

Let, as in Section 2.1, the wedge under study occupy in a cylindrical coordinate system (r, φ, z) the region given by $\Phi \leq |\varphi| \leq \pi$ with $0 < \Phi < \pi$ and $0 \leq |z| < \infty$. Introduce, without loss of generality, an arbitrarily oriented Hertzian dipole in the $r\varphi$-plane, that is, with the coordinates $(r_0, \varphi_0, 0)$, or what is the same, $(x_0, y_0, 0)$ in the related Cartesian system (Fig. 3.1). Of interest in this chapter is the electromagnetic field $\boldsymbol{E}$ and $\boldsymbol{H}$ excited by the Hertzian dipole outside the wedge.

The time-harmonic, that is, with an assumed time-dependence $\exp(-ikct)$ not shown explicitly, electromagnetic field outside the wedge satisfies Maxwell's equations (1.22)–(1.23), or in a slightly different form,

$$\operatorname{curl} \boldsymbol{H} = -\mathrm{i}k\boldsymbol{E} + \boldsymbol{J}, \quad \operatorname{curl} \boldsymbol{E} = \mathrm{i}k\boldsymbol{H}. \tag{3.1}$$

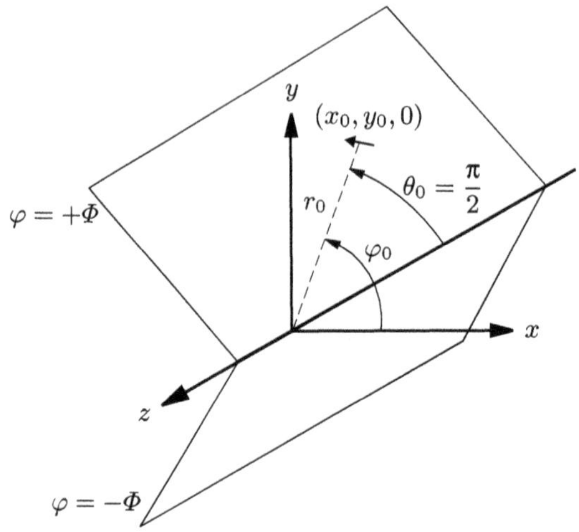

Figure 3.1 A wedge illuminated by a Hertzian dipole.

Here $\boldsymbol{H}$ and $\boldsymbol{J}$ stand for the magnetic field strength and current density multiplied by Z_0, the intrinsic impedance of free space. As usual, k denotes the wavenumber in free space. For a Hertzian dipole, $\boldsymbol{J}$ is given by

$$\boldsymbol{J} = -\mathrm{i}k\boldsymbol{P}\,\delta(x - x_0)\,\delta(y - y_0)\,\delta(z), \tag{3.2}$$

where $\delta(x)$ is the Dirac delta function and $\boldsymbol{P}$ a constant vector with $\boldsymbol{P} = \left(P_x, P_y, P_z\right)$.

On the surfaces of the wedge, the tangential components of the electric and magnetic field depend upon each other according to (2.1):

$$\begin{aligned} E_r(r, \pm\Phi, z) &= \mp\eta^{\pm} H_z(r, \pm\Phi, z), \\ E_z(r, \pm\Phi, z) &= \pm\eta^{\pm} H_r(r, \pm\Phi, z). \end{aligned} \tag{3.3}$$

For simplicity, this chapter deals with a scalar impedance wedge, that is, with $a_{12}^{\pm} = a_{21}^{\pm} = \eta^{\pm}$. Remember that $\eta^{\pm}$ are the (in terms of Z_0) normalized surface impedances.

Furthermore, the total electromagnetic field obeys Meixner's conditions at the edge of the wedge and the radiation conditions at infinity

$$\int_{S_R} |\boldsymbol{E} + \boldsymbol{e}_R \times \boldsymbol{H}|^2\,\mathrm{d}S \to 0, \quad \int_{S_R} |\boldsymbol{H} - \boldsymbol{e}_R \times \boldsymbol{E}|^2\,\mathrm{d}S \to 0, \quad R \to \infty, \tag{3.4}$$

where S_R denotes the exterior part of a sphere of radius R centered at the origin of the Cartesian system, $R = \|\boldsymbol{R}\|$ is the distance from the origin, and $\boldsymbol{e}_R = \boldsymbol{R}/R$ denotes the unit vector in the radial direction.

Remark: The radiation conditions in the form of (3.4) are valid for $\operatorname{Re}\eta^{\pm} > 0$, that is, the surface waves in their traditional definition attenuate exponentially on the wedge's surfaces with $R \to \infty$. Provided that $\operatorname{Re}\eta^{\pm} = 0$ holds good, the surface waves propagate actually from the edge of the wedge to infinity and do vanish as $R \to \infty$ but as $\mathrm{O}(1/\sqrt{R})$ on the wedge's faces. In this case, the radiation conditions (3.4) must be appropriately modified to be also met by the outgoing surface waves. The modification implies the addition of special integral terms in (3.4) for such surface waves in the close vicinities of the wedge's faces; see also Section 4.4.

3.1.2 The Hertz vector and plane-wave expansion of the incident field

Let us represent the electric and magnetic field vectors $\boldsymbol{E}$ and $\boldsymbol{H}$ in terms of the Hertz vector $\boldsymbol{\Pi}$ according to (1.25), (1.26), and (1.28); then the first of Maxwell's equations (3.1) amounts to

$$\left(\Delta + k^2\right) \boldsymbol{\Pi} = -\boldsymbol{P}\, \delta\,(x - x_0)\; \delta\,(y - y_0)\; \delta\,(z)\,. \tag{3.5}$$

Therefore, the Hertz vector is given in an explicit form

$$\boldsymbol{\Pi} = \boldsymbol{P} \exp\left(\mathrm{i}k R_0\right) / \left(4\pi R_0\right), \quad R_0 = \sqrt{(x - x_0)^2 + (y - y_0)^2 + z^2}. \tag{3.6}$$

Hence, the field strengths of the incident electromagnetic wave $\boldsymbol{E}^{\mathrm{i}}$ and $\boldsymbol{H}^{\mathrm{i}}$ read, in line with (1.30) and (1.31),

$$\boldsymbol{H}^{\mathrm{i}} = -\mathrm{i}k \operatorname{curl} \boldsymbol{\Pi}, \quad \boldsymbol{E}^{\mathrm{i}} = \operatorname{grad} \operatorname{div} \boldsymbol{\Pi} + k^2 \boldsymbol{\Pi}. \tag{3.7}$$

$\boldsymbol{H}^{\mathrm{i}}$ stands for the incident magnetic field strength multiplied by Z_0.

To make use of the results of the previous chapter, it is necessary to express the incident field in terms of plane waves in a suitable way (see [3.1]). In view of the explicit form of the Hertz vector $\boldsymbol{\Pi}$, let us study at first the Green function in free space $g(x, y, z; x_0, y_0) = \exp(\mathrm{i}k R_0) / (4\pi R_0)$, which solves a scalar partial differential equation related to (3.5)

$$\left(\operatorname{div} \operatorname{grad} + k^2\right) g(x, y, z; x_0, y_0) = -\delta\,(x - x_0)\; \delta\,(y - y_0)\; \delta\,(z)\,.$$

This equation, after having multiplied on both sides with $\exp(-\mathrm{i}\zeta z)$ and integrated from $z = -\infty$ to $z = +\infty$, takes on the form

$$\left[\frac{\partial^2}{\partial x^2} + \frac{\partial^2}{\partial y^2} + \left(k^2 - \zeta^2\right)\right] G(x, y; x_0, y_0; \zeta) = -\delta\,(x - x_0)\; \delta\,(y - y_0)\,,$$

where $G(x, y; x_0, y_0; \zeta)$ is the Fourier transform of $g(x, y, z; x_0, y_0)$ with respect to z:

$$G(x, y; x_0, y_0; \zeta) = \int_{\mathbb{R}} g(x, y, z; x_0, y_0) \exp(-\mathrm{i}\zeta z)\, \mathrm{d}z.$$

It is well-known that the two-dimensional Green function $G(x, y; x_0, y_0; \zeta)$ is proportional to the Hankel function of the first kind and the zeroth order, namely,

$$G(x, y; x_0, y_0; \zeta) = (\mathrm{i}/4) \mathrm{H}_0^{(1)}\left(\varrho\sqrt{k^2 - \zeta^2}\right), \quad \varrho = \sqrt{(x - x_0)^2 + (y - y_0)^2},$$

under the conditions

$$0 \leq \arg \sqrt{k^2 - \zeta^2} < \pi, \quad 0 \leq \arg k < \pi\,. \tag{3.8}$$

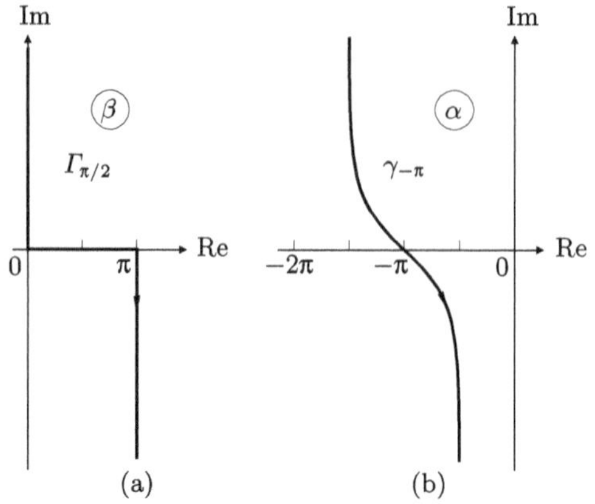

Figure 3.2 Contours of integration $\Gamma_{\pi/2}$ (a) and $\gamma_{-\pi}$ (b).

On use of the inverse Fourier transform, we get the Green function $g(x, y, z; x_0, y_0)$

$$g(x, y, z; x_0, y_0) = \frac{\mathrm{i}}{8\pi} \int_{\mathbb{R}} \exp(\mathrm{i}\zeta z)\, \mathrm{H}_0^{(1)}\left(\varrho\sqrt{k^2 - \zeta^2}\right) \mathrm{d}\zeta \tag{3.9}$$

under the same conditions (3.8). See (6.616/3) of [3.2].

Now let us introduce a new variable β via $\zeta = k\cos\beta$, this relation can be written as

$$\frac{\exp(\mathrm{i}kR_0)}{4\pi R_0} = \frac{\mathrm{i}k}{8\pi} \int_{\Gamma_{\pi/2}} \sin\beta \exp(\mathrm{i}kz\cos\beta)\, \mathrm{H}_0^{(1)}(k\varrho\sin\beta)\, \mathrm{d}\beta.$$

The integration contour is given by $\Gamma_{\pi/2} = (\mathrm{i}\infty, 0] \cup [0, \pi] \cup [\pi, \pi - \mathrm{i}\infty)$, depicted in Fig. 3.2(a).

In the next step, replace the Hankel function in the previous equation by the Sommerfeld integral expression [3.3]:

$$\mathrm{H}_0^{(1)}(\rho) = \frac{1}{\pi} \int_{\gamma_{-\pi}} \exp\left(-\mathrm{i}\rho\cos\alpha'\right) \mathrm{d}\alpha',$$

with $\gamma_{-\pi} = (-\delta_+ - \pi + \mathrm{i}\infty, \delta_- - \pi - \mathrm{i}\infty)$ and $0 < \delta_\pm < \pi$, shown in Fig. 3.2(b), and get therefore

$$\frac{\exp(\mathrm{i}kR_0)}{4\pi R_0} = \frac{\mathrm{i}k}{8\pi^2} \int_{\Gamma_{\pi/2}} \int_{\gamma_{-\pi}} \sin\beta \exp\left[\mathrm{i}k\left(z\cos\beta - \varrho\sin\beta\cos\alpha'\right)\right] \mathrm{d}\alpha' \mathrm{d}\beta.$$

To warrant $\operatorname{Im} k\varrho\sin\beta\cos\alpha' < 0$, the admissible range of $\delta_\pm$ must be narrowed to $0 < \delta_\pm < \pi/2$.

Notice that there is

$$\varrho\cos\alpha' = \varrho\cos(\alpha - \psi) = (x - x_0)\cos\alpha + (y - y_0)\sin\alpha, \quad 0 \le \psi < 2\pi;$$

hence, Green's function in free space can lastly be expanded in terms of plane waves

$$\frac{\exp(ikR_0)}{4\pi R_0} = \frac{ik}{8\pi^2} \int_{\Gamma_{\pi/2}} \int_{\gamma_{\psi-\pi}} \sin\beta \, e^{ik[z\cos\beta-(x-x_0)\sin\beta\cos\alpha-(y-y_0)\sin\beta\sin\alpha]} d\alpha \, d\beta. \tag{3.10}$$

Here, the new contour of integration $\gamma_{\psi-\pi}$ is related to the previous one $\gamma_{-\pi}$ via $\gamma_{\psi-\pi} = \gamma_{-\pi} + \psi$.

From (3.6) and (3.10), the plane-wave expansion of the Hertz vector $\boldsymbol{\Pi}$ turns out

$$\boldsymbol{\Pi} = \int_{\Gamma_{\pi/2}} \int_{\gamma_{\psi-\pi}} e^{-ik\sin\beta(x\cos\alpha+y\sin\alpha)+ikz\cos\beta} \boldsymbol{\mathcal{F}}(\alpha, \beta) \, d\alpha \, d\beta, \tag{3.11}$$

$$\boldsymbol{\mathcal{F}}(\alpha, \beta) = \left(\mathcal{F}_x, \mathcal{F}_y, \mathcal{F}_z\right) = \frac{ik}{8\pi^2} \boldsymbol{P} \sin\beta \, e^{ik\sin\beta(x_0\cos\alpha+y_0\sin\alpha)}. \tag{3.12}$$

Expression (3.11) is related to Weyl's formula [3.4], and, written in cylindrical coordinates, it reads

$$\boldsymbol{\Pi} = \int_{\Gamma_{\pi/2}} \int_{\gamma_{\psi-\pi}} e^{-ikr\sin\beta\cos(\alpha-\varphi)+ikz\cos\beta} \boldsymbol{\mathcal{F}}(\alpha, \beta) d\alpha \, d\beta. \tag{3.13}$$

Therefore, the sought-for plane-wave expansion of the incident field generated by a Hertzian dipole follows from (3.7) and (3.11)

$$\begin{bmatrix} H_z^{\mathrm{i}} \\ E_z^{\mathrm{i}} \end{bmatrix} = \frac{ik^3}{8\pi^2} \int_{\Gamma_{\pi/2}} \int_{\gamma_{\psi-\pi}} e^{-ikr\sin\beta\cos(\alpha-\varphi)+ikz\cos\beta} \times \sin^2\beta \, e^{ikr_0\sin\beta\cos(\alpha-\varphi_0)} \, \overline{U}_0(\alpha, \beta) \, d\alpha \, d\beta, \tag{3.14}$$

with

$$\begin{aligned} \overline{U}_0(\alpha, \beta) &= \left[U_{10}(\alpha, \beta), \; U_{20}(\alpha, \beta)\right]^{\mathrm{T}}, \\ U_{10}(\alpha, \beta) &= \sin\alpha \, P_x - \cos\alpha \, P_y, \\ U_{20}(\alpha, \beta) &= \cos\beta \left(\cos\alpha \, P_x + \sin\alpha \, P_y\right) + \sin\beta \, P_z. \end{aligned}$$

On use of a multidimensional saddle-point method summarized in Section 3.7.1, the asymptotic expression as $k \to \infty$ for the incident field, or more precisely, its z-components, for a point O on the edge with the Cartesian coordinates $(0, 0, z_O)$ is given by

$$\begin{bmatrix} H_z^{\mathrm{i}}(O) \\ E_z^{\mathrm{i}}(O) \end{bmatrix} \approx \frac{e^{ik\ell^{\mathrm{i}}}}{4\pi \ell^{\mathrm{i}}} k^2 \sin\beta_0 \, \overline{U}_0(\varphi_0, \beta_0), \tag{3.15}$$

where $\ell^{\mathrm{i}} = \sqrt{r_0^2 + z_O^2}$ measures the distance between the source point and the edge point O, and $\beta_0 = \arctan(r_0/z_O)$ denotes the angle subtended between the z-axis (the edge) and ℓ^{i}.

It is worth mentioning that as a special case of the present problem, diffraction of a dipole-field by a perfectly conducting wedge has been dealt with by many researchers (for example [3.1, 3.5, 3.6]), with its explicit solution given by Malyuzhinets and Tuzhilin [3.6]. (The readers are referred to a well-documented collection of known solutions to wave scattering by canonical bodies [3.7].) For comparison purposes, a uniform expression for the far field derived from the aforementioned exact solution will be called upon in Section 3.6.

3.2 The integral representation of the total field

3.2.1 Integral formulation

In this section, we build up the sought-for solution to the problem under study from the plane-wave solution detailed in Chapter 2: to superpose in a suitable way the fields generated by skew-incident plane waves under different angles α and β. Contrary to Chapter 2, however, these angles must take on, in addition to real values, also complex values. As exhibited by (3.14), the complex amplitude of a plane wave arriving under the angles α and β is given by

$$\mathrm{i}k^3\left(8\pi^2\right)^{-1}\sin^2\!\beta\,\exp\left[\mathrm{i}kr_0\sin\beta\cos(\alpha-\varphi_0)\right]\overline{U}_0(\alpha,\beta);$$

therefore, the integral representation of the z-components of the total field reads

$$\begin{bmatrix} H_z \\ E_z \end{bmatrix} = \frac{\mathrm{i}k^3}{8\pi^2}\int\limits_{\Gamma_{\pi/2}}\int\limits_{\gamma_{\varphi_0}} \mathrm{e}^{\mathrm{i}kz\cos\beta} \times \sin^2\!\beta\,\mathrm{e}^{\mathrm{i}kr_0\sin\beta\cos(\alpha-\varphi_0)}\,\overline{U}(r,\varphi,\alpha,\beta)\,\mathrm{d}\alpha\,\mathrm{d}\beta, \tag{3.16}$$

with the unknown functions $\overline{U}(r,\varphi,\alpha,\beta) = [U_1(r,\varphi,\alpha,\beta),\ U_2(r,\varphi,\alpha,\beta)]^{\mathrm{T}}$ which are assumed to be meromorphic with respect to (α,β) in some domain $\mathbb{C}\times\mathbb{C}$ for $r > 0$ and $|\varphi| \leq \Phi$.

In expression (3.16), the integration in the α-plane is to be performed along γ_{φ_0}, implying that the angle ψ equals $\pi+\varphi_0$, the value at the wedge's edge. Therefore, the integral expression for the incident wave (3.14) converges for $|\varphi_0| < \Phi - \pi/2$.[1] This notwithstanding, the asymptotic expressions to be derived later in this chapter hold good for $|\varphi_0| < \Phi$. Furthermore, the integration contour $\Gamma_{\pi/2}$ must not traverse branch cuts related to complex angles $\theta^{\pm}$ and $\chi^{\pm}$ to be discussed later in Section 3.3.2.

3.2.2 Formulation of the problem for $\overline{U}(r,\varphi,\alpha,\beta)$

Let us formulate the problem for $\overline{U}(r,\varphi,\alpha,\beta)$ as functions of (r,φ) and notice that the problem depends analytically on the parameters (α,β). We also postulate an analytic dependence of $\overline{U}(r,\varphi,\alpha,\beta)$ on these parameters at least in some appropriate subdomain of $\mathbb{C}\times\mathbb{C}$.[2] This, in particular, implies the possibility to deform appropriately the contour of the iterated integration. On the other hand, recent works [3.8, 3.9] (see also Chapter 2) enable us to develop an efficient and self-consistent procedure to determine $\overline{U}(r,\varphi,\alpha,\beta)$ for all $\alpha\in(-\Phi,\Phi)$, $\beta\in(0,\pi)$. This procedure, an appropriate analytic extension as well as asymptotic evaluation of the resulting integrals, offers a way to deduce efficient nonuniform expressions of the far field and specify domains of their applicability.

---

[1] In this way it is warranted that the wedge under study is fully situated inside the half-space in $\mathbb{R}^3$ $r\cos(\varphi-\varphi_0) < r_0$, inside which the Weyl-type integral (3.14) exhibits the stated property.

[2] From the mathematical point of view, the description of analytic continuation to a maximal domain is a complex problem.

The unknown functions are sought in the form of the Sommerfeld integrals[3] with integration to be carried out along the Sommerfeld double-loop contour γ

$$\overline{U}(r,\varphi,\alpha,\beta)=\frac{1}{2\pi\mathrm{i}}\int\limits_{\gamma}\overline{\overline{f}}(s+\varphi,\alpha,\beta)\cdot\overline{U}_0(\alpha,\beta)\,\mathrm{e}^{-ikr\sin\beta\cos s}\,\mathrm{d}s, \tag{3.17}$$

they meet the two-dimensional Helmholtz equation outside the wedge with $|\varphi|<\Phi$ and $r>0$:

$$\left[\frac{1}{r}\frac{\partial}{\partial r}\left(r\frac{\partial}{\partial r}\right)+\frac{1}{r^2}\frac{\partial^2}{\partial\varphi^2}+(k\sin\beta)^2\right]U_{1,2}(r,\varphi,\alpha,\beta)=0. \tag{3.18}$$

The radial components of the electromagnetic field are expressed in terms of $U_{1,2}(r,\varphi)$[4]

$$\begin{bmatrix}H_r\\E_r\end{bmatrix}=-\frac{k^2}{8\pi^2}\int\limits_{\Gamma_{\pi/2}}\mathrm{d}\beta\int\limits_{\gamma_{\varphi_0}}\mathrm{d}\alpha\,\mathrm{e}^{ikz\cos\beta}\,\mathrm{e}^{ikr_0\sin\beta\cos(\alpha-\varphi_0)}$$
$$\times\left[\cos\beta\frac{\partial U_1(r,\varphi)}{\partial r}-\frac{\partial U_2(r,\varphi)}{r\partial\varphi},\ \frac{\partial U_1(r,\varphi)}{r\partial\varphi}+\cos\beta\frac{\partial U_2(r,\varphi)}{\partial r}\right]^{\mathrm{T}}, \tag{3.19}$$

and satisfy the equations

$$\begin{aligned}\left(k^2+\frac{\partial^2}{\partial z^2}\right)E_r&=\frac{\partial}{\partial z}\left(\frac{\partial E_z}{\partial r}\right)+\mathrm{i}\frac{k}{r}\frac{\partial H_z}{\partial\varphi},\\ \left(k^2+\frac{\partial^2}{\partial z^2}\right)H_r&=\frac{\partial}{\partial z}\left(\frac{\partial H_z}{\partial r}\right)-\mathrm{i}\frac{k}{r}\frac{\partial E_z}{\partial\varphi}.\end{aligned} \tag{3.20}$$

It is remarkable that, provided $U_{1,2}(r,\varphi)$ solves the Helmholtz equation, then H_r and E_r in (3.19) fulfill (3.20), which turn out directly from Maxwell's equations. The φ-components of the field can be found from

$$\mathrm{i}kH_\varphi=\frac{\partial E_r}{\partial z}-\frac{\partial E_z}{\partial r},\quad -\mathrm{i}kE_\varphi=\frac{\partial H_r}{\partial z}-\frac{\partial H_z}{\partial r}. \tag{3.21}$$

From the boundary conditions for the electromagnetic field (3.3) we get the respective ones for the vector $\overline{U}(r,\varphi)$:

$$\overline{\overline{\mathcal{I}}}\cdot\left.\frac{\partial\overline{U}(r,\varphi)}{kr\partial\varphi}\right|_{\varphi=\pm\Phi}=\mp\sin^2\!\beta\,\overline{\overline{\mathcal{A}}}^{\pm}\cdot\overline{U}(r,\pm\Phi)+\cos\beta\,\overline{\overline{\mathcal{B}}}\cdot\frac{\mathrm{i}\partial\overline{U}(r,\pm\Phi)}{k\partial r}, \tag{3.22}$$

with the constant matrices $\overline{\overline{\mathcal{I}}}$, $\overline{\overline{\mathcal{A}}}^{\pm}$, and $\overline{\overline{\mathcal{B}}}$ defined by

$$\overline{\overline{\mathcal{I}}}=\begin{bmatrix}1&0\\0&1\end{bmatrix},\quad \overline{\overline{\mathcal{A}}}^{\pm}=\begin{bmatrix}\eta^{\pm}&0\\0&1/\eta^{\pm}\end{bmatrix},\quad \overline{\overline{\mathcal{B}}}=\begin{bmatrix}0&-1\\1&0\end{bmatrix}. \tag{3.23}$$

---

[3]It is obvious that the function $\overline{\overline{f}}(s,\alpha,\beta)$ should be analytically extended also with respect to (α,β).

[4]We shall omit the dependence of $U_{1,2}(r,\varphi,\alpha,\beta)$ on α and β if it does not lead to misunderstanding.

We can assert that, provided $\overline{U}(r, \varphi, \alpha, \beta)$ obeys the boundary conditions (3.22), the resulting electromagnetic field fulfills the boundary conditions (3.3).

As was mentioned already, the problem at hand—the problem of determining $\overline{U}(r, \varphi, \alpha, \beta)$—is nothing else than the problem of a skew-incident plane wave on an impedance wedge provided $\alpha \in (-\Phi, \Phi)$ and $\beta \in (0, \pi)$. Having resolved the problem for such real values of α and β, we then exploit the argumentation of a global analytic extension to all complex values including those on the contours $\beta \in \Gamma_{\pi/2}$ and $\alpha \in \gamma_{\varphi_0}$. It should be remarked that such a continuation is generally not a trivial task.

The fulfillment of Meixner's conditions and radiation condtions can be directly verified using the expressions for the total field (3.16), (3.19), and (3.21).

3.2.3 Representation for the spectral functions

To complete the solution to the problem under study, we turn now our attention to the spectral functions $\overline{\overline{f}}(s, \alpha, \beta)$. The procedure described in detail in [3.8, 3.9] and Chapter 2 allows us to write down $\overline{\overline{f}}(s, \alpha, \beta)$ in a rigorous form, although implicitly as a solution to some Fredholm integral equations of the second kind. As can be shown, these spectral functions meet the following identities:

$$\begin{aligned}\overline{\overline{f}}(s, \alpha, \beta) = F_0(s, \beta) \Bigg\{ & \frac{\nu \overline{\overline{\mathcal{I}}}}{F_0(\alpha, \beta) \sin[\nu(s-\alpha)]} \\ & + \frac{\nu \mathrm{H}(\operatorname{Re}\alpha) \overline{\overline{R}}^{+}(\alpha, \beta)}{F_0(2\Phi-\alpha, \beta) \sin[\nu(s-(2\Phi-\alpha))]} \\ & + \frac{\nu \mathrm{H}(-\operatorname{Re}\alpha) \overline{\overline{R}}^{-}(\alpha, \beta)}{F_0(-2\Phi-\alpha, \beta) \sin[\nu(s-(-2\Phi-\alpha))]} + \overline{\overline{C}}^{+}(\beta) \mathrm{e}^{-\mathrm{i}\nu s} \\ & + \overline{\overline{C}}^{-}(\beta) \mathrm{e}^{\mathrm{i}\nu s} - \frac{\mathrm{i}}{8\Phi} \int_{\mathrm{i}\mathbb{R}} \frac{\overline{\overline{Q}}(-t, \beta) \cdot \overline{\overline{F}}(t, \alpha, \beta)}{\cos[\nu(s+t)]} \mathrm{d}t \Bigg\},\end{aligned} \tag{3.24}$$

with

$$\nu = \pi/(4\Phi), \quad \overline{\overline{Q}}(t, \beta) = \operatorname{diag}[Q_1(t, \beta), Q_2(t, \beta)],$$

$$\overline{\overline{C}}^{\pm}(\beta) = \begin{bmatrix} C_{11}^{\pm}(\beta) & C_{12}^{\pm}(\beta) \\ C_{21}^{\pm}(\beta) & C_{22}^{\pm}(\beta) \end{bmatrix}, \quad \overline{\overline{F}}(s, \alpha, \beta) = \begin{bmatrix} F_{11}(s, \alpha, \beta) & F_{12}(s, \alpha, \beta) \\ F_{21}(s, \alpha, \beta) & F_{22}(s, \alpha, \beta) \end{bmatrix}.$$

For $\Phi > \pi/2$, matrices $\overline{\overline{C}}^{\pm}(\beta)$ differ from zero; hence, the following constraint must be imposed ($\ell = 1, 2$) [3.8]:

$$\begin{aligned} f_{1\ell}(\pm\Phi - \pi/2) &= b_1^{\mp}(\mp\Phi - \pi/2) f_{1\ell}(\pm\Phi + \pi/2), \\ f_{2\ell}(\pm\Phi \pm \pi/2) &= b_4^{\mp}(\mp\Phi \pm \pi/2) f_{1\ell}(\pm\Phi \mp \pi/2). \end{aligned} \tag{3.25}$$

Other functions and parameters are defined in Chapter 2, especially in Sections 2.3–2.5. The integral term is regular in the strip

$$\Pi(-2\Phi, 2\Phi) := \{s : -2\Phi < \mathrm{Re}\, s < 2\Phi\}.$$

The functions $\overline{\overline{F}}(s, \alpha, \beta)$ solve the following integral equations for $s \in \mathrm{i}\mathbb{R}$

$$\overline{\overline{F}}(s, \alpha, \beta) = \frac{\nu\overline{\overline{\mathcal{I}}}(\alpha, \beta)}{F_0(\alpha, \beta)\sin[\nu(s-\alpha)]} + \frac{\nu\mathrm{H}(\mathrm{Re}\,\alpha)\overline{\overline{R}}^{+}(\alpha, \beta)}{F_0(2\Phi-\alpha, \beta)\sin[\nu(s-(2\Phi-\alpha))]}$$
$$+\frac{\nu\mathrm{H}(-\mathrm{Re}\,\alpha)\overline{\overline{R}}^{-}(\alpha, \beta)}{F_0(-2\Phi-\alpha, \beta)\sin[\nu(s-(-2\Phi-\alpha))]} + \overline{\overline{C}}^{+}(\beta)\mathrm{e}^{-\mathrm{i}\nu s}$$
$$+\overline{\overline{C}}^{-}(\beta)\mathrm{e}^{\mathrm{i}\nu s} - \frac{\mathrm{i}}{8\Phi}\int_{\mathrm{i}\mathbb{R}} \frac{\overline{\overline{Q}}(-t, \beta)\cdot\overline{\overline{F}}(t, \alpha, \beta)}{\cos[\nu(s+t)]}\mathrm{d}t, \tag{3.26}$$

and the constraints (3.25).

After having solved the previous integral equations for a given pair of angles (α, β), the spectral functions $\overline{\overline{f}}(s, \alpha, \beta)$ are efficiently computed by means of (3.24), extrapolating $\overline{\overline{f}}(s, \alpha, \beta)$ along the imaginary axis of the complex s-plane to the strip $\Pi(-2\Phi, 2\Phi)$ in that plane. To extend s beyond this strip, the matrix functional difference equation

$$\overline{\overline{A}}^{\pm}(s, \beta)\cdot\overline{\overline{f}}(s \pm \Phi, \alpha, \beta) = \overline{\overline{A}}^{\pm}(-s, \beta)\cdot\overline{\overline{f}}(-s \pm \Phi, \alpha, \beta) \tag{3.27}$$

is to be employed, if necessary, repeatedly. In (3.27), the matrix $\overline{\overline{A}}^{\pm}(s, \beta)$ is defined as

$$\overline{\overline{A}}^{\pm}(s, \beta) = \sin s\,\overline{\overline{\mathcal{I}}} \pm \sin\beta\,\overline{\overline{\mathcal{A}}}^{\pm} - \cos\beta\cos s\,\overline{\overline{\mathcal{B}}}.$$

3.3 Deformation of the contours of integration and the geometrical-optics (GO) field

3.3.1 Saddle points, polar singularities, and residues

To deduce far-field asymptotics, the very first step lies in determining the saddle points, poles, and branch points of the integrands of (3.16) and (3.17). (It is noted that poles are not isolated singularities but some singular lines or surfaces in a multidimentional case. Therefore, we are going to speak of polar singularities instead of poles.) Hence, let us begin with the determination of the saddle points of the integrand of (3.16) and (3.17).

According to Section 3.7.1, in the integrand of (3.16) and (3.17), given explicitly in (3.32), k plays the part of a large parameter whereas function S is given by

$$S = \mathrm{i}\{z\cos\beta + \sin\beta\,[r_0\cos(\alpha-\varphi_0) - r\cos s]\}. \tag{3.28}$$

Furthermore, at a simple saddle point (α_0, β_0, s_0) it is required that

$$\mathrm{grad}\, S(\alpha, \beta, s)\big|_{\alpha=\alpha_0, \beta=\beta_0, s=s_0} = 0, \quad \det S''(\alpha, \beta, s)\big|_{\alpha=\alpha_0, \beta=\beta_0, s=s_0} \neq 0$$

be met. As can be shown with ease, the following two saddle points are found from (3.28)

$$(\varphi_0, \beta_0, -\pi), \quad (\varphi_0, \beta_0, \pi) \tag{3.29}$$

with $\beta_0 = \arctan[(r + r_0)/z]$ and $r = \sqrt{x^2 + y^2}$. (The field point is characterized by its Cartesian coordinates (x, y, z).)

The moduli of det S'' at these two saddle points are identical and given by

$$\left|\det S''\right|_{\alpha=\varphi_0, \beta=\beta_0, s=\pm\pi} = r_0 r \left(\ell^{\mathrm{i}} + \ell\right) \sin^2\beta_0, \tag{3.30}$$

with the trigonometric relations

$$\ell^{\mathrm{i}} = r_0 / \sin\beta_0, \quad \ell = r/\sin\beta_0, \quad \sqrt{(r + r_0)^2 + z^2} = \ell^{\mathrm{i}} + \ell.$$

For the moduli of det S'' differ from zero, the two points given in (3.29) are indeed the sought-for simple saddle points.

In the next step, we deform the contours of integration $F_{\pi/2}$, γ_{φ_0}, and γ to those of steepest descents in one dimension $\mathrm{S}(\beta_0)$, $\mathrm{S}(\varphi_0)$, $\mathrm{S}(-\pi)$, and $\mathrm{S}(\pi)$; thereby polar singularities of different nature might be captured. Based on a similar study described in Section 2.6.1, such polar singularities and their respective residues of the spectral function $\overline{\overline{f}}(s + \varphi, \alpha, \beta)$ for the present case with $\pi/2 < \Phi < \pi$ are listed in Table 3.1

In Table 3.1, the notations are those adopted in [3.8, 3.9]; see also Chapter 2. Especially, the reflection coefficients $\overline{\overline{R}}^{\pm}$ are defined in (2.32), whereas the excitation coefficients for surface waves $\overline{\overline{R}}^{\mathrm{s}}_{\theta^\pm, \chi^\pm}$ are given in (2.67)–(2.71) with $a_{12}^{\pm} = a_{21}^{\pm} = \eta^{\pm}$.

Therefore, the integral expressions for the z-components of the total field (3.16) can be put into an equivalent form

$$\begin{bmatrix} H_z \\ E_z \end{bmatrix} = \begin{bmatrix} H_z^{\mathrm{e}} \\ E_z^{\mathrm{e}} \end{bmatrix} + \sum_{n=1}^{7} \frac{\mathrm{i}k^3}{8\pi^2} \int_{\mathrm{S}(\beta_0)} \int_{\mathrm{S}(\varphi_0)} \mathrm{e}^{\mathrm{i}kz\cos\beta}$$
$$\times \sin^2\beta \, \mathrm{e}^{\mathrm{i}kr_0 \sin\beta\cos(\alpha-\varphi_0)} \, \overline{r}_n \, \mathrm{d}\alpha \, \mathrm{d}\beta, \tag{3.31}$$

Table 3.1 Polar singularities and residues of $\overline{\overline{f}}(s + \varphi, \alpha, \beta)$

n	Polar Singularities s_n	Residue $\overline{\overline{R}}_n$
1	$\alpha - \varphi$	$\overline{\overline{I}}$
2	$2\Phi - \alpha - \varphi$	$\overline{\overline{R}}^{+}(\alpha, \beta)$
3	$-2\Phi - \alpha - \varphi$	$\overline{\overline{R}}^{-}(\alpha, \beta)$
4	$\pi + \Phi + \theta^+(\beta) - \varphi$	$\overline{\overline{R}}^{\mathrm{s}}_{\theta^+}(\alpha, \beta) \cdot \overline{\overline{f}}(\Phi - \pi - \theta^+(\beta), \alpha, \beta)$
5	$-\pi - \Phi - \theta^-(\beta) - \varphi$	$\overline{\overline{R}}^{\mathrm{s}}_{\theta^-}(\alpha, \beta) \cdot \overline{\overline{f}}(-\Phi + \pi + \theta^-(\beta), \alpha, \beta)$
6	$\pi + \Phi + \chi^+(\beta) - \varphi$	$\overline{\overline{R}}^{\mathrm{s}}_{\chi^+}(\alpha, \beta) \cdot \overline{\overline{f}}(\Phi - \pi - \chi^+(\beta), \alpha, \beta)$
7	$-\pi - \Phi - \chi^-(\beta) - \varphi$	$\overline{\overline{R}}^{\mathrm{s}}_{\chi^-}(\alpha, \beta) \cdot \overline{\overline{f}}(-\Phi + \pi + \chi^-(\beta), \alpha, \beta)$

where

$$\begin{bmatrix} H_z^e \\ E_z^e \end{bmatrix} = \frac{k^3}{16\pi^3} \int\limits_{S(\beta_0)} \int\limits_{S(\varphi_0)} \int\limits_{S(-\pi)+S(\pi)} \sin^2\!\beta\, \overline{\overline{f}}(s+\varphi,\alpha,\beta)\cdot \overline{U}_0(\alpha,\beta) \times e^{ik\{z\cos\beta+\sin\beta[r_0\cos(\alpha-\varphi_0)-r\cos s]\}}\, ds\, d\alpha\, d\beta, \tag{3.32}$$

and

$$\overline{r}_n = \mathcal{H}_\sigma(s_n)\,\overline{\overline{R}}_n(\alpha,\beta)\cdot\overline{U}_0(\alpha,\beta)\exp(-ikr\sin\beta\cos s_n)\,. \tag{3.33}$$

Here $\mathcal{H}_\sigma(s)$ denotes a characteristic function for a domain σ on the complex s-plane. If $s \in \sigma$, there is $\mathcal{H}_\sigma(s) = 1$; otherwise $\mathcal{H}_\sigma(s) = 0$. In the present case, we assume that σ is the domain between the steepest-descent paths $S(-\pi)$ and $S(\pi)$ of the Sommerfeld contours on the complex s-plane.

The edge-diffracted field $\left[H_z^e,\ E_z^e\right]^T$ (3.32) can be rewritten in the following form:

$$\begin{bmatrix} H_z^e \\ E_z^e \end{bmatrix} = \frac{k^3}{16\pi^3} \int\limits_{S(\beta_0)} \int\limits_{S(\varphi_0)} \int\limits_{S(0)} \sin^2\!\beta\, \overline{\overline{D}}(s+\varphi,\alpha,\beta)\cdot \overline{U}_0(\alpha,\beta) \times e^{ik\{z\cos\beta+\sin\beta[r_0\cos(\alpha-\varphi_0)+r\cos s]\}}\, ds\, d\alpha\, d\beta\,, \tag{3.34}$$

with

$$\overline{\overline{D}}(s+\varphi,\alpha,\beta) = \overline{\overline{f}}(s+\pi+\varphi,\alpha,\beta) - \overline{\overline{f}}(s-\pi+\varphi,\alpha,\beta). \tag{3.35}$$

3.3.2 Branch cuts for auxiliary angles

As outlined in Chapter 2 and made conspicuous in Table 3.1, auxiliary angles $\theta^{\pm}$ and $\chi^{\pm}$ occur in the procedure of solution. They depend upon the angle of incidence θ_0 and the normalized surface impedances $\eta^{\pm}$. In the present case, they are given by

$$\sin\theta^{\pm}(\beta) = \frac{\eta^{\pm}}{\sin\beta}, \qquad \sin\chi^{\pm}(\beta) = \frac{1}{\eta^{\pm}\sin\beta}. \tag{3.36}$$

Therefore, branches of the roots $\theta^{\pm}(\beta)$ and $\chi^{\pm}(\beta)$ of equations (3.36) need be fixed. It suffices to confine, for example, our attention to $\theta^{+}(\beta)$ with $\operatorname{Im}\eta^{+} > 0$.

Equations (3.36) suggest as the starting point of our study the branch cuts of the inverse trigonometric function arcsin w, which lie on the real axis of the complex w-plane with $w = \operatorname{Re} w = p$ and $|p| > 1$. From the first equation of (3.36) it turns out that, along the branch cuts on the complex β-plane, $\sin\beta = \eta^{+}/p$ must hold good. It implies that one branch cut in the complex β-plane consists of an arc connecting the point arcsin η^{+} with the origin and a mirror image of this arc with respect to the origin. Owing to the periodicity of $\sin\beta$, this branch cut, shifted by an integral number of π along both the positive and negative real axis in the complex β-plane, are also branch cuts of $\theta^{+}(\beta)$ (cf. Fig. 3.3). If $\operatorname{Re}\eta^{+} = 0$, then the cuts run parallel to the imaginary axis of the complex β-plane. Likewise, branch cuts of $\theta^{+}(\beta)$ in the case of $\operatorname{Im}\eta^{+} < 0$ and branch cuts of $\theta^{-}(\beta)$ and $\chi^{\pm}(\beta)$ can be constructed.

As a consequence, the contours of integration in the complex β-plane $\Gamma_{\pi/2}$ and $S(\beta_0)$ must be modified in such a way that they are led around the branch cuts of $\theta^{\pm}(\beta)$ and $\chi^{\pm}(\beta)$.

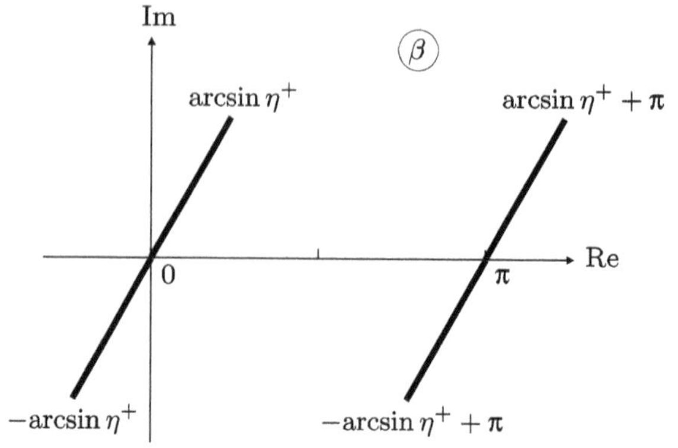

Figure 3.3 Branch cuts (thick lines) of $\theta^+(\beta)$ in the complex β-plane ($\mathrm{Im}\,\eta^+ > 0$).

3.3.3 The geometrical-optics field

In the present case, the geometrical-optics field may comprise the incident part, the parts bounced off the upper and lower faces of the wedge. These three ingredients can be obtained from Section 3.3.1, or more precisely from Table 3.1, (3.31), and (3.33). Because $s_1 = \alpha - \varphi$ and the integration in the complex α-plane is executed along $\mathrm{S}(\varphi_0)$, we have $\mathcal{H}_\sigma(s_1) = \mathrm{H}(\pi - |\varphi_0 - \varphi|)$. Using (3.14), we obtain the incident part of the geometrical-optics field:

$$\begin{bmatrix} H_z^{\mathrm{inc}} \\ E_z^{\mathrm{inc}} \end{bmatrix} = \begin{cases} \begin{bmatrix} H_z^{\mathrm{i}} \\ E_z^{\mathrm{i}} \end{bmatrix}, & |\varphi_0 - \varphi| < \pi \\ 0, & \text{otherwise} \end{cases} \tag{3.37}$$

The far-field expression for $\left[H_z^{\mathrm{i}}, E_z^{\mathrm{i}}\right]^{\mathrm{T}}$ turns out from an asymptotic evaluation of (3.14):

$$\begin{bmatrix} H_z^{\mathrm{i}} \\ E_z^{\mathrm{i}} \end{bmatrix} \approx \sin\beta_{\mathrm{i}} k^2 \frac{\exp(ikR_0)}{4\pi R_0} \overline{U}_0(\alpha_{\mathrm{i}}, \beta_{\mathrm{i}}), \tag{3.38}$$

with R_0 denoting the distance between the source and the field points, cf. (3.6), and

$$\alpha_{\mathrm{i}} = \arctan\frac{y_0 - y}{x_0 - x}, \quad \beta_{\mathrm{i}} = \arctan\frac{(x_0 - x)\cos\alpha_{\mathrm{i}} + (y_0 - y)\sin\alpha_{\mathrm{i}}}{z}.$$

These results can easily be verified using (3.6) and (3.7).

Analogously, the reflected waves can be obtained:

$$\begin{bmatrix} H_z^{\mathrm{r}} \\ E_z^{\mathrm{r}} \end{bmatrix} = \sum_{n=2}^{3} \int_{\mathrm{S}(\beta_0)} \mathrm{d}\beta \int_{\mathrm{S}(\varphi_0)} \mathrm{d}\alpha\, \mathrm{e}^{ikz\cos\beta} \frac{ik^3}{8\pi^2} (\sin\beta)^2 \mathrm{e}^{ikr_0 \sin\beta\cos(\alpha - \varphi_0)}$$
$$\times \mathcal{H}_\sigma(s_n) \overline{\overline{R}}^{\pm}(\alpha, \beta) \cdot \overline{U}_0(\alpha, \beta) \mathrm{e}^{-ikr\sin\beta\cos s_n}. \tag{3.39}$$

Because of $s_{2,3} = \pm 2\Phi - \alpha - \varphi$ (cf. Table 3.1) and of the integration in the complex α-plane being carried out along $\mathrm{S}(\varphi_0)$, there is $\mathcal{H}_\sigma(s_{2,3}) = \mathrm{H}(\pi - |\pm 2\Phi - \varphi_0 - \varphi|)$. The matrix reflection coefficients $\overline{\overline{R}}^{\pm}(\alpha, \beta)$, defined in (2.23)–(2.28) and (2.32), are reproduced for convenience:

$$\overline{\overline{R}}^{\pm}(\alpha,\beta) = -\frac{\overline{\overline{N}}^{\pm}(-\alpha,\beta)}{D^{\pm}(-\alpha,\beta)}, \tag{3.40}$$

with

$$\begin{aligned}
\overline{\overline{N}}^{\pm}(\alpha,\beta) &= \begin{bmatrix} N_1^{\pm}(\alpha,\beta) & N_2^{\pm}(\alpha,\beta) \\ N_3^{\pm}(\alpha,\beta) & N_4^{\pm}(\alpha,\beta) \end{bmatrix}, \\
N_1^{\pm}(\alpha,\beta) &= -2\cot^2\beta\sin^2(\alpha\pm\Phi) \\
&\quad - \left[\sin(\alpha\pm\Phi)\mp\sin\theta^{\pm}(\beta)\right]\left[\sin(\alpha\pm\Phi)\pm\sin\chi^{\pm}(\beta)\right], \\
N_2^{\pm}(\alpha,\beta) &= 2\cot\beta\csc\beta\sin(\alpha\pm\Phi)\cos(\alpha\pm\Phi), \\
N_3^{\pm}(\alpha,\beta) &= -N_2^{\pm}(\alpha,\beta), \\
N_4^{\pm}(\alpha,\beta) &= -2\cot^2\beta\sin^2(\alpha\pm\Phi) \\
&\quad - \left[\sin(\alpha\pm\Phi)\pm\sin\theta^{\pm}(\beta)\right]\left[\sin(\alpha\pm\Phi)\mp\sin\chi^{\pm}(\beta)\right], \\
D^{\pm}(\alpha,\beta) &= \left[\sin(\alpha\pm\Phi)\pm\sin\theta^{\pm}(\beta)\right]\left[\sin(\alpha\pm\Phi)\pm\sin\chi^{\pm}(\beta)\right].
\end{aligned} \tag{3.41}$$

Using $\overline{\overline{R}}^{\pm}(\alpha_0^{\pm},\beta_0^{\pm})$ in place of $\overline{\overline{R}}^{\pm}(\alpha,\beta)$, where $\alpha_0^{\pm}$, $\beta_0^{\pm}$ denote the simple saddle points of (3.39) in the respective cases, an approximation for the reflected parts of the geometrical-optics field follows:

$$\begin{aligned}
\begin{bmatrix} H_z^{\mathrm{r}} \\ E_z^{\mathrm{r}} \end{bmatrix} &\approx \sum_{n=2}^{3} \int_{S(\beta_0)} \mathrm{d}\beta \int_{S(\varphi_0)} \mathrm{d}\alpha\, \mathrm{e}^{\mathrm{i}kz\cos\beta}\, \frac{\mathrm{i}k^3}{8\pi^2}(\sin\beta)^2\, \mathrm{e}^{\mathrm{i}kr_0\sin\beta\cos(\alpha-\varphi_0)} \\
&\quad\times \mathcal{H}_\sigma(s_n)\, \overline{\overline{R}}^{\pm}(\alpha_0^{\pm},\beta_0^{\pm})\cdot\overline{U}_0(\alpha,\beta)\,\mathrm{e}^{-\mathrm{i}kr\sin\beta\cos[\alpha-(\pm 2\Phi-\varphi)]} \\
&= \begin{bmatrix} H_z^{\mathrm{r}+} \\ E_z^{\mathrm{r}+} \end{bmatrix} + \begin{bmatrix} H_z^{\mathrm{r}-} \\ E_z^{\mathrm{r}-} \end{bmatrix}
\end{aligned} \tag{3.42}$$

with

$$\begin{bmatrix} H_z^{\mathrm{r}\pm} \\ E_z^{\mathrm{r}\pm} \end{bmatrix} = \begin{cases} \overline{\overline{R}}^{\pm}(\alpha_0^{\pm},\beta_0^{\pm})\cdot\begin{bmatrix} H_z^{\mathrm{i}}(r,\pm 2\Phi-\varphi,z) \\ E_z^{\mathrm{i}}(r,\pm 2\Phi-\varphi,z) \end{bmatrix}, & |\pm 2\Phi-\varphi_0-\varphi|<\pi \\ 0, & \text{otherwise} \end{cases} \tag{3.43}$$

Evidently, the points with the cylindrical coordinates $(r,\pm 2\Phi-\varphi,z)$ are the images of the point of observation located at (r,φ,z) with respect to the upper and lower faces of the wedge. Therefore, the respective simple saddle points $\alpha_0^{\pm}$ and $\beta_0^{\pm}$ turn out geometrically from the point of observation (its image) and the source point:

$$\alpha_0^{\pm} = \arctan\frac{y_0 - r\sin(\pm 2\Phi-\varphi)}{x_0 - r\cos(\pm 2\Phi-\varphi)}, \quad \beta_0^{\pm} = \arccos\frac{z}{\psi^{\mathrm{r}\pm}}, \tag{3.44}$$

where $\psi^{\mathrm{r}\pm}$ denote the distances (eikonals) that the reflected waves travel from the source point via the respective points of reflection to the point of observation:

$$\psi^{\mathrm{r}\pm} = \sqrt{r^2 + r_0^2 - 2rr_0\cos(\pm 2\Phi-\varphi-\varphi_0) + z^2}. \tag{3.45}$$

These results for the reflected parts of the geometrical-optics field are traditional and hence could be directly obtained by the ray method without evaluating the double integral (3.39) for large arguments. This option also enables us to consider the appropriate branch of the square root of det S'' in the formula for the leading term of the method of steepest descents (3.63), see [3.10, Chapter 4, Section 1].

3.4 The diffracted wave from the edge of the wedge

3.4.1 Nonuniform expression

It is well-known that simple saddle points are responsible for edge-diffracted waves. By evaluating the multiple integral (3.34) by means of the many-dimensional steepest-descent method, and using the relations given in Section 3.3.1, we arrive at the nonuniform expression for the edge-diffracted space wave

$$\begin{bmatrix} H_z^{\mathrm{e}} \\ E_z^{\mathrm{e}} \end{bmatrix} \approx -\frac{\mathrm{e}^{ik\ell^{\mathrm{i}}}}{4\pi\ell^{\mathrm{i}}}\frac{\mathrm{e}^{i\pi/4}k^2}{\sqrt{2\pi k}}\sqrt{\frac{\ell^{\mathrm{i}}}{\ell\,(\ell+\ell^{\mathrm{i}})}}\,\mathrm{e}^{ik\ell}\,\overline{\overline{D}}(\varphi,\varphi_0,\beta_0)\cdot\overline{U}_0(\varphi_0,\beta_0).$$

Here the matrix diffraction coefficient $\overline{\overline{D}}(\varphi,\varphi_0,\beta_0)$ is defined in (3.35); the distance from the source point to the point of diffraction ℓ^{i} and the distance from the point of diffraction to the point of observation ℓ are given in Section 3.3.1.

In view of (3.15), this expression contains $\left[H_z^{\mathrm{i}}(O),\ E_z^{\mathrm{i}}(O)\right]^{\mathrm{T}}$, the z-components of the incident dipole field at a point O on the edge whose Cartesian coordinates are $(0, 0, z_O)$ with $z_O = r_0 z/(r_0 + r)$. The point O is usually called the point of edge diffraction, which is at the same time the apex of the so-called cone of edge diffraction.

Hence, the z-components of the nonuniform edge-diffracted field are expressible as

$$\begin{bmatrix} H_z^{\mathrm{e}} \\ E_z^{\mathrm{e}} \end{bmatrix} \approx -\frac{\mathrm{e}^{i\pi/4}}{\sqrt{2\pi k}\,\sin\beta_0}\sqrt{\frac{\ell^{\mathrm{i}}}{\ell\,(\ell+\ell^{\mathrm{i}})}}\,\mathrm{e}^{ik\ell}\,\overline{\overline{D}}(\varphi,\varphi_0,\beta_0)\cdot\begin{bmatrix} H_z^{\mathrm{i}}(O) \\ E_z^{\mathrm{i}}(O) \end{bmatrix}. \tag{3.46}$$

As is well-known, the diffracted field by a wedge under a point-source illumination exhibits the structure of an astigmatic ray because of the different radii of curvature ℓ and $\ell + \ell^{\mathrm{i}}$ of its wavefront.

Now move the Hertzian dipole into infinity with $\ell^{\mathrm{i}} \to \infty$ in such a way that the point of edge diffraction remains fixed and the incident field at that point $\left[H_z^{\mathrm{i}}(O), E_z^{\mathrm{i}}(O)\right]^{\mathrm{T}}$ is kept finite. Under this circumstance, the spreading factor becomes simply

$$\lim_{\ell^{\mathrm{i}}\to\infty}\sqrt{\frac{\ell^{\mathrm{i}}}{\ell\,(\ell+\ell^{\mathrm{i}})}} = \frac{1}{\sqrt{\ell}}.$$

By recalling the geometric relation $r = \ell\sin\beta_0$, equation (3.46) reduces to

$$\begin{bmatrix} H_z^{\mathrm{e}} \\ E_z^{\mathrm{e}} \end{bmatrix} \approx -\frac{\mathrm{e}^{i\pi/4}}{\sqrt{2\pi k\sin\beta_0}}\frac{\mathrm{e}^{ikr/\sin\beta_0}}{\sqrt{r}}\,\overline{\overline{D}}(\varphi,\varphi_0,\beta_0)\cdot\begin{bmatrix} H_z^{\mathrm{i}}(O) \\ E_z^{\mathrm{i}}(O) \end{bmatrix},$$

fully agreeing with the nonuniform part of the third term of (2.77) (see also (2.76)), the first-order uniform expression for the edge-diffracted field excited by an incident plane wave. Thereby it has been taken into account that in the latter case the point of edge diffraction is located at $z_O = z - r \cot \beta_0$.

The matrix diffraction coefficient $\overline{\overline{D}}(\varphi, \varphi_0, \beta_0)$ is related to the matrix spectral function $\overline{\overline{f}}(s, \alpha, \beta)$ in accordance with (3.35) at $s = \varphi \pm \pi$, $\alpha = \varphi_0$, and $\beta = \beta_0$. As indicated clearly by (3.35), the matrix diffraction coefficient for an incident spherical wave is identical with that for an incident plane wave with real-valued incident angles φ_0 and β_0. This fact renders the nonuniform expression for the edge-diffracted field (3.46), as well as its uniform expression (3.49) given in the next subsection, numerically efficient.

Departing from the expressions for both the geometrical-optics field (3.37), (3.42), and (3.43) and the edge-diffracted field (3.46), we are capable of deriving a uniform formulation in the framework of the uniform asymptotic theory of diffraction (UAT). This is a uniform version of the geometrical theory of diffraction, which modifies the discontinuous geometrical-optics field and keeps at the same time the nonuniform diffraction coefficient, and hence nonuniform edge-diffracted field, unchanged [3.11–3.13]. It is worth noting that such a procedure for devising a uniform expression out of nonuniform ones is the simplest case of the so-called parabolic cylinder-function Ansatz (PC Ansatz) (see the review [3.14]).

3.4.2 The UAT formulation

Let us recall at first the Fresnel integral $\mathrm{F}(\zeta)$ for a real-valued variable ζ defined by

$$\mathrm{F}(\zeta) = \frac{\mathrm{e}^{-\mathrm{i}\pi/4}}{\sqrt{\pi}} \int_{-\infty}^{\zeta} \mathrm{e}^{\mathrm{i}s^2} \mathrm{d}s. \tag{3.47}$$

Of particular interest is its asymptotic expression for $|\zeta| \gg 1$, namely,

$$\mathrm{F}(\zeta) \sim \mathrm{H}(\zeta) - \frac{\mathrm{e}^{\mathrm{i}\pi/4}}{2\zeta\sqrt{\pi}} \mathrm{e}^{\mathrm{i}\zeta^2} + \mathrm{O}\left(\frac{1}{\zeta^2}\right), \tag{3.48}$$

where $\mathrm{H}(\zeta)$ denotes the Heaviside unit-step function. It is reminded that the Fresnel integral $\mathrm{F}(\zeta)$ defined in (3.47) is related to the widely used cosine and sine Fresnel integrals $\mathrm{C}(x)$ and $\mathrm{S}(x)$ as follows:

$$\mathrm{F}(\zeta) = \frac{1}{2} + \frac{1-\mathrm{i}}{2}\left[\mathrm{C}\left(\sqrt{\frac{2}{\pi}}\zeta\right) + \mathrm{i}\,\mathrm{S}\left(\sqrt{\frac{2}{\pi}}\zeta\right)\right],$$

where

$$\mathrm{C}(x) = \int_0^x \cos\left(\frac{\pi}{2}\xi^2\right) \mathrm{d}\xi, \quad \mathrm{S}(x) = \int_0^x \sin\left(\frac{\pi}{2}\xi^2\right) \mathrm{d}\xi.$$

The properties of the Fresnel integral allow us to devise a first-order uniform expression out of the discontinuous geometrical-optics field (3.37, 3.42) and the nonuniform expression of the edge-diffracted field (3.46):

$$\begin{bmatrix} H_z \\ E_z \end{bmatrix} = \begin{bmatrix} H_z^{\mathrm{e}} \\ E_z^{\mathrm{e}} \end{bmatrix} + \left\{ \mathrm{F}\left[\sqrt{k\left(\psi^{\mathrm{e}} - \psi^{\mathrm{i}}\right)}\right] + \frac{\mathrm{e}^{\mathrm{i}\pi/4}\mathrm{e}^{\mathrm{i}k(\psi^{\mathrm{e}}-\psi^{\mathrm{i}})}}{2\sqrt{\pi}\sqrt{k\left(\psi^{\mathrm{e}} - \psi^{\mathrm{i}}\right)}} \right\} \begin{bmatrix} H_z^{\mathrm{i}} \\ E_z^{\mathrm{i}} \end{bmatrix}$$
$$+ \left\{ \mathrm{F}\left[\sqrt{k\left(\psi^{\mathrm{e}} - \psi^{\mathrm{r}_+}\right)}\right] + \frac{\mathrm{e}^{\mathrm{i}\pi/4}\mathrm{e}^{\mathrm{i}k(\psi^{\mathrm{e}}-\psi^{\mathrm{r}_+})}}{2\sqrt{\pi}\sqrt{k\left(\psi^{\mathrm{e}} - \psi^{\mathrm{r}_+}\right)}} \right\} \begin{bmatrix} H_z^{\mathrm{r}_+} \\ E_z^{\mathrm{r}_+} \end{bmatrix}$$
$$+ \left\{ \mathrm{F}\left[\sqrt{k\left(\psi^{\mathrm{e}} - \psi^{\mathrm{r}_-}\right)}\right] + \frac{\mathrm{e}^{\mathrm{i}\pi/4}\mathrm{e}^{\mathrm{i}k(\psi^{\mathrm{e}}-\psi^{\mathrm{r}_-})}}{2\sqrt{\pi}\sqrt{k\left(\psi^{\mathrm{e}} - \psi^{\mathrm{r}_-}\right)}} \right\} \begin{bmatrix} H_z^{\mathrm{r}_-} \\ E_z^{\mathrm{r}_+} \end{bmatrix}. \tag{3.49}$$

Here, ψ^{e}, ψ^{i}, and $\psi^{\mathrm{r}_\pm}$ are the eikonals of the edge-diffracted, incident, and (from the upper and lower faces of the wedge) reflected rays with

$$\psi^{\mathrm{e}} = \ell^{\mathrm{i}} + \ell, \quad \psi^{\mathrm{i}} = \sqrt{(x-x_0)^2 + (y-y_0)^2 + z^2}, \tag{3.50}$$

and $\psi^{\mathrm{r}_\pm}$ defined in (3.45).

The sign of the square roots contained in (3.49) depends upon the location of the point of observation. For example, $\sqrt{k\left(\psi^{\mathrm{e}} - \psi^{\mathrm{i}}\right)}$ is taken to be positive, that is $\left|\sqrt{k\left(\psi^{\mathrm{e}} - \psi^{\mathrm{i}}\right)}\right|$, in the lit region and negative, namely, $-\left|\sqrt{k\left(\psi^{\mathrm{e}} - \psi^{\mathrm{i}}\right)}\right|$, in the shadow of the incident part of the geometrical-optics field; the remaining square roots of (3.49) are to be dealt with accordingly. The so-defined arguments of the Fresnel integral in (3.49) are called detour parameters [3.13]. Expression (3.49) implies that the incident and reflected parts of the geometrical-optics field have to be extended into their respective shadows.

That the uniform expression (3.49) reduces to the nonuniform one is easily seen by replacing the respective Fresnel integral in (3.49) with its asymptotic formula (3.48). To show that (3.49) is indeed finitely valued at and continuous across different shadow boundaries, let us consider as an example a small vicinity of the shadow boundary of the incident wave at $\varphi = \varphi_0 - \pi$.

To this end, we recall that the first term on the right-hand side of (3.24) corresponds to the incident part of the spectral functions. Together with (3.35), the principal part of the matrix diffraction coefficient can be written as

$$\overline{\overline{D}} = \overline{\overline{I}}/\delta\varphi + \cdots, \quad \delta\varphi = \varphi - (\varphi_0 - \pi).$$

Hence, the diffracted part tends to infinity as $1/\delta\varphi$, that is, inversely proportional to the angular deviation from the shadow boundary under consideration.

On the other side, by means of simple geometric relations and in line with the definition of the detour parameters there is

$$\sqrt{\psi^{\mathrm{e}} - \psi^{\mathrm{i}}} \approx \delta\varphi\sqrt{r_0 r/\left(2\psi^{\mathrm{e}}\right)},$$

causing the singular part of the second term of (3.49) to grow at the same speed as the diffracted part does, but with an opposite sign, hence canceling exactly the singularity of the latter. See also Section 3.6.

3.5 Expressions for surface waves

3.5.1 Surface waves excited directly by the dipole

When extracting the reflected waves from the rigorous expression (3.39), the method of steepest descents can also be applied to the evaluation of the integration there in α, hence shifting the contour of integration $\mathrm{S}(\varphi_0)$ to the required one $\mathrm{S}(\alpha_0^{\pm})$. At the end, one arrives again at (3.42) for the reflected waves.

During the deformation of the path of integration, however, polar singularities might be traversed under certain conditions. As indicated by (3.39), these polar singularities are the zeros of $D^{\pm}(-\alpha, \beta)$, the common denominators of the matrix reflection coefficients $\overline{\overline{R}}^{\pm}(\alpha, \beta)$ given explicitly in (3.40). From (3.41), the definition for $D^{\pm}(\alpha, \beta)$, it turns out that the singular polarities that need be taken into consideration are the following:

$$\alpha_{\zeta^{\pm}}(\beta) = \pm\left(\Phi - \zeta^{\pm} - \pi\right), \quad \zeta^{\pm} \in \left\{\theta^{\pm}(\beta), \chi^{\pm}(\beta)\right\}. \tag{3.51}$$

They can be captured if they lie between the contours $\mathrm{S}(\varphi_0)$ and $\mathrm{S}(\alpha_0^{\pm})$. The corresponding waves are denoted by $\left[H_z^{\mathrm{s}}, E_z^{\mathrm{s}}\right]^{\mathrm{T}}$ and read

$$\begin{bmatrix} H_z^{\mathrm{s}} \\ E_z^{\mathrm{s}} \end{bmatrix} = \begin{bmatrix} H_z^{\mathrm{s},\theta^+} \\ E_z^{\mathrm{s},\theta^+} \end{bmatrix} + \begin{bmatrix} H_z^{\mathrm{s},\theta^-} \\ E_z^{\mathrm{s},\theta^-} \end{bmatrix} + \begin{bmatrix} H_z^{\mathrm{s},\chi^+} \\ E_z^{\mathrm{s},\chi^+} \end{bmatrix} + \begin{bmatrix} H_z^{\mathrm{s},\chi^-} \\ E_z^{\mathrm{s},\chi^-} \end{bmatrix} \tag{3.52}$$

with

$$\begin{aligned} \begin{bmatrix} H_z^{\mathrm{s},\zeta^{\pm}} \\ E_z^{\mathrm{s},\zeta^{\pm}} \end{bmatrix} = \pm\frac{k^3}{4\pi} \int\limits_{\mathrm{S}(\beta_0)} & \mathrm{d}\beta\,(\sin\beta)^2 \mathrm{e}^{\mathrm{i}kz\cos\beta} \mathcal{H}_\sigma(s_{2,3}) \\ & \times \mathcal{H}_\sigma(\alpha_{\zeta^{\pm}}(\beta))\, \mathrm{e}^{\mathrm{i}k\sin\beta\left[r\cos(\Phi\mp\varphi+\zeta^{\pm}) - r_0\cos(\Phi\mp\varphi_0-\zeta^{\pm})\right]} \\ & \times \overline{\overline{R}}^{\mathrm{s}}_{\zeta^{\pm}} \cdot \overline{U}_0\left(\alpha_{\zeta^{\pm}}(\beta), \beta\right). \end{aligned} \tag{3.53}$$

The factors $\overline{\overline{R}}^{\mathrm{s}}_{\zeta^{\pm}}$ in the excitation coefficients of surface waves are defined in (2.67–2.71), but now with $a_{12}^{\pm} = a_{21}^{\pm} = \eta^{\pm}$.

Carrying out the integration in (3.53) by means of the method of steepest descents, one arrives at an asymptotic expression for this kind of surface waves. The in general complex-valued saddle points $\beta_{\mathrm{s}}^{\zeta^{\pm}}$ solve the transcendental equation

$$-z\sin\beta + \frac{\cos\beta}{\cos\zeta^{\pm}(\beta)}\left[r\cos\left(\Phi\mp\varphi\right) - r_0\cos\left(\Phi\mp\varphi_0\right)\right] = 0,$$

where use has been made of the relation $\mathrm{d}\zeta^{\pm}(\beta)/\mathrm{d}\beta = -\tan\zeta^{\pm}(\beta)\cot\beta$. It is reminded that the range of the saddle points $\beta_{\mathrm{s}}^{\zeta^{\pm}}$ is fixed by the two characteristic functions contained in (3.53).

At the saddle points we have

$$\begin{aligned} S\left(z^0\right) &= \mathrm{i}\left\{z\cos\beta_{\mathrm{s}}^{\zeta^{\pm}} + \sin\beta_{\mathrm{s}}^{\zeta^{\pm}}\left[r\cos\left(\Phi\mp\varphi+\zeta^{\pm}\left(\beta_{\mathrm{s}}^{\zeta^{\pm}}\right)\right)\right.\right. \\ &\qquad \left.\left. - r_0\cos\left(\Phi\mp\varphi_0-\zeta^{\pm}\left(\beta_{\mathrm{s}}^{\zeta^{\pm}}\right)\right)\right]\right\}, \\ \det S''\left(z^0\right) &= -\mathrm{i}z\left[1+\tan^2\zeta^{\pm}\left(\beta_{\mathrm{s}}^{\zeta^{\pm}}\right)\cos^2\beta_{\mathrm{s}}^{\zeta^{\pm}}\right]/\cos\beta_{\mathrm{s}}^{\zeta^{\pm}}. \end{aligned}$$

Therefore, the asymptotic expression for this kind of surface waves arises from evaluating the contour integral of (3.53) by means of the one-dimensional saddle-point method:

$$\begin{bmatrix} H_z^{s,\zeta\pm} \\ E_z^{s,\zeta\pm} \end{bmatrix} \approx \pm\sqrt{\frac{2\pi}{k}} \left[\det S''\left(z^0\right)\right]^{-1/2} \mathrm{e}^{kS(z^0)} \frac{k^3}{4\pi} \sin^2 \beta_s^{\zeta^\pm} \mathcal{H}_\sigma\left(s_{2,3}\right)$$
$$\times \mathcal{H}_\sigma\left(\alpha_{\zeta^\pm}\left(\beta_s^{\zeta^\pm}\right)\right) \overline{\overline{R}}{}^{s}_{\zeta^\pm} \cdot \overline{U}_0\left(\alpha_{\zeta^\pm}\left(\beta_s^{\zeta^\pm}\right), \beta_s^{\zeta^\pm}\right). \tag{3.54}$$

The right branch of the square root in this formula is chosen in a standard manner, see for instance [3.10, 3.15] and [3.16].

3.5.2 Surface waves excited at the edge by an incident space wave

The surface waves described in the previous subsection are precisely of the same type as those excited by a dipole above an infinitely large impedance plane. In this subsection, we discuss a kind of wedge-specific surface waves. This kind of surface waves is excited by the incident space wave and "born" at the edge of the wedge and then outgoes to infinity along the wedge's faces. Like the surface waves (3.53), they depend upon the value of the imaginary part of $\chi^\pm$ and of $\theta^\pm$, and hence of the normalized surface impedances $\eta^\pm$.

Let us look now at this kind of surface waves related to $\zeta^\pm$ whose definition is given in (3.51). In accordance with Table 3.1 and (3.31, 3.33), they are given by

$$\begin{bmatrix} H_z^{sd} \\ E_z^{sd} \end{bmatrix} = \begin{bmatrix} H_z^{sd,\theta^+} \\ E_z^{sd,\theta^+} \end{bmatrix} + \begin{bmatrix} H_z^{sd,\theta^-} \\ E_z^{sd,\theta^-} \end{bmatrix} + \begin{bmatrix} H_z^{sd,\chi^+} \\ E_z^{sd,\chi^+} \end{bmatrix} + \begin{bmatrix} H_z^{sd,\chi^-} \\ E_z^{sd,\chi^-} \end{bmatrix} \tag{3.55}$$

with

$$\begin{bmatrix} H_z^{sd,\zeta^\pm} \\ E_z^{sd,\zeta^\pm} \end{bmatrix} = \frac{ik^3}{8\pi^2} \int\limits_{S(\beta_0)} d\beta \int\limits_{S(\varphi_0)} d\alpha\, (\sin\beta)^2$$
$$\times \mathrm{e}^{ik\{z\cos\beta + \sin\beta[r_0\cos(\alpha-\varphi_0) + r\cos(\Phi\mp\varphi+\zeta^\pm)]\}}$$
$$\times \mathrm{H}\left(\pm\varphi - \Phi - \mathrm{Re}\,\zeta^\pm - \mathrm{gd}\left(\mathrm{Im}\,\zeta^\pm\right)\right)$$
$$\times \overline{\overline{R}}{}^{s}_{\zeta^\pm}(\alpha,\beta) \cdot \overline{f}\left(\pm\left(\Phi - \pi - \zeta^\pm\right), \alpha, \beta\right). \tag{3.56}$$

Here $\mathrm{H}(x)$ is the Heaviside function. Remember that the complex angles $\zeta^\pm$ depend upon β and $\eta^\pm$ according to (3.36).

To evaluate this expression we use the method of steepest descents. The saddle points of interest are located at $\alpha = \varphi_0$, $\beta = \beta_{sd}^{\zeta^\pm}$, where $\beta_{sd}^{\zeta^\pm}$ are roots of the transcendental equation

$$-z\sin\beta + \cos\beta\left[r_0 + r\frac{\cos(\Phi\mp\varphi)}{\cos\zeta^\pm(\beta)}\right] = 0.$$

and are limited in their range by the Heaviside function contained in (3.56). That the saddle points have real-valued $\alpha = \varphi_0$ explains why the related surface waves are regarded as excited at the edge by an incident space wave radiated by the dipole.

At these saddle points the following relations hold:

$$\det S''\left(z^0\right) = -r_0\left[z\tan\beta_{\rm sd}^{\zeta^\pm} + r\cos^2\beta_{\rm sd}^{\zeta^\pm}\cos(\Phi\mp\varphi)\frac{\sin^2\beta_{\rm sd}^{\zeta^\pm}}{\cos^3\beta_{\rm sd}^{\zeta^\pm}}\right],$$

$$S\left(z^0\right) = \mathrm{i}\left\{z\cos\beta_{\rm sd}^{\zeta^\pm} + \sin\beta_{\rm sd}^{\zeta^\pm}\left[r_0 + r\cos\left(\Phi\mp\varphi+\zeta^\pm\left(\beta_{\rm sd}^{\zeta^\pm}\right)\right)\right]\right\}.$$

By using the many-dimensional method of steepest descents we get

$$\begin{aligned}\begin{bmatrix} H_z^{\mathrm{sd},\zeta^\pm} \\ E_z^{\mathrm{sd},\zeta^\pm}\end{bmatrix} &\approx \frac{ik^2}{4\pi}\left(\sin\beta_{\rm sd}^{\zeta^\pm}\right)^2\left[\det S''\left(z^0\right)\right]^{-1/2} \mathrm{e}^{kS(z^0)} \\ &\times \mathrm{H}\left(\pm\varphi-\Phi-\mathrm{Re}\,\zeta^\pm\left(\beta_{\rm sd}^{\zeta^\pm}\right)-\mathrm{gd}\left(\mathrm{Im}\,\zeta^\pm\left(\beta_{\rm sd}^{\zeta^\pm}\right)\right)\right) \\ &\times \overline{\overline{R}}^{\,\mathrm{S}}_{\zeta^\pm}\left(\varphi_0,\beta_{\rm sd}^{\zeta^\pm}\right)\cdot\overline{f}\left(\pm\left(\Phi-\pi-\zeta^\pm\left(\beta_{\rm sd}^{\zeta^\pm}\right)\right),\varphi_0,\beta_{\rm sd}^{\zeta^\pm}\right).\end{aligned} \tag{3.57}$$

For a successive application of the one-dimensional saddle-point method to (3.56) leads to the same result, the right branch of the square root in the above formula is accordingly chosen.

As $kr_0 \to \infty$,[5] in a standard manner the expressions reduce to those in Chapter 2 for the case of the plane-wave incidence. The branch of the square root in (3.57) is also verified by this limiting procedure.

3.6 Numerical results

As pointed out in Section 3.4, because of the real-valued angles φ_0 and θ_0, the numerical computation of either the nonuniform expression for the edge-diffracted space wave (3.46) or the first-order uniform expression for the total space field (3.49) is to be performed in the same manner as in the case of plane-wave diffraction by an impedance wedge discussed in detail in Section 2.7. Therefore, on use of the experiences garnered there, a similarly efficient, that is accurate and fast, solution procedure has been obtained.

To show the attained accuracy in a special case, let us take a PEC wedge as an example, whose UAT formulation has been put forward in [3.13] with the respective diffraction coefficient matrix $\overline{\overline{D}}^{\,\mathrm{PEC}}$ as follows:

$$\overline{\overline{D}}^{\,\mathrm{PEC}}(\varphi,\varphi_0,\beta_0) = \mathrm{diag}\left[D_{\rm h}(\varphi,\varphi_0,\beta_0),\; D_{\rm s}(\varphi,\varphi_0,\beta_0)\right]^{\mathrm{T}}. \tag{3.58}$$

Here, $D_{\rm h,s}(\varphi,\varphi_0,\beta_0)$ stand for the diffraction coefficients for an acoustically hard (soft) wedge:

$$D_{\rm h,s}(\varphi,\varphi_0,\beta_0) = \frac{\mu\sin(\mu\pi)}{\cos\mu(\varphi-\varphi_0)-\cos(\mu\pi)} \mp \frac{\mu\sin(\mu\pi)}{\cos\mu(\varphi+\varphi_0)+\cos(\mu\pi)}. \tag{3.59}$$

As displayed in Fig. 3.4, the numerical results based on the procedure described in this chapter for $\eta_\pm = 10^{-20}$ agrees fully with those obtained from the UAT formulas for a PEC wedge, that is, using $\overline{\overline{D}}^{\,\mathrm{PEC}}$ in place of $\overline{\overline{D}}$ in the nonuniform expression (3.46) and hence in the first-order uniform expression (3.49). In addition, the well-known fact that a PEC wedge

[5] In this case $\beta_{\rm sd}^{\zeta^\pm}$ becomes purely real and is no longer dependent upon either Φ, φ, or $\zeta^\pm$.

does not cause depolarization of the incident wave has also been called upon to confirm the accuracy of the present procedure in this case.

For the numerical results depicted in Fig. 3.4, the modulus of $\boldsymbol{P}$ has been chosen in such a way that the maximum field strength generated by the Hertzian dipole at a normalized distance $kR_0 = 1000$ equals 1 mV/m. The orientation of the Hertzian dipole and hence vector $\boldsymbol{P}$ is determined by the angles θ_s and φ_s, in line with

$$P_x = P \sin\theta_s \cos\varphi_s, \quad P_y = \sin\theta_s \sin\varphi_s, \quad P_z = -P\cos\theta_s. \tag{3.60}$$

Here P denotes the complex amplitude of $\boldsymbol{P}$.

As in the case of diffraction of a skew-incident plane electromagnetic wave at an impedance wedge (see for instance [3.8, 3.9] and Chapter 2), not only copolarized field components but also cross-polarized field components exist in the case of diffraction of a spherical electromagnetic wave by an impedance wedge.[6] Such an example is depicted in Fig. 3.5.

Shown in this figure and also in Fig. 3.6 are the nonuniform diffracted fields given by (3.46): the first term on the right-hand side of (3.49), the "modified" geometrical-optics (GO) parts described by the remaining terms on the right-hand side of (3.49), and their sums. As is evident, both the nonuniform diffracted fields and the modified GO parts grow without bound at the shadow boundaries of incidence (in this example located at $\varphi = \varphi_0 - \pi = -120°$) and reflection (located at $\varphi = 2\Phi - \pi - \varphi_0 = 90°$); but their singularities cancel each other precisely to give continuous and smooth total fields across these shadow boundaries.

As usual, the cross-polarized diffracted fields are mainly due to the (on the spindle-like wave fronts of the edge-diffracted rays [3.17]) transversal diffusion of waves from the "sources" distributed along the shadow boundary of reflection. See in particular the nonuniform diffracted fields in Fig. 3.5. (A fine exposition of the concept of transversal wave diffusion can be found in [3.18].)

For the sake of completeness, Fig. 3.6 shows the respective copolarized fields as a function of the azimuthal angle φ. It is noted that, as indicated by the nonuniform diffracted parts displayed in Fig. 3.6, the copolarized components of diffraction are fed by "sources" located at both the shadow boundary of reflection and the shadow boundary of incidence.

As made conspicuous by Figs. 3.4–3.6, the obtained results are everywhere continuous and smooth, especially around the shadow boundaries. This is a well-known advantage of the uniform asymptotic theory of diffraction compared with the uniform geometrical theory of diffraction (UTD), an alternative formulation that employs only the incident wave field at the point of diffraction for the diffracted field (for a thorough comparison between these two widely used uniform versions of Keller's GTD the reader is referred to [3.17]).

In the case of diverse surface-waves ingredients in the total field, a part of which has been studied in Section 3.5, however, a careful extension of the procedure described in Section 2.7 is to be performed in the future.

---

[6]This can be inferred from the fact that the matrix diffraction coefficient for a PEC wedge is always diagonal, whereas that for an impedance wedge has in general nonvanishing antidiagonal entries.

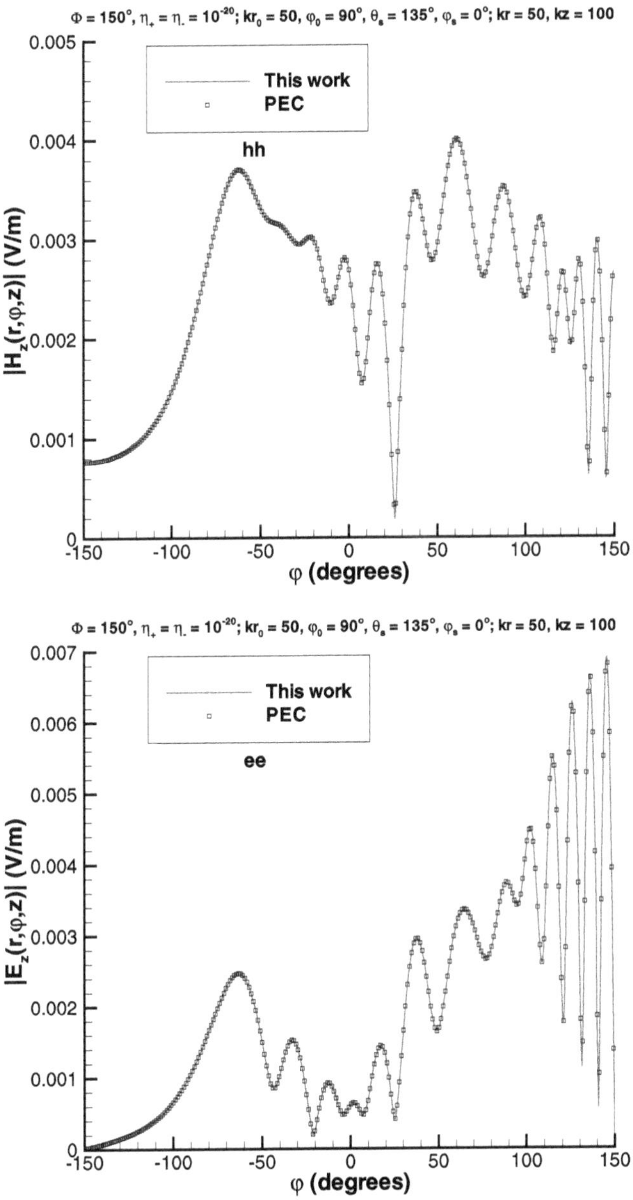

Figure 3.4 Total fields excited by a Hertzian dipole in the presence of a PEC wedge. Comparison with the UAT solution for a PEC wedge [3.13].

3.7 Appendices

3.7.1 Appendix A. Multidimensional saddle-point method

As can be seen in Chapters 1 and 2, the exact solutions to plane-wave diffraction by an impedance wedge under either normal or skew incidence are expressed in terms of the

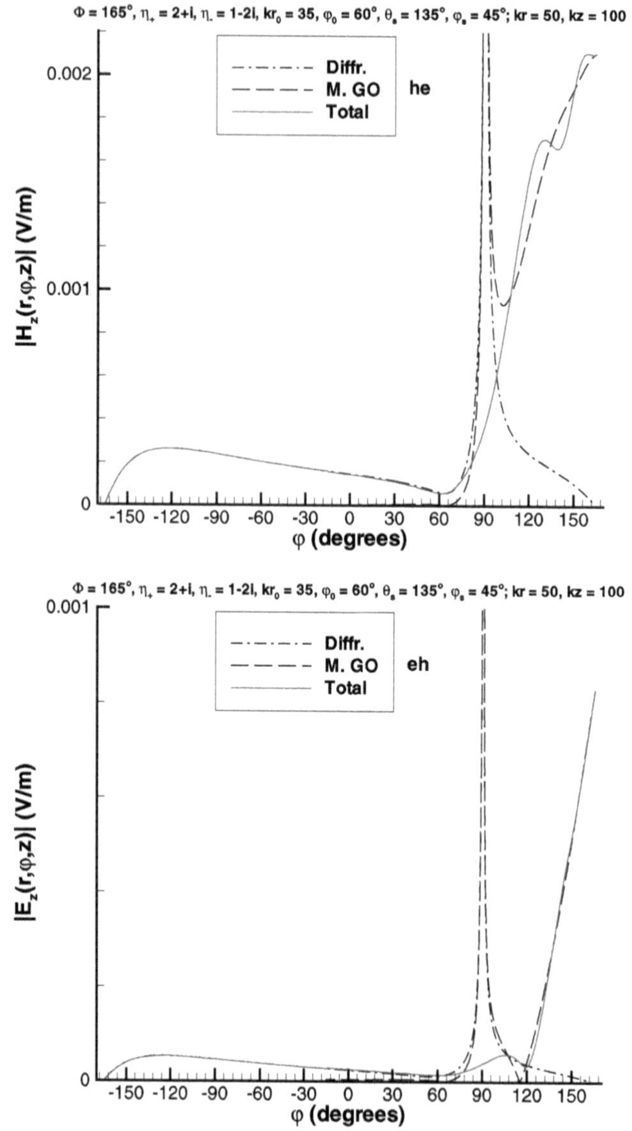

Figure 3.5 Cross-polarized field components excited by a Hertzian dipole in the presence of an impedance wedge (dash-dot line: nonuniform diffracted fields, long-dash line: modified GO fields, solid line: total fields).

Sommerfeld integrals, that is integrals along a contour in the complex α-plane. To deduce the far-field behavior, we have carried out an asymptotic evaluation of the Sommerfeld integrals by means of the classical saddle-point technique (see for instance [3.19–3.21]).

To express either the wave-field generated by a Hertzian dipole or the exact solution to diffraction of a dipole-field by an impedance wedge, however, integration over two and three complex variables needs be carried out, see (3.14), and (3.31)–(3.32). To compute

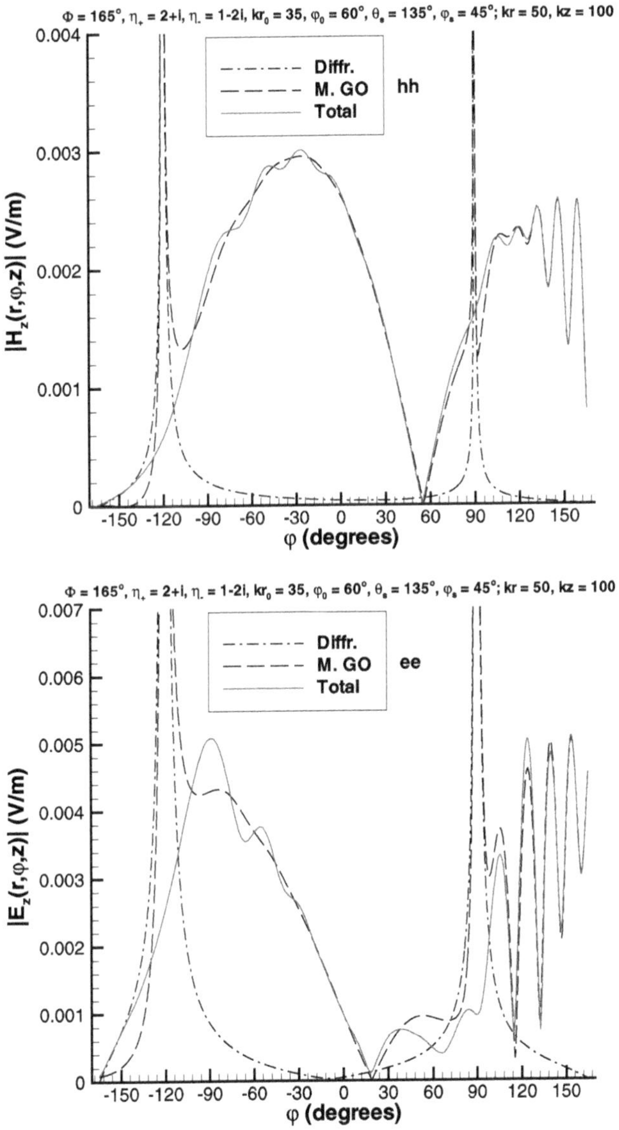

Figure 3.6 Copolarized field components excited by a Hertzian dipole in the presence of an impedance wedge (dash-dot line: nonuniform diffracted fields, long-dash line: modified GO fields, solid line: total fields).

asymptotically these integrals is a formidable task. As rightly observed by Ursell in his 1980 paper [3.22], "much less is known about" double complex integrals, let alone treble or higher-order ones. Even today one can find mainly treatment of asymptotic evaluation of multidimensional real integrals in the literature, see for instance [3.23–3.25]. The sole exception seems to be the late Fedoryuk, whose 1977 book contains a chapter on multidimensional saddle-point technique [3.26]. See also [3.10, 3.15].

We give the main results of the multidimensional saddle-point method, based on [3.10, 3.15]. The multiple complex integral in question is given by

$$F(\lambda) = \int_{\gamma^n} f(z) \exp\{\lambda S(z)\} \, dz. \tag{3.61}$$

Here λ is a large positive parameter, γ^n denotes an n-dimensional contour, $f(z)$ and $S(z)$ are analytic functions of the n complex variables $z = (z_1, \ldots, z_n)$ in some domain containing γ^n, and there is $dz = dz_1 \ldots dz_n$.

As in the more familiar one-dimensional case, the main contributions are expected to stem from the critical points of the exponent $S(z)$, the so-called saddle points. In a similar manner, a multidimensional saddle point z^0 is a point at which there is

$$\text{grad}\, S(z)\big|_{z=z^0} = 0. \tag{3.62}$$

If in addition the Hessian matrix at such a point z^0 is not singular, that is $\det S''\left(z^0\right) \neq 0$, then one speaks of a simple saddle point.

In case the end points of the contour γ^n are located in different "valleys" with $\text{Re}\, S(z) < \text{Re}\, S\left(z^0\right)$, then γ^n can be deformed to a saddle contour $\widetilde{\gamma}^n$ defined by the minimax property; see [3.15]. As shown by Fedoryuk in [3.26], if the maximum of $\text{Re}\, S(z)$ is attained at such a simple saddle point z^0 on the saddle contour $\widetilde{\gamma}^n$, then as $\lambda \to \infty$, the first term of an asymptotic expansion reads

$$F(\lambda) \sim \left(\frac{2\pi}{\lambda}\right)^{n/2} \left[\det S''\left(z^0\right)\right]^{-1/2} \exp\left\{\lambda S\left(z^0\right)\right\} f\left(z^0\right). \tag{3.63}$$

According to Fedoryuk, the branch of the root has to be chosen in accordance with the orientation of the contour.

Because no result is known yet for the possible coalesce of a polar singularity, that is, the pole of the complex amplitude $f(z)$, and a saddle point, the uniform expressions given in Section 3.4.2 have been constructed from the nonuniform expressions with the aid of a kind of asymptotic matching.

For recent works on extending the classical saddle-point technique to multidimensional integrals over several complex variables, the readers are referred to [3.27, 3.28].

3.7.2 Appendix B. The reciprocity principle

Sometimes the reciprocity principle enables one to extend the range of applicability of deduced formulas for the fields including those with the point-source illumination. For example, having a formula derived for $r > r_0$ (or $r < r_0$) then by means of the reciprocity, one can easily obtain the corresponding extension of the result for $r < r_0$ ($r > r_0$) by simply interchanging the locations of the source and of the observation point. In this appendix we consider the reciprocity principle in the problem for a point source illuminating an impedance wedge.

Let the electromagnetic field satisfy Maxwell's equations outside the wedge (see also (3.1)),

$$\mathrm{curl}\boldsymbol{E} = \mathrm{i}k\boldsymbol{H} - \boldsymbol{M}\,, \quad \mathrm{curl}\boldsymbol{H} = -\mathrm{i}k\boldsymbol{E} + \boldsymbol{J}\,, \tag{3.64}$$

we recall that $\boldsymbol{H}$ and $\boldsymbol{J}$ denote the magnetic field strength and electric current density multiplied with Z_0, the intrinsic impedance of free space. As usual, k is the wavenumber. The wave field is generated by compact sources, like an electric dipole

$$\boldsymbol{J} = -\mathrm{i}\,k\boldsymbol{P}_{\mathrm{e}}\,\delta\,(x - x_0)\,\delta\,(y - y_0)\,\delta\,(z - z_0) \tag{3.65}$$

and a fictitious magnetic dipole

$$\boldsymbol{M} = -\mathrm{i}\,k\boldsymbol{P}_{\mathrm{m}}\,\delta\,(x - x_0)\,\delta\,(y - y_0)\,\delta\,(z - z_0)\,. \tag{3.66}$$

On the wedge's faces the impedance boundary conditions are satisfied (3.3) ($\mathrm{Re}\,\eta > 0$):

$$\boldsymbol{E} - (\boldsymbol{E}\cdot\boldsymbol{n})\,\boldsymbol{n} = -\eta\,\boldsymbol{n}\times\boldsymbol{H} \tag{3.67}$$

and the radiation conditions at infinity ($R = \sqrt{x^2 + y^2 + z^2} \to \infty$) written in the form

$$\boldsymbol{E},\,\boldsymbol{H} \sim \mathrm{O}\left(R^{-1}\right), \tag{3.68}$$

$$\boldsymbol{E} + \boldsymbol{e}_R \times \boldsymbol{H} = \mathrm{o}\left(R^{-1}\right)\,, \quad \boldsymbol{H} - \boldsymbol{e}_R \times \boldsymbol{E} = \mathrm{o}\left(R^{-1}\right)\,. \tag{3.69}$$

The Meixner conditions are assumed in the form of the estimates ($r = \sqrt{x^2 + y^2} \to 0$)

$$E_z,\, H_z \sim C_0(z) + C(z)r^{\delta}, \quad E_r,\, H_r,\, E_\varphi,\, H_\varphi \sim C(z)r^{\delta-1}, \quad \delta > 0 \tag{3.70}$$

which are valid provided that $C_0(z)$, $C(z)$ are uniformly bounded with respect to z, where (r, φ, z) are the cylindrical coordinates associated with the edge of the wedge (see Fig. 3.1).

Let $\boldsymbol{E}'$, $\boldsymbol{H}'$ denote another electromagnetic field generated by the currents (compact sources) $\boldsymbol{J}'$, $\boldsymbol{M}'$ and this field be governed by the "primed" Maxwell's equations and the conditions (3.64)–(3.70). We multiply the equations (3.64) by $\boldsymbol{H}'$ and $\boldsymbol{E}'$, subtract them, and then integrate the result over the domain $\Omega_{\varepsilon,R}$. The domain $\Omega_{\varepsilon,R}$ is the region outside the wedge and the cylindrical surface $r = \varepsilon$ and inside the sphere $x^2 + y^2 + z^2 = R^2$.

Thus, we obtain[7]

$$\int\limits_{\Omega_{\varepsilon,R}} \left(\boldsymbol{M}\cdot\boldsymbol{H}' - \boldsymbol{J}\cdot\boldsymbol{E}'\right)\mathrm{d}V = \int\limits_{\Omega_{\varepsilon,R}} \left[(\mathrm{i}k\boldsymbol{H} - \mathrm{curl}\boldsymbol{E})\cdot\boldsymbol{H}' - (\mathrm{i}k\boldsymbol{E} + \mathrm{curl}\boldsymbol{H})\cdot\boldsymbol{E}'\right]\mathrm{d}V$$

$$= \int\limits_{\Omega_{\varepsilon,R}} \left[\left(\mathrm{i}k\boldsymbol{H}' - \mathrm{curl}\boldsymbol{E}'\right)\cdot\boldsymbol{H} - \left(\mathrm{i}k\boldsymbol{E}' + \mathrm{curl}\boldsymbol{H}'\right)\cdot\boldsymbol{E}\right]\mathrm{d}V$$

---

[7]The identity $\boldsymbol{A}\cdot\mathrm{curl}\boldsymbol{B} - \boldsymbol{B}\cdot\mathrm{curl}\boldsymbol{A} = \mathrm{div}(\boldsymbol{B}\times\boldsymbol{A})$ will be exploited in the following derivation.

$$+\int\limits_{\Omega_{\varepsilon,R}} \left[(\boldsymbol{H}\cdot \mathrm{curl}\boldsymbol{E}' - \boldsymbol{E}'\cdot \mathrm{curl}\boldsymbol{H}) - (\boldsymbol{H}'\cdot \mathrm{curl}\boldsymbol{E} - \boldsymbol{E}\cdot \mathrm{curl}\boldsymbol{H}') \right] \mathrm{d}V$$

$$= \int\limits_{\Omega_{\varepsilon,R}} \left(\boldsymbol{M}'\cdot\boldsymbol{H} - \boldsymbol{J}'\cdot\boldsymbol{E}\right)\mathrm{d}V + \int\limits_{\Omega_{\varepsilon,R}} \mathrm{div}\left(\boldsymbol{E}'\times\boldsymbol{H} - \boldsymbol{E}\times\boldsymbol{H}'\right)\mathrm{d}V$$

$$= \int\limits_{\Omega_{\varepsilon,R}} \left(\boldsymbol{M}'\cdot\boldsymbol{H} - \boldsymbol{J}'\cdot\boldsymbol{E}\right)\mathrm{d}V + \int\limits_{\partial\Omega_{\varepsilon,R}} \left(\boldsymbol{E}'\times\boldsymbol{H} - \boldsymbol{E}\times\boldsymbol{H}'\right)\cdot\boldsymbol{n}\,\mathrm{d}S$$

$$= \int\limits_{\Omega_{\varepsilon,R}} \left(\boldsymbol{M}'\cdot\boldsymbol{H} - \boldsymbol{J}'\cdot\boldsymbol{E}\right)\mathrm{d}V + \int\limits_{S_R} \left(\boldsymbol{E}'\times\boldsymbol{H} - \boldsymbol{E}\times\boldsymbol{H}'\right)\cdot\boldsymbol{e}_R\,\mathrm{d}S$$

$$+ \int\limits_{S^+_{\varepsilon,R}\cup S^-_{\varepsilon,R}} \left(\boldsymbol{E}'\times\boldsymbol{H} - \boldsymbol{E}\times\boldsymbol{H}'\right)\cdot\boldsymbol{n}\,\mathrm{d}S + \int\limits_{C_{\varepsilon,R}} \left(\boldsymbol{E}'\times\boldsymbol{H} - \boldsymbol{E}\times\boldsymbol{H}'\right)\cdot\boldsymbol{n}\,\mathrm{d}S$$

$$\to \int\limits_{\Omega} \left(\boldsymbol{M}'\cdot\boldsymbol{H} - \boldsymbol{J}'\cdot\boldsymbol{E}\right)\mathrm{d}V \quad \text{as } \varepsilon\to 0,\ R\to\infty\,. \tag{3.71}$$

We remark that the integral over the spherical surface S_R vanishes because of the radiation conditions (3.69). The integral over the cylindrical surface $C_{\varepsilon,R}$

$$\lim_{R\to\infty}\lim_{\varepsilon\to 0} \int\limits_{C_{\varepsilon,R}} \left(\boldsymbol{E}'\times\boldsymbol{H} - \boldsymbol{E}\times\boldsymbol{H}'\right)\cdot\boldsymbol{n}\,\mathrm{d}S = 0$$

disappears in accordance with Meixner's conditions in the form (3.70), whereas the integral over the parts of the wedge's faces $S^{\pm}_{\varepsilon,R}$ cancels in view of the boundary conditions (3.67) and of the obvious equalities

$$\begin{aligned}
&\left(\boldsymbol{E}'\times\boldsymbol{H}\right)\cdot\boldsymbol{n} = -\left(\boldsymbol{n}\times\boldsymbol{H}\right)\cdot\boldsymbol{E}' = \eta^{-1}\left[\boldsymbol{E} - (\boldsymbol{E}\cdot\boldsymbol{n})\,\boldsymbol{n}\right]\cdot\boldsymbol{E}' \\
&= \eta^{-1}\left[\boldsymbol{E}\cdot\boldsymbol{E}' - (\boldsymbol{E}\cdot\boldsymbol{n})\left(\boldsymbol{n}\cdot\boldsymbol{E}'\right)\right].
\end{aligned}$$

Hence the reciprocity principle is expressed by the equality[8]

$$\int\limits_{\Omega} \left(\boldsymbol{M}\cdot\boldsymbol{H}' - \boldsymbol{J}\cdot\boldsymbol{E}'\right)\mathrm{d}V = \int\limits_{\Omega} \left(\boldsymbol{M}'\cdot\boldsymbol{H} - \boldsymbol{J}'\cdot\boldsymbol{E}\right)\mathrm{d}V. \tag{3.72}$$

All known particular forms follow from this general form of the reciprocity principle.

[8]We note that this equality seems to be more general than the corresponding symmetry of the diffraction coefficients of the scattered wave from the edge.

CHAPTER 4

Diffraction of a TM surface wave by an angular break of an impedance sheet

In the present chapter,[1] we study the diffraction of an electromagnetic surface wave by an impedance sheet[2] with an angular break. The vector of the magnetic field strength $\boldsymbol{H}$ of the wave is directed along the edge of the break. The angular break divides $\mathbb{R}^3$ into two wedge-shaped domains, Ω_1 and Ω_2 (see Fig. 4.1). The axis OZ coincides with the edge of the structure, whereas the transverse magnetic (TM) surface wave is incident along the side Γ_+ in the direction orthogonal to the edge. In this case, a solution depends on the coordinates x and y. The principal problem consists of computing the scattering diagram (diffraction coefficient) of the cylindrical wave propagating from the edge. As in previous chapters, the approach is also based on the Sommerfeld–Malyuzhinets technique. Note that excitation of the wave field by an incident surface wave leads to some novelties in formulating the radiation condition for the problem at hand.

4.1 Formulation of the problem

For the problem at hand, the electromagnetic field can be expressed in terms of the H_z-component of the magnetic field exploiting Maxwell's equations. Denote the H_z-component in the domain

$$\Omega_1 = \{(r, \varphi);\ r > 0,\ |\varphi| < \Phi\},\quad \pi/2 < \Phi < \pi,$$

by $u_1(r, \varphi)$ and in the domain

$$\Omega_2 = \left\{(r, \varphi);\ r > 0,\ |\overline{\varphi}| < \overline{\Phi}\right\},\quad \overline{\varphi} = \pi - \varphi,$$

by $u_2(r, \varphi)$, $\overline{\Phi} = \pi - \Phi$; $\Gamma_\pm$ are the boundaries of the domains Ω_1 and Ω_2 coinciding with the location of the broken impedance sheet.

---

[1]This chapter is based on the work [4.1].

[2]Such an idealized structure, while termed by Harrington and Mautz [4.2] as an impedance sheet, is also known under the name resistive sheet; see for instance [4.3].

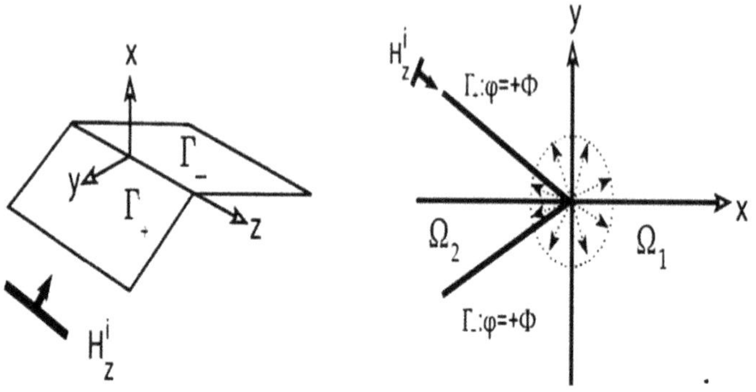

Figure 4.1 Diffraction of a surface wave by an angular break.

The wave field

$$u_{1,2} = u^{\mathrm{i}}_{1,2} + v_{1,2} \tag{4.1}$$

is represented by the sum of the surface wave incident along Γ_+,

$$\begin{aligned} u^{\mathrm{i}}_1(r,\varphi) &= \mathrm{e}^{-ikr\cos[(\Phi-\varphi)-\theta]}, \quad |\varphi| \le \Phi, \\ u^{\mathrm{i}}_2(r,\varphi) &= -\mathrm{e}^{-ikr\cos[(\overline{\Phi}-\overline{\varphi})-\theta]}, \quad |\overline{\varphi}| \le \overline{\Phi}, \end{aligned} \tag{4.2}$$

where θ is a parameter dependent on the impedance of the sheet, k is the wavenumber, and the scattered field is $v_{1,2}(r,\varphi)$.

From physical arguments it is obvious that the scattered field consists of surface waves reflected from the edge and transmitted across the edge as well as of a cylindrical wave propagating from the edge of the structure and having the asymptotics

$$v^{\mathrm{c}}_{1,2}(r,\varphi) = \frac{\mathrm{e}^{ikr+\mathrm{i}\pi/4}}{\sqrt{2\pi kr}} D_{1,2}(\varphi)\left[1 + \mathrm{O}\left(\frac{1}{kr}\right)\right] \tag{4.3}$$

as $kr \to \infty$, where $D_{1,2}(\varphi)$ is the sought-for-scattering diagram.

The total field satisfies the Helmholtz equation in $\Omega_{1,2}$

$$(\operatorname{div}\operatorname{grad} + k^2)u_{1,2} = 0 \tag{4.4}$$

(the dependence on time e^{-ikct} is omitted where c denotes the speed of light in the ambient medium), and the boundary conditions on the broken sheet

$$\left[\frac{1}{r}\frac{\partial u_1}{\partial \varphi} - \frac{1}{r}\frac{\partial u_2}{\partial \varphi}\right]\Bigg|_{\Gamma_\pm} = 0, \tag{4.5}$$

$$\left[\frac{1}{r}\frac{\partial u_1}{\partial \varphi} + \frac{1}{r}\frac{\partial u_2}{\partial \varphi} \mp ik\sin\theta(u_1 - u_2)\right]\Bigg|_{\Gamma_\pm} = 0, \tag{4.6}$$

which describe an impedance sheet characterized by the (normalized) surface admittance $2/\sin\theta$. In what follows we also assume that

$$\operatorname{Re}\theta = 0, \quad \operatorname{Im}\theta < 0. \tag{4.7}$$

Restrictions (4.7) ensure existence of the surface wave (4.2) propagating along Γ_+ to the edge. The wave field satisfies Meixner's condition on the edge ([4.4, p. 19] and Chapter 1)

$$\int_{S_\varepsilon} \frac{\partial u}{\partial n} \overline{u}\, \mathrm{d}s \to 0, \quad \text{as } \varepsilon \to 0, \tag{4.8}$$

(S_ε is a circumference of a circle of a small radius ε centered at the origin) and the radiation condition (see also (4.3) and Chapter 1). The radiation condition for the problem at hand requires some additional commenting. It can be written in a weak integral form

$$\int_{S_R^c} \left| \frac{\partial (u - u^{\mathrm{i}})}{\partial r} - ik\,(u - u^{\mathrm{i}}) \right|^2 \mathrm{d}s + \int_{S_R^b} \left| \frac{\partial (u - u^{\mathrm{i}})}{\partial r} - ik\cos\theta\,(u - u^{\mathrm{i}}) \right|^2 \mathrm{d}s \to 0 \tag{4.9}$$

as $R \to \infty$, where S_R^b is asymptotically small (as $R \to \infty$) parts of the circumference $S_R = \{(r, \varphi) : \ r = R, \ -\pi \le \varphi < \pi\}$ described by

$$S_R^b = \left\{(r, \varphi) : \ r = R, \ -Cr^{-\delta} \le \varphi + \Phi < Cr^{-\delta} \text{ and } -Cr^{-\delta} \le \varphi - \Phi < Cr^{-\delta}\right\},$$

where $C > 0$, $\delta > 0$ are some positive constants, u^{i} is the incident surface wave (4.2). In these parts of the circumference the main contribution to the far-field asymptotics is given by the surface waves outgoing from the edge.

The rest part of the circumference, that is, $S_R^c = S_R \setminus S_R^b$, corresponds to the domain, where only the cylindrical wave propagating at infinity exists and the contribution of the surface waves to the far-field asymptotics is exponentially small. So the first summand in (4.9) is responsible for the cylindrical wave, whereas the second summand extracts the surface waves reflected from and transmitted across the edge.

It is useful to comment that in Chapter 2 we also considered surface waves propagating from the edge of an impedance wedge; however, in that case we assumed some absorption ($\operatorname{Re}\theta > 0$) in the wedge's faces. The outgoing surface waves then attenuated with r tending to infinity. The surface waves were really propagating at infinity as $\operatorname{Re}\theta \to 0$; that is, we have actually exploited a version of the limiting-absorption principle to extract the outgoing surface waves. It is remarkable that so-defined outgoing surface waves satisfy a radiation condition analogous to the second summand in (4.9).

4.2 Functional equations and reduction to integral equations

As in Chapters 2 and 3, the Sommerfeld–Malyuzhinets technique plays a key role for the problem at hand. The solution is sought in the form of the Sommerfeld integrals

$$\begin{aligned} u_1(r, \varphi) &= \frac{1}{2\pi \mathrm{i}} \int_\gamma \mathrm{e}^{-ikr\cos\alpha} s_1(\alpha + \varphi)\, \mathrm{d}\alpha, \quad |\varphi| \le \Phi, \\ u_2(r, \varphi) &= \frac{1}{2\pi \mathrm{i}} \int_\gamma \mathrm{e}^{-ikr\cos\alpha} s_2(\alpha - \overline{\varphi})\, \mathrm{d}\alpha, \quad |\overline{\varphi}| \le \overline{\Phi}, \end{aligned} \tag{4.10}$$

where γ is the Sommerfeld double-loop contour (see [4.5] and Chapter 1), and the spectral functions $s_{1,2}(\alpha)$ are sought in a class of meromorphic functions. The boundary conditions are valid, provided that the unknown spectral functions satisfy a system of functional

(Malyuzhinets-type) equations

$$\begin{aligned} R_+(\alpha)s_1(\alpha+\Phi)+R_-(\alpha)s_1(-\alpha+\Phi) &= -R_-(\alpha)s_2(\alpha-\overline{\Phi})-R_+(\alpha)s_2(-\alpha-\overline{\Phi}),\\ s_1(\alpha+\Phi)+s_1(-\alpha+\Phi) &= s_2(\alpha-\overline{\Phi})+s_2(-\alpha-\overline{\Phi}),\\ R_-(\alpha)s_1(\alpha-\Phi)+R_+(\alpha)s_1(-\alpha-\Phi) &= -R_+(\alpha)s_2(\alpha+\overline{\Phi})-R_-(\alpha)s_2(-\alpha+\overline{\Phi}),\\ s_1(\alpha-\Phi)+s_1(-\alpha-\Phi) &= s_2(\alpha+\overline{\Phi})+s_2(-\alpha+\overline{\Phi}), \end{aligned} \tag{4.11}$$

where $R_\pm(\alpha)=\sin\alpha\pm\sin\theta$. A solution of the system satisfies the requirements

$$s_{1,2}(\alpha)=C^{\pm}_{1,2}+\mathrm{O}(\mathrm{e}^{\pm\mathrm{i}\delta_0\alpha}) \text{ as } \alpha\to\pm\mathrm{i}\infty, \tag{4.12}$$

uniformly with respect to $|\mathrm{Re}\,\alpha|<$ const. $\delta_0\,(>0)$ and $C^{\pm}_{1,2}$ are some constants. Condition (4.12) ensures Meixner's condition on the edge.

Moreover, $s_1(\alpha)-1/(\alpha-\alpha_1)$ is regular in the basic strip $|\mathrm{Re}\,\alpha|<\Phi$, $\alpha_1=\Phi-\theta$, and $s_2(\alpha)+1/(\alpha-\alpha_2)$ is regular in the basic strip $|\mathrm{Re}\,\alpha|<\overline{\Phi}$, $\alpha_2=-\overline{\Phi}+\theta$. These requirements are inherited from the radiation condition and guarantee that the only wave coming to the edge from infinity is the incident surface wave. The spectral functions have the prescribed residues at the poles $\alpha_{1,2}$ being responsible for the incident wave. The situation with the prescribed poles is completely analogous to that in Chapter 2 for the plane-wave illumination. However, contrary to the case of the plane-wave incidence, the corresponding poles in the basic strips are complex valued.

In the general case, we cannot expect that the problem for the functional equations (4.11) is explicitly solvable; however, we intend to reduce it, first, to that for a functional difference equation of the second order then to an integral equation.

4.2.1 Reduction to a second-order functional equation

It is useful to introduce a notation $r(\alpha)=R_-(\alpha)/R_+(\alpha)$ and to note that $r(\alpha)=1/r(-\alpha)$. Combining the first two equations in (4.11), we obtain

$$s_2(\alpha)=\frac{2s_1(\alpha+[\Phi+\overline{\Phi}])}{1-r(\alpha+\overline{\Phi})}+\frac{1+r(\alpha+\overline{\Phi})}{1-r(\alpha+\overline{\Phi})}s_1(-\alpha+[\Phi-\overline{\Phi}]). \tag{4.13}$$

In a similar manner, the last two equations in (4.11) lead to

$$s_2(\alpha)=\frac{2s_1(\alpha-[\Phi+\overline{\Phi}])}{1-r^{-1}(\alpha-\overline{\Phi})}+\frac{1+r^{-1}(\alpha-\overline{\Phi})}{1-r^{-1}(\alpha-\overline{\Phi})}s_1(-\alpha-[\Phi-\overline{\Phi}]). \tag{4.14}$$

As a consequence of equations (4.13) and (4.14), we have the equation for the spectral function $s_1(\alpha)$

$$\begin{aligned} &[\sin(\alpha+\overline{\Phi})+\sin\theta]\,s_1(\alpha+\pi)+[\sin(\alpha-\overline{\Phi})-\sin\theta]\,s_1(\alpha-\pi)\\ &=-\left[\sin(\alpha+\overline{\Phi})s_1(-\alpha+\Phi-\overline{\Phi})+\sin(\alpha-\overline{\Phi})s_1(-\alpha-\Phi+\overline{\Phi})\right]. \end{aligned} \tag{4.15}$$

Provided that a solution to equation (4.15) is found from an appropriate class of functions, the second spectral function $s_2(\alpha)$ is determined from equations (4.13) or (4.14). Although formally equation (4.15) is not a functional difference equation of the second order, its properties prove to be very similar.

It is convenient to transform equation (4.15) to a simpler form with constant coefficients on the left-hand side. To this end, we consider a special solution $F_0(\alpha)$ (recall a similar trick in Chapter 2) of the auxiliary equation

$$[\sin(\alpha+\overline{\Phi})+\sin\theta]\,F_0(\alpha+\pi) = -[\sin(\alpha-\overline{\Phi})-\sin\theta]\,F_0(\alpha-\pi).$$

It is not difficult to demonstrate that the desired solution of the auxiliary equation can be represented in the form

$$F_0(\alpha) = \frac{\chi_{\pi/2}(\alpha+\Phi-\theta)\,\chi_{\pi/2}(\alpha-\overline{\Phi}+\theta)}{\chi_{\pi/2}(\alpha-\Phi+\theta)\,\chi_{\pi/2}(\alpha+\overline{\Phi}-\theta)},$$

where $\chi_{\Phi_0}(\alpha)$ is the generalized Malyuzhinets function (see Section 2.4.2). The auxiliary solution $F_0(\alpha)$ has the asymptotics

$$F_0(\alpha) \sim c_1^{\pm}\exp\left[\mp\frac{\mathrm{i}}{2\pi}(\Phi-\overline{\Phi})\alpha\right],\quad \alpha\to\pm\mathrm{i}\infty.$$

We seek a solution of equation (4.15) in the form

$$s_1(\alpha) = F_0(\alpha)\,\widetilde{\Psi}(\alpha), \tag{4.16}$$

where $\widetilde{\Psi}(\alpha)$ is a new unknown function

$$\widetilde{\Psi}(\alpha) \sim C^{\pm}\exp\left[\pm\frac{\mathrm{i}}{2\pi}(\Phi-\overline{\Phi})\alpha\right],\quad \alpha\to\pm\mathrm{i}\infty,$$

satisfying the equation

$$\begin{aligned}&\widetilde{\Psi}(\alpha+\pi)-\widetilde{\Psi}(\alpha-\pi)\\ &=-\left\{a_+(\alpha)\widetilde{\Psi}(-\alpha+[\Phi-\overline{\Phi}])-a_-(\alpha)\widetilde{\Psi}(-\alpha-[\Phi-\overline{\Phi}])\right\},\end{aligned} \tag{4.17}$$

where

$$a_\pm(\alpha) = \frac{\sin(\alpha\pm\overline{\Phi})F_0(-\alpha\pm[\Phi-\overline{\Phi}])}{[\sin(\alpha\pm\overline{\Phi})\pm\sin\theta]F_0(\alpha\pm\pi)},\quad a_\pm(\alpha)\sim a_\pm^0 = \mathrm{const}_\pm \text{ as } \alpha\to\pm\mathrm{i}\infty.$$

It is useful to separate the singularity corresponding to the incident surface wave. To that end, we represent the required function $\widetilde{\Psi}(\alpha)$ in the form

$$\widetilde{\Psi}(\alpha) = \Psi(\alpha)+\Psi_{\mathrm{i}}(\alpha),\quad \Psi_{\mathrm{i}}(\alpha) = \frac{1}{2F_0(\alpha_1)}\,\frac{1}{\sin[(\alpha-\alpha_1)/2]}, \tag{4.18}$$

where $\Psi(\alpha)$ is regular in the strip $|\mathrm{Re}\,\alpha|<\pi$. We then verify easily that $\Psi(\alpha)$ solves the equation

$$\begin{aligned}&\Psi(\alpha+\pi)-\Psi(\alpha-\pi)\\ &=-\left\{a_+(\alpha)\Psi(-\alpha+[\Phi-\overline{\Phi}])-a_-(\alpha)\Psi(-\alpha-[\Phi-\overline{\Phi}])\right\}+f_{\mathrm{i}}(\alpha),\end{aligned} \tag{4.19}$$

where

$$f_{\mathrm{i}}(\alpha) = -\Psi_{\mathrm{i}}(\alpha+\pi) + \Psi_{\mathrm{i}}(\alpha-\pi) - \left\{a_+(\alpha)\Psi_{\mathrm{i}}(-\alpha+[\Phi-\overline{\Phi}]) - a_-(\alpha)\Psi_{\mathrm{i}}(-\alpha-[\Phi-\overline{\Phi}])\right\}.$$

The next step is to transform equation (4.19) to an integral equation of the second kind "inverting" the difference operator on the left-hand side.

4.2.2 An integral equation of the second kind

Assuming that $|\mathrm{Re}\,\alpha| < \pi$, we transform equation (4.19), first, to an integral representation for $\Psi(\alpha)$ in that strip. Equation (4.19) can be written in the form

$$\Psi(\alpha+\pi) - \Psi(\alpha-\pi) = g(\alpha). \tag{4.20}$$

Representing $\Psi(\alpha) = \Psi_{\mathrm{e}}(\alpha) + \Psi_{\mathrm{o}}(\alpha)$ as a sum of even and odd summands, we write the equation (4.20) with the help of two systems

$$\begin{cases} \Psi_{\mathrm{e}}(\alpha+\pi) - \Psi_{\mathrm{e}}(-\alpha+\pi) = g_{\mathrm{o}}(\alpha), \\ \Psi_{\mathrm{e}}(\alpha-\pi) - \Psi_{\mathrm{e}}(-\alpha-\pi) = -g_{\mathrm{o}}(\alpha) \end{cases} \tag{4.21}$$

and

$$\begin{cases} \Psi_{\mathrm{o}}(\alpha+\pi) + \Psi_{\mathrm{o}}(-\alpha+\pi) = g_{\mathrm{e}}(\alpha), \\ \Psi_{\mathrm{o}}(\alpha-\pi) + \Psi_{\mathrm{o}}(-\alpha-\pi) = -g_{\mathrm{e}}(\alpha). \end{cases} \tag{4.22}$$

Systems (4.21) and (4.22) can be explicitly solved with respect to $\Psi_{\mathrm{e,o}}(\alpha)$, exploiting the Fourier transform (along the imaginary axis) or an equivalent technique of the $\mathcal{S}$-integrals.

Omitting standard calculations (see Chapters 1 and 2), we arrive at an integral representation in the strip $|\mathrm{Re}\,\alpha| < \pi$

$$\begin{aligned} \Psi(-\alpha) = {} & \frac{\mathrm{i}}{4\pi}\int_{\mathrm{i}\mathbb{R}} \mathrm{d}\tau \left(\tan\frac{\alpha}{2} - \tan\frac{\alpha+\tau}{2}\right) \\ & \times \left\{a_+(\tau)\Psi(-\tau+[\Phi-\overline{\Phi}]) - a_-(\tau)\Psi(-\tau-[\Phi-\overline{\Phi}])\right\} \\ & - \frac{\mathrm{i}}{4\pi}\int_{\mathrm{i}\mathbb{R}} \mathrm{d}\tau \left(\tan\frac{\alpha}{2} - \tan\frac{\alpha+\tau}{2}\right) f_{\mathrm{i}}(\tau). \end{aligned} \tag{4.23}$$

Representation (4.23) enables us to determine $\Psi(\alpha)$ in the strip $|\mathrm{Re}\,\alpha| < \pi$ provided that the values $\Psi(-\tau \pm [\Phi-\overline{\Phi}])$ for $\tau \in \mathrm{i}\mathbb{R} = (-\mathrm{i}\infty, \mathrm{i}\infty)$ are known.

Introduce the vector $p(\alpha) = [p_+(\alpha), p_-(\alpha)]^{\mathrm{T}}$, $p_\pm(\alpha) = \Psi(-\alpha \pm [\Phi-\overline{\Phi}])$. Taking $\alpha = t - [\Phi-\overline{\Phi}]$ and $\alpha = t + [\Phi-\overline{\Phi}]$, $t \in \mathrm{i}\mathbb{R}$ in the representation (4.23), we arrive at the integral equation

$$p(t) = \int_{\mathrm{i}\mathbb{R}} \mathcal{K}(t,\tau)\, p(\tau)\, \mathrm{d}\tau + q(t), \tag{4.24}$$

with

$$q(\alpha) = [q_+(\alpha), q_-(\alpha)]^{\mathrm{T}}$$

$$\mathcal{K}(t,\tau) = \frac{\mathrm{i}}{4\pi}\begin{bmatrix} K_+(t,\tau)a_+(\tau), & -K_+(t,\tau)a_-(\tau) \\ K_-(t,\tau)a_+(\tau), & -K_-(t,\tau)a_-(\tau)\end{bmatrix},$$

$$K_\pm(t,\tau) = \left(\tan\frac{\alpha}{2} - \tan\frac{\alpha+\tau}{2}\right)\Bigg|_{\alpha=t\mp[\Phi-\overline{\Phi}]},$$

$$q_\pm(t) = \frac{\mathrm{i}}{4\pi}\int\limits_{\mathrm{i}\mathbb{R}} K_\pm(t,\tau)\, f_{\mathrm{i}}(\tau)\,\mathrm{d}\tau.$$

Vector integral equation (4.24) can naturally be studied in $L_2(-\mathrm{i}\infty, \mathrm{i}\infty)$. The kernel $\mathcal{K}(t,\tau)$, however, is not of the Hilbert–Schmidt class, so the Fredholm property cannot be directly verified. One could proceed in a more tricky way and develop an integral equation of the second kind, equivalent to (4.24), with a compact operator. We omit the corresponding details because equation (4.24) is, nevertheless, Fredholm and is completely efficient for numerical computation.

4.3 Analytic continuation of the spectral functions and scattering diagram

Equation (4.24) can be numerically solved, and the results turn out to be stable and withstand testing with respect to internal convergence. With the help of the representation (4.23), the values of $\Psi(\alpha)$, as well as those of $s_1(\alpha)$, are found in the strip $|\mathrm{Re}\,\alpha| < \pi$. We need to describe a procedure of analytic continuation of $s_{1,2}$ to any strip parallel to the imaginary axis and thus to the whole complex plane. Actually, it is sufficient to determine the analytic continuation to the strip $|\mathrm{Re}\,\alpha| < \pi + \Phi$, because, in computing the asymptotics of the far field, only the saddle points and poles of the integrand are essential that are captured in deforming the Sommerfeld contour γ into the steepest-descent paths going through the saddle points $\pm\pi$.

If α belongs to the strip $-\overline{\Phi} < \mathrm{Re}\,\alpha < 0$, then for calculation of $s_2(\alpha)$ in that strip, we use equation (4.13) because $s_1(\alpha)$ has already been computed in $\mathrm{i}\mathbb{R} \pm [\Phi - \overline{\Phi}]$ and $|\mathrm{Re}\,\alpha| < \pi$. To continue $s_2(\alpha)$ into the strip $0 < \mathrm{Re}\,\alpha < \overline{\Phi}$ we use equation (4.14). As a consequence of the functional equations (4.11), we write formulas that enable us to obtain an analytic continuation of $s_{1,2}(\alpha)$ to wider strips

$$\begin{aligned} s_1(\alpha+2\Phi) &= -\frac{\sin(\alpha+\Phi)}{\sin(\alpha+\Phi)+\sin\theta}\, s_1(-\alpha) + \frac{\sin\theta}{\sin(\alpha+\Phi)+\sin\theta}\, s_2(\alpha+[\Phi+\overline{\Phi}]), \\ s_1(\alpha-2\Phi) &= -\frac{\sin(\alpha-\Phi)}{\sin(\alpha-\Phi)-\sin\theta}\, s_1(-\alpha) - \frac{\sin\theta}{\sin(\alpha-\Phi)-\sin\theta}\, s_2(\alpha-[\Phi-\overline{\Phi}]). \end{aligned} \tag{4.25}$$

If α on the right-hand side of the first equation in (4.25) belongs to the strip $-\Phi < \mathrm{Re}\,\alpha < \Phi$, then $\alpha + 2\Phi$ lies in the strip $\Phi < \mathrm{Re}\,\alpha < 3\Phi$. Analogously, if α on the right-hand side of the second equation in (4.25) belongs to the strip $-\Phi < \mathrm{Re}\,\alpha < \Phi$, then the argument $\alpha - 2\Phi$ on the left-hand side lies in the strip $-3\Phi < \mathrm{Re}\,\alpha < -\Phi$. Moreover, it may be necessary to compute

$s_2(\alpha)$ on the right-hand side of (4.25) in the strips $\overline{\Phi} < \operatorname{Re}\alpha < 3\overline{\Phi}$ and $-3\overline{\Phi} < \operatorname{Re}\alpha < -\overline{\Phi}$ adjacent to $-\overline{\Phi} < \operatorname{Re}\alpha < \overline{\Phi}$. Such continuation is performed by means of the formulas

$$s_2(\alpha + 2\overline{\Phi}) = -\frac{\sin(\alpha + \overline{\Phi})}{\sin(\alpha + \overline{\Phi}) + \sin\theta} s_1(-\alpha) + \frac{\sin\theta}{\sin(\alpha + \overline{\Phi}) + \sin\theta} s_1(\alpha - [\Phi - \overline{\Phi}]),$$

$$s_2(\alpha - 2\overline{\Phi}) = -\frac{\sin(\alpha - \overline{\Phi})}{\sin(\alpha - \overline{\Phi}) - \sin\theta} s_1(-\alpha) - \frac{\sin\theta}{\sin(\alpha - \overline{\Phi}) - \sin\theta} s_1(\alpha + [\Phi - \overline{\Phi}]).$$

If the parameter $\overline{\Phi}$ is sufficiently small, further continuation of $s_2(\alpha)$ to the strips $3\overline{\Phi} < \operatorname{Re}\alpha < 5\overline{\Phi}$ and $-5\overline{\Phi} < \operatorname{Re}\alpha < -3\overline{\Phi}$ may be required. Assuming that α belongs to the strip $-\overline{\Phi} < \operatorname{Re}\alpha < \overline{\Phi}$, such a continuation is performed with the help of the formulas[3]

$$\begin{aligned} s_2(\alpha + 4\overline{\Phi}) = &\frac{\sin(\alpha + 3\overline{\Phi})}{\sin(\alpha + 3\overline{\Phi}) + \sin\theta} \frac{\sin(\alpha + \overline{\Phi})}{\sin(\alpha + \overline{\Phi}) + \sin\theta} s_2(\alpha) \\ &- \frac{\sin(\alpha + 3\overline{\Phi})}{\sin(\alpha + 3\overline{\Phi}) + \sin\theta} \frac{\sin\theta}{\sin(\alpha + \overline{\Phi}) + \sin\theta} s_1(-\alpha + [\Phi - \overline{\Phi}]) \\ &+ \frac{\sin\theta}{\sin(\alpha + 3\overline{\Phi}) + \sin\theta} s_1(\alpha - [\Phi - 3\overline{\Phi}]), \quad -\overline{\Phi} < \operatorname{Re}\alpha < \overline{\Phi}, \end{aligned}$$

$$\begin{aligned} s_2(\alpha - 4\overline{\Phi}) = &\frac{\sin(\alpha - 3\overline{\Phi})}{\sin(\alpha - 3\overline{\Phi}) - \sin\theta} \frac{\sin(\alpha - \overline{\Phi})}{\sin(\alpha - \overline{\Phi}) - \sin\theta} s_2(\alpha) \\ &+ \frac{\sin(\alpha - 3\overline{\Phi})}{\sin(\alpha - 3\overline{\Phi}) - \sin\theta} \frac{\sin\theta}{\sin(\alpha - \overline{\Phi}) - \sin\theta} s_1(-\alpha - [\Phi - \overline{\Phi}]) \\ &- \frac{\sin\theta}{\sin(\alpha - 3\overline{\Phi}) - \sin\theta} s_1(\alpha + [\Phi - 3\overline{\Phi}]), \quad -\overline{\Phi} < \operatorname{Re}\alpha < \overline{\Phi}. \end{aligned}$$

It is obvious that formulas for further continuation can also be easily written, although they become more and more cumbersome.

4.3.1 Scattering diagram

The most important characteristic of the cylindrical wave from the edge is the scattering diagram $D_{1,2}(\varphi)$ in (4.3). If the saddle-point technique is applied to the Sommerfeld integrals (4.10), then, deforming contour γ into the steepest-descent paths (see also Chapters 1 and 2), we obtain the expression for the scattering diagram

$$\begin{aligned} D_1(\varphi) &= s_1(-\pi + \varphi) - s_1(\pi + \varphi), \quad |\varphi| < \Phi, \\ D_2(\varphi) &= s_2(-\pi - \overline{\varphi}) - s_2(\pi - \overline{\varphi}), \quad |\overline{\varphi}| < \overline{\Phi}. \end{aligned} \tag{4.26}$$

In Fig. 4.2 the results of calculating the absolute value of the scattering diagram $|D_{1,2}(\varphi)|$ for two different angles of the break and two impedances $\sin\theta/2$ are given. The computations[4]

[3] Actually, these formulas are iterations of those previously given.

[4] The numerical results were obtained by V. E. Grikurov (deceased in 2008).

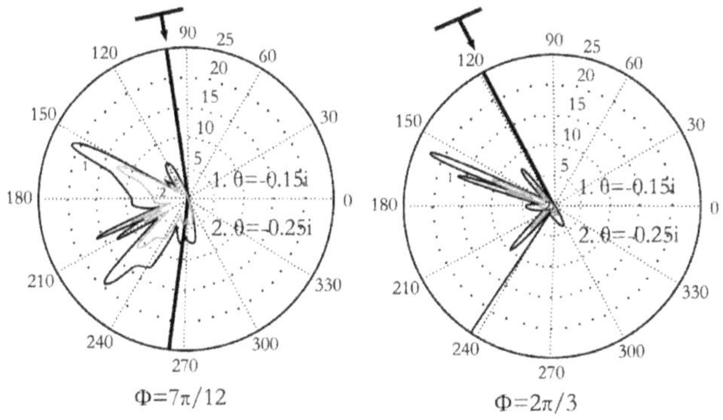

Figure 4.2 Scattering diagrams for two different angles of the break and for two impedances.

are based on the solution of the integral equation (4.24) and on the procedure of analytic continuation described already.

4.3.2 Reflected and transmitted surface waves

As an example, we consider the case, where $\pi/2 < \Phi < \pi$ and $\pi/4 < \overline{\Phi} < \pi/2$, then in deforming the Sommerfeld contour into the steepest-descent paths, two poles $\alpha_1^{\pm} = \pm(\pi + \Phi + \theta)$ of the spectral function $s_1(\alpha)$ and two poles $\alpha_2^{\pm} = \pm(\pi + \overline{\Phi} + \theta)$ of the spectral function $s_2(\alpha)$ can be captured, in addition to the poles corresponding to the incident wave. By the theorem on residues, the poles $\alpha_{1,2}^{+}$ specify the surface waves reflected from the edge

$$u_1^s(r,\varphi) = \mathrm{res}_{\alpha_1^+} s_1(\alpha) e^{ikr\cos[(\Phi-\varphi)+\theta]}, \quad 0 \le \Phi - \varphi \le -\mathrm{gd}(\mathrm{Im}\,\theta) - \mathrm{Re}\,\theta,$$

$$u_2^s(r,\varphi) = \mathrm{res}_{\alpha_2^+} s_2(\alpha) e^{ikr\cos[(\overline{\Phi}-\overline{\varphi})+\theta]}, \quad 0 \le \overline{\Phi} - \overline{\varphi} \le -\mathrm{gd}(\mathrm{Im}\,\theta) - \mathrm{Re}\,\theta,$$

and the transmitted surface waves

$$u_1^s(r,\varphi) = \mathrm{res}_{\alpha_1^-} s_1(\alpha) e^{ikr\cos[(-\Phi-\varphi)-\theta]}, \quad 0 \le \Phi + \varphi \le -\mathrm{gd}(\mathrm{Im}\,\theta) - \mathrm{Re}\,\theta,$$

$$u_2^s(r,\varphi) = \mathrm{res}_{\alpha_2^-} s_2(\alpha) e^{ikr\cos[(\overline{\Phi}+\overline{\varphi})+\theta]}, \quad 0 \le \overline{\Phi} + \overline{\varphi} \le -\mathrm{gd}(\mathrm{Im}\,\theta) - \mathrm{Re}\,\theta,$$

where $\mathrm{gd}(x)$ is the Gudermann function defined in (1.149), and there is $\mathrm{Re}\,\theta \ge 0$.

The surface waves ($\mathrm{Re}\,\theta = 0$) propagate along $\Gamma_{\pm}$ from the edge with the velocity $c^\star = c/\cosh|\theta|$, where c is the velocity of the electromagnetic waves in $\Omega_{1,2}$, and are exponentially damping with distance from $\Gamma_{\pm}$. If the value of $\overline{\Phi}$ is sufficiently small, the number of captured poles increases and the study of reflected and transmitted surface waves becomes more complicated. In particular, for a sufficiently large number of poles captured a specific interference picture arises in Ω_2, which requires additional study.

Solution of the analogous problem of diffraction of a plane incident wave encounters no difficulties. It is useful to remark that the scattering of a surface wave by a conical singularity is studied in the paper [4.6]; however, the mathematical technique of that work is, generally speaking, different.

4.4 Discussion of uniqueness

In the present section we consider uniqueness of a (classical) solution given by the Sommerfeld integral, that is, we intend to demonstrate that $u \equiv 0$ as $u^{\mathrm{i}} = 0$, for a slightly more general formulation of the problem at hand. Our aim is to study sufficient conditions in which the solution obtained in the present chapter is unique. The latter is, in an indirect way, responsible for the stable numerical solvability of the integral equation, which is Fredholm, and offers a deeper insight in the particular roles played by each condition in the formulation of the problem.

Namely, we assume here that the sides of the impedance sheet have different surface impedances, so we have

$$\left[\frac{1}{r}\frac{\partial u_1}{\partial \varphi} - \frac{1}{r}\frac{\partial u_2}{\partial \varphi}\right]\Bigg|_{\Gamma_\pm} = 0, \tag{4.27}$$

$$\left[\frac{1}{r}\frac{\partial u_1}{\partial \varphi} + \frac{1}{r}\frac{\partial u_2}{\partial \varphi} \mp ik\sin\theta_\pm(u_1 - u_2)\right]\Bigg|_{\Gamma_\pm} = 0, \tag{4.28}$$

which describe the conditions on the sides of the broken impedance sheet characterized by the surface admittances $2/\sin\theta_\pm$. In proving the uniqueness we also assume that

$$k > 0,\ \ \mathrm{Re}\,\theta_+ = 0,\ \ \mathrm{Im}\,\theta_+ < 0,\ \ \pi/2 > \mathrm{Re}\,\theta_- > 0,\ \ \mathrm{Im}\,\theta_- < 0, \tag{4.29}$$

that is, the lower side of the broken sheet is absorbing, whereas the upper side is lossless.[5]

The radiation condition reads

$$\int_{S_R^c}\left|\frac{\partial u}{\partial r} - ik\,u\right|^2 ds + \int_{S_{R,b}^+}\left|\frac{\partial u}{\partial r} - ik\cos\theta_+\,u\right|^2 ds \to 0 \tag{4.30}$$

as $R \to \infty$, where $S_{R,b}^+ = \{(r,\varphi):\ r = R,\ \ -Cr^{-\delta} \le \varphi - \Phi < Cr^{-\delta}\}$ and $S_R^c = S_R \backslash S_{R,b}^+$.

We exploit the Helmholtz equation (4.4) for u_1 and u_2 and the Green theorem thus obtain

$$\begin{aligned} -\int_{\Omega^1_{\varepsilon,R}} |\nabla u_1|^2\,dxdy + k^2\int_{\Omega^1_{\varepsilon,R}} |u_1|^2\,dxdy + \int_{\partial\Omega^1_{\varepsilon,R}} \frac{\partial u_1}{\partial n}\,\overline{u}_1 ds &= 0,\\ -\int_{\Omega^2_{\varepsilon,R}} |\nabla u_2|^2\,dxdy + k^2\int_{\Omega^2_{\varepsilon,R}} |u_2|^2\,dxdy + \int_{\partial\Omega^2_{\varepsilon,R}} \frac{\partial u_2}{\partial n}\,\overline{u}_1 ds &= 0, \end{aligned} \tag{4.31}$$

where $\Omega^1_{\varepsilon,R}$ and $\Omega^2_{\varepsilon,R}$ are, respectively, the parts of the domains Ω_1 and Ω_2 located outside the circumference S_ε of small radius ε centered at the origin and inside the circumference S_R of large radius R, $\partial\Omega^{1,2}_{\varepsilon,R}$ are the boundaries of the domains and n is the normal vector directed into the exterior of the corresponding domain.

[5]No proof can be given when both sides are lossfree, i.e., $\mathrm{Re}\,\theta_+ = 0$ and $\mathrm{Re}\,\theta_- = 0$.

We sum up the equalities (4.31), use the continuity of the normal derivatives (4.27), and take the imaginary part of the result

$$\mathrm{Im}\left\{\int\limits_{S_R^c}\frac{\partial u}{\partial n}\overline{u}\mathrm{d}s+\int\limits_{S_\varepsilon}\frac{\partial u}{\partial n}\overline{u}\mathrm{d}s+\int\limits_{S_{R,b}^+}\frac{\partial u}{\partial n}\overline{u}\mathrm{d}s+\right.$$
$$\left.\int\limits_{l_{R,\varepsilon}^+}\frac{\partial u}{\partial n}(\overline{u}_1-\overline{u}_2)\mathrm{d}s+\int\limits_{l_{R,\varepsilon}^-}\frac{\partial u}{\partial n}(\overline{u}_1-\overline{u}_2)\mathrm{d}s\right\}=0, \tag{4.32}$$

where $l_{R,\varepsilon}^+$ and $l_{R,\varepsilon}^-$ are the parts of the sheet sides Γ_+ and Γ_-, located between the circumferences S_ε and S_R. Note that $u=u_1$ in Ω_1, $u=u_2$ in Ω_2. We take into account the boundary conditions (4.28) and write

$$\mathrm{Im}\int\limits_{l_{R,\varepsilon}^+}\frac{\partial u}{\partial n}(\overline{u}_1-\overline{u}_2)\mathrm{d}s=2\mathrm{Im}\left\{\frac{\mathrm{i}\sin\theta_+}{k|\sin\theta_+|^2}\right\}\int\limits_{l_{R,\varepsilon}^+}\left|\frac{\partial u}{\partial n}\right|^2\mathrm{d}s=0,$$
$$\mathrm{Im}\int\limits_{l_{R,\varepsilon}^-}\frac{\partial u}{\partial n}(\overline{u}_1-\overline{u}_2)\mathrm{d}s=2\mathrm{Im}\left\{\frac{\mathrm{i}\sin\theta_-}{k|\sin\theta_-|^2}\right\}\int\limits_{l_{R,\varepsilon}^-}\left|\frac{\partial u}{\partial n}\right|^2\mathrm{d}s\geq 0, \tag{4.33}$$

where we have used the restrictions (4.29). We then exploit the correlations (4.33), Meixner's condition (4.8) (see also [4.4, Section 2.2, p. 19]) and from (4.32) obtain

$$-\mathrm{Im}\left\{\int\limits_{S_R^c}\frac{\partial u}{\partial n}\overline{u}\,\mathrm{d}s+\int\limits_{S_{R,b}^+}\frac{\partial u}{\partial n}\overline{u}\,\mathrm{d}s\right\}=2\mathrm{Im}\left\{\frac{\mathrm{i}\sin\theta_-}{k|\sin\theta_-|^2}\right\}\int\limits_{l_{R,0}^-}\left|\frac{\partial u}{\partial n}\right|^2\mathrm{d}s\geq 0. \tag{4.34}$$

Further estimates are straightforward. We have

$$0\leq-\mathrm{Im}\left\{\int\limits_{S_R^c}\frac{\partial u}{\partial n}\overline{u}\,\mathrm{d}s+\int\limits_{S_{R,b}^+}\frac{\partial u}{\partial n}\overline{u}\,\mathrm{d}s\right\}$$
$$=-\mathrm{Im}\left\{\int\limits_{S_R^c}\left(\frac{\partial u}{\partial r}-\mathrm{i}ku\right)\overline{u}\,\mathrm{d}s+\mathrm{i}k\int\limits_{S_R^c}|u|^2\,\mathrm{d}s\right. \tag{4.35}$$
$$\left.+\int\limits_{S_{R,b}^+}\left(\frac{\partial u}{\partial r}-\mathrm{i}k\cos\theta_+u\right)\overline{u}\,\mathrm{d}s+\mathrm{i}k\cos\theta_+\int\limits_{S_{R,b}^+}|u|^2\,\mathrm{d}s\right\}.$$

From (4.35), we conclude that

$$
\begin{aligned}
0 &\leq k \int\limits_{S_R^c} |u|^2 \, ds + k\mathrm{Re}(\cos\theta_+) \int\limits_{S_{R,b}^+} |u|^2 \, \mathrm{d}s \\
&\leq -\mathrm{Im} \left\{ \int\limits_{S_R^c} \left(\frac{\partial u}{\partial r} - iku \right) \bar{u} \, \mathrm{d}s + \int\limits_{S_{R,b}^+} \left(\frac{\partial u}{\partial r} - ik\cos\theta_+ u \right) \bar{u} \, \mathrm{d}s \right\} \\
&\leq \left(\int\limits_{S_R^c} \left| \frac{\partial u}{\partial r} - iku \right|^2 \mathrm{d}s \right)^{1/2} \left(\int\limits_{S_R^c} |u|^2 \, \mathrm{d}s \right)^{1/2} \\
&\quad + \left(\int\limits_{S_{R,b}^+} \left| \frac{\partial u}{\partial r} - ik\cos\theta_+ u \right|^2 \mathrm{d}s \right)^{1/2} \left(\int\limits_{S_{R,b}^+} |u|^2 \, \mathrm{d}s \right)^{1/2},
\end{aligned}
\tag{4.36}
$$

where we have also exploited the Cauchy inequality. The chain of inequalities (4.36) enables us to write

$$
\begin{aligned}
&\left(k \int\limits_{S_R^c} |u|^2 \, \mathrm{d}s + k\mathrm{Re}(\cos\theta_+) \int\limits_{S_{R,b}^+} |u|^2 \, \mathrm{d}s \right)^{1/2} \\
&\leq \left(\int\limits_{S_R^c} \left| \frac{\partial u}{\partial r} - iku \right|^2 \mathrm{d}s \right)^{1/2} \left(\frac{\int_{S_R^c} |u|^2 \, \mathrm{d}s}{k \int_{S_R^c} |u|^2 \, \mathrm{d}s + k\mathrm{Re}(\cos\theta_+) \int_{S_{R,b}^+} |u|^2 \, \mathrm{d}s} \right)^{1/2} \\
&+ \left(\int\limits_{S_{R,b}^+} \left| \frac{\partial u}{\partial r} - ik\cos\theta_+ u \right|^2 \mathrm{d}s \right)^{1/2} \left(\frac{\int_{S_{R,b}^+} |u|^2 \, \mathrm{d}s}{k \int_{S_R^c} |u|^2 \, \mathrm{d}s + k\mathrm{Re}(\cos\theta_+) \int_{S_{R,b}^+} |u|^2 \, \mathrm{d}s} \right)^{1/2} \to 0,
\end{aligned}
\tag{4.37}
$$

as $R \to \infty$, in view of the radiation condition (4.30).

Remark: From the estimates (4.36) and (4.37) we also have

$$
-\mathrm{Im} \left\{ \int\limits_{S_R^c} \left(\frac{\partial u}{\partial r} - iku \right) \bar{u} \, \mathrm{d}s + \int\limits_{S_{R,b}^+} \left(\frac{\partial u}{\partial r} - ik\cos\theta_+ u \right) \bar{u} \, \mathrm{d}s \right\} \to 0 \text{ as } R \to \infty,
\tag{4.38}
$$

and from (4.38), (4.37) and (4.35) and (4.34) conclude that

$$
\int\limits_{l_{\infty,0}^-} \left| \frac{\partial u}{\partial n} \right|^2 \mathrm{d}s = 0 \text{ and } \left. \frac{\partial u}{\partial n} \right|_{l_{\infty,0}^-} = 0.
$$

From (4.37) we have

$$\int_{S_R^c} |u|^2\, ds + \mathrm{Re}(\cos\theta_+) \int_{S_{R,b}^+} |u|^2\, ds \to 0, \quad \text{as } R \to \infty. \tag{4.39}$$

The latter limit means that the scattering diagram $D_{1,2}(\varphi) \equiv 0$, $\varphi \in [0, 2\pi)$ because of the asymptotic behavior (4.3). In particular,

$$D_1(\varphi) = s_1(-\pi + \varphi) - s_1(\pi + \varphi) = 0, \quad |\varphi| < \Phi$$

and, in view of analyticity,

$$s_1(-\pi + \alpha) - s_1(\pi + \alpha) = 0, \quad \alpha \in \mathbb{C},$$

$s_1(\alpha)$ is regular and bounded in the strip $|\mathrm{Re}\,\alpha| \le \pi$. (Note that from the second summand in the limiting correlation (4.39) $s_1(\alpha)$ has no poles in the strip $|\mathrm{Re}\,\alpha| \le \pi$, because the amplitudes of surface waves, specified by the captured poles, are zero.) The 2π−periodic function $s_1(\alpha)$ is then constant, which is followed (see (4.10)) by $u_1 \equiv 0$.[6]

[6]Then it is obvious that there is also $u_2 \equiv 0$.

CHAPTER 5

Acoustic scattering of a plane wave by a circular impedance cone

In the present chapter we consider diffraction of a plane acoustic wave illuminating completely the surface of a right-circular impedance cone. Formulation of the problem is also given for an arbitrary convex cone and uniqueness of the solution is discussed. By separating the radial variable the problem at hand is then reduced to that for a spectral function. Further separation of the angular variables for the circular cone leads to integral equations for the Fourier coefficients. The use of the Sommerfeld integral enables one to describe the far-field asymptotics for, in particular, the reflected wave, the spherical wave from the vertex of the cone, and the surface waves.

5.1 Formulation of the problem and uniqueness

5.1.1 Formulation of the problem

Let a conical surface C cut out a domain Σ on the unit sphere S^2 centered at the point O, which is the vertex of the cone (Fig. 5.1). The points ω_0 and ω on the unit sphere are related to the direction, from which the incident wave comes, and to the observation point $\boldsymbol{r}$, $\boldsymbol{\omega} = \boldsymbol{r}/r$ correspondingly. We assume that the plane wave illuminates completely the conical surface C.

Let $\hat{\theta}(\omega, \omega_0)$ be the geodesic distance between ω and ω_0 on the unit sphere, $\cos\hat{\theta}(\omega, \omega_0) = \cos\theta\cos\theta_0 + \sin\theta\sin\theta_0\cos(\varphi - \varphi_0)$, $\omega = (\theta, \varphi)$ be the traditional spherical coordinates on the unit sphere, $\omega_0 = (\theta_0, \varphi_0)$, and k be the wavenumber; then the incident plane wave is[1]

$$U^{\mathrm{i}} = \mathrm{e}^{-ikr\cos\hat{\theta}(\omega,\omega_0)}. \tag{5.1}$$

We construct a classical solution of the problem. The scattered field $U(kr, \omega, \omega_0)$ satisfies the Helmholtz wave equation

$$\left(\Delta + k^2\right) U = 0, \tag{5.2}$$

[1]Recall that the time dependence e^{-ikct} is omitted throughout this chapter.

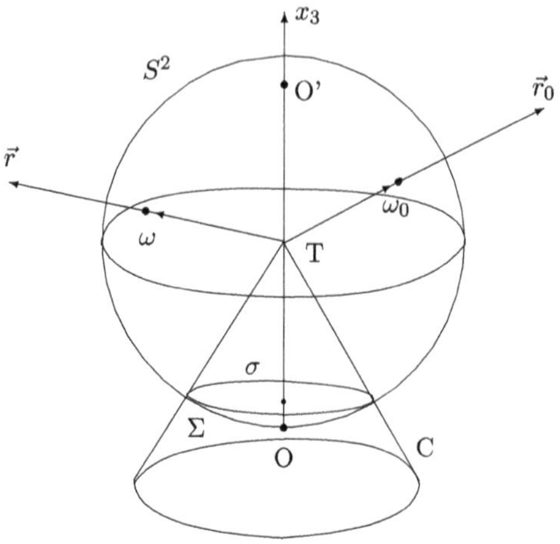

Figure 5.1 Diffraction by an impedance cone.

and is subject, in the sum with the incident wave, to the boundary condition

$$\left.\frac{\partial\left(U+U^{\mathrm{i}}\right)}{\partial n}\right|_{C}-\left.ik\eta\left(U+U^{\mathrm{i}}\right)\right|_{C}=0, \tag{5.3}$$

where n is the normal vector directed to the interior of C, and η is the normalized and k-independent surface impedance. We assume that $\operatorname{Re}\eta>0$, which means that the conical surface is absorbing provided the opposite was not declared. Sometimes we shall also use the complex angle ζ with $\eta=\sin\zeta,\ \ 0<\operatorname{Re}\zeta<\pi/2$.

The solution U satisfies Meixner's condition

$$\operatorname{Im}\left(ik\int_{S_\varepsilon}\frac{\partial U}{\partial r}\overline{U}\mathrm{d}s\right)\to 0,\ \ \varepsilon\to 0, \tag{5.4}$$

which implies that the energy flux through the sphere S_ε of a small radius ε tends to zero as $\varepsilon\to 0$. The latter condition is definitely valid as

$$|U|\le \operatorname{const} r^h,\quad r\,|\nabla U|\le \operatorname{const} r^h,\quad h>-1/2, \tag{5.5}$$

uniformly with respect to the angular variables. The density of energy is locally integrable.

Assume that the cone is acute and convex, that is, Σ (see Fig 5.1) belongs to a hemisphere of S^2, and is convex.

We turn to the radiation condition. Introduce

$$\hat{\theta}'(\omega,\omega_0)=\min_{s\in\sigma}\left(\hat{\theta}(\omega,s)+\hat{\theta}(s,\omega_0)\right)$$

the length of the broken geodesic on the unit sphere [5.10]. The magnitude $\hat{\theta}'(\omega,\omega_0)$ has a simple geometrical meaning. If a ray (geodesic) is emanated from ω_0 on $S^2\setminus\Sigma$, it reflects from σ in accordance with the geometrical-optics law and then arrives at the point ω so that $\hat{\theta}'(\omega,\omega_0)$ coincides with the minimal length of such a broken geodesic on $S^2\setminus\Sigma$ connecting ω and ω_0. The domain $\hat{\theta}'(\omega,\omega_0)>\pi$ is usually called "oasis" and denoted by M. (We point out that this domain $M\subset S^2$ is, generally speaking, not small.) It turns out that in this domain the far field U (as $kr \to \infty$) is described by a spherical wave propagating from the vertex of the cone. The surface L defined by $\hat{\theta}'(\omega, \omega_0) = \pi$ is called the surface of singular directions. The diffraction coefficient of the spherical wave from the vertex has singularities as $\omega \in L \cap S^2$. Consider a small vicinity (of $O((kr)^{-1/2+\varepsilon})$, $\varepsilon > 0$ small) of the conical surface L. We call it transition region to the domain illuminated by the reflected rays. As $\omega \in M$ the wave field has the asymptotics

$$U(kr, \omega, \omega_0) = D(\omega, \omega_0, kr) \frac{e^{ikr}}{-ikr}, \tag{5.6}$$

where $D(\omega, \omega_0, kr)$ is an asymptotic series with respect to the powers $(kr)^{-m}$, $m \geq 0$. The coefficient $D(\omega, \omega_0)$ at the leading term with $m = 0$ is called the scattering diagram (diffraction coefficient) of the spherical wave.

The rays reflected from the cone illuminate the domain $\hat{\theta}'|_\sigma \leq \hat{\theta}' \leq \pi$. The wave field scattered at infinity outside the transition zone (i.e., in the domain $N \subset S^2$ defined by $\hat{\theta}'|_\sigma \leq \hat{\theta}' < \pi$) is the sum of the reflected, spherical and surface waves

$$U(kr, \omega, \omega_0) = D_r(\omega, \omega_0, kr)\, e^{-ikr \cos \hat{\theta}'(\omega, \omega_0)} + D(\omega, \omega_0, kr) \frac{e^{ikr}}{-ikr} \\ + D_s(\omega, \omega_0, kr)\, e^{ikr \cos \theta_s(\omega, \omega_0)}, \tag{5.7}$$

where the last summand in (5.7) with complex $\theta_s(\omega, \omega_0)$ describes the so-called surface wave, $D_s(\omega, \omega_0, kr) = O((kr)^{-1/2})$. This wave is bounded on the conical surface C and vanishes exponentially as the observation point goes away from the surface C. As $\operatorname{Re} \eta > 0$, that is, for an absorbing surface, this wave is also exponentially decreasing as kr increases.

Such a behavior (as $\operatorname{Re} \eta > 0$) is, in some sense, different from conventional definition of the surface wave; nevertheless, we call such waves with the characteristic asymptotic behavior surface waves. It is worth noting that the corresponding summand appears in the asymptotics in some specific conditions. In particular, the necessary conditions of excitation of the acoustic surface wave are the inequalities $\operatorname{Im} \zeta < 0$, $\eta = \sin \zeta$, $0 \leq \operatorname{Re} \zeta \leq \pi/2$.

In the transition region the wave field is described by the parabolic cylinder function, which is an analog of the Fresnel integral in the problems of diffraction by a wedge.

5.1.2 On uniqueness of the classical solution

Let $k > 0$, $\operatorname{Re} \eta > 0$. We consider the homogeneous problem ($U^i = 0$) (5.2)–(5.5). Assume that its solution satisfies the Sommerfeld radiation conditions in the integral form

$$\int_{S_R} \left| \frac{\partial U}{\partial r} - ikU \right|^2 ds \to 0, \quad R \to \infty, \tag{5.8}$$

where S_R is the part of a sphere of radius R outside the cone and centered at the vertex T. Note that, provided $\operatorname{Re} \eta = 0$, a surface wave may propagate from the vertex, which leads to the modification of the radiation condition (5.8) by inclusion of a specific term describing the outgoing surface wave (see also [5.42]).

Exploit the equation and the Green theorem in the domain $\Omega_{\varepsilon,R}$ (which is in the exterior of the cone and of the sphere S_ε and in the interior of S_R) with the boundary $\partial\Omega_{\varepsilon,R} = S_\varepsilon \cup C_{\varepsilon,R} \cup S_R$, where S_ε is the part of the sphere of small radius ε in the exterior of the cone, $C_{\varepsilon,R}$ is the part of the conical surface between S_ε, S_R, and the boundary condition. Compute the imaginary part and thus obtain

$$k\,\mathrm{Re}(\eta)\int\limits_{C_{\varepsilon,R}} |U|^2 \mathrm{d}s + \mathrm{Im}\int\limits_{S_\varepsilon} \frac{\partial U}{\partial n}\bar{U}\mathrm{d}s + \mathrm{Im}\int\limits_{S_R} \frac{\partial U}{\partial n}\bar{U}\mathrm{d}s = 0,$$

where n is the normal directed to the exterior of the domain. In the latter equality we let ε tend to zero. We have

$$\begin{aligned}
0 \le k\,\mathrm{Re}\,\eta \int\limits_{C_R} |U|^2 \mathrm{d}s &= -\mathrm{Im}\int\limits_{S_R} \frac{\partial U}{\partial R}\bar{U}\mathrm{d}s \\
= -\mathrm{Im}\int\limits_{S_R}\left(\frac{\partial U}{\partial R} - ikU\right)\bar{U}\mathrm{d}s &- \mathrm{Im}(ik)\int\limits_{S_R} |U|^2 \mathrm{d}s \\
\le \left[\int\limits_{S_R}\left|\frac{\partial U}{\partial r} - ikU\right|^2 ds\right]^{1/2}\left[\int\limits_{S_R}|U|^2\mathrm{d}s\right]^{1/2} &- k\int\limits_{S_R}|U|^2\mathrm{d}s.
\end{aligned} \tag{5.9}$$

Therefore, we obtain

$$\mathrm{Im}(ik)\int\limits_{S_R}|U|^2\mathrm{d}s \le \left[\int\limits_{S_R}\left|\frac{\partial U}{\partial r} - ikU\right|^2 \mathrm{d}s\right]^{1/2}\left[\int\limits_{S_R}|U|^2\mathrm{d}s\right]^{1/2}.$$

Then, exploiting the condition (5.8), we verify that

$$\mathrm{Im}(ik)\int\limits_{S_R}|U|^2\mathrm{d}s \to 0,\ R\to\infty.$$

Applying the previous relation to (5.9), we find that $U|_C = 0$, and, hence, in view of the boundary condition, $\partial_n U|_C = 0$. Finally, we use the fact that a solution of a Cauchy problem for the Helmholtz (which means elliptic) equation with the trivial initial data on C is unique then conclude $U = 0$.

We note that the radiation condition (5.8) is valid in the case of illumination by a compact source and is not satisfied for a plane-wave illumination.

For the plane-wave illumination with the conditions

$$k = k_1 + \mathrm{i}\kappa,\ k_1 > 0,$$

and

$$k_1\mathrm{Re}\,\eta \ - \kappa\,\mathrm{Im}\eta \ge 0,\quad \kappa > 0,$$

for the scattered field satisfying the estimates

$$|U| < \text{const}\exp(-a\kappa r),\ \ |\partial_r U| < \text{const}\exp(-a\kappa r),\ a > 0 \tag{5.10}$$

uniformly with respect to the angular variables, the solution of the homogeneous ($U^{\mathrm{i}} = 0$) problem (5.2)–(5.5) is also trivial, $U = 0$.

Indeed, the proof follows directly from the identity

$$2k_1\kappa \int\limits_{\Omega_{\varepsilon,R}} |U|^2 \mathrm{d}V + (k_1 \mathrm{Re}\,\eta - \kappa\,\mathrm{Im}\,\eta) \int\limits_{C_{\varepsilon,R}} |U|^2 \mathrm{d}s$$

$$+\mathrm{Im}\left(\mathrm{i}k \int\limits_{S_\varepsilon} \frac{\partial U}{\partial n} \overline{U} \mathrm{d}s \right) + \mathrm{Im}\,\mathrm{i}k \int\limits_{S_R} \frac{\partial U}{\partial n} \overline{U} \mathrm{d}s = 0,$$

where the last summand is estimated by use of (5.10).

5.2 Kontorovich–Lebedev (KL) transform and incomplete separation of variables

5.2.1 Integral representation of the solution

Actually the basic idea of transformations lies in the incomplete separation of variables: the radial variable is specific in view of the geometry of the scatterer. However, in our case the separation of the radial variable by means of the Kontorovich–Lebedev transform leads, contrary to the ideal boundary conditions, to a boundary-value problem on S^2 with a (with respect to the spectral variable) nonlocal boundary condition on σ.

The plane incident wave admits the integral representation (see, e.g., [5.29])

$$U^{\mathrm{i}}(kr, \omega, \omega_0) = \frac{4}{\mathrm{i}\sqrt{2\pi}} \int\limits_{\mathrm{i}\mathbb{R}} \nu \sin \pi\nu\, u^{\mathrm{i}}_{\nu}(\omega, \omega_0) \frac{\mathrm{K}_\nu(-\mathrm{i}kr)}{\sqrt{-\mathrm{i}kr}} \mathrm{d}\nu, \tag{5.11}$$

where

$$u^{\mathrm{i}}_{\nu}(\omega, \omega_0) = -\frac{\mathrm{P}_{\nu-1/2}\left(-\cos\hat{\theta}(\omega, \omega_0)\right)}{4\cos(\pi\nu)}, \tag{5.12}$$

$\mathrm{P}_{\nu-1/2}(x)$ is the Legendre function, $\mathrm{K}_\nu(z)$ is the modified Bessel function (Macdonald function)

$$\mathrm{K}_\nu(z) = -\mathrm{i}\,(\pi/2)\exp(-\mathrm{i}\nu\pi/2)\,\mathrm{H}^{(2)}_\nu(-\mathrm{i}z),$$

where the integral (5.11) converges exponentially provided

$$\hat{\theta}(\omega, \omega_0) > \pi/2 + |\arg(-\mathrm{i}k)|, \quad |\arg(-\mathrm{i}k)| \le \pi/2. \tag{5.13}$$

The convergence is easily verified by use of the asymptotics

$$\mathrm{K}_\nu(z) \sim \mathrm{const}\, \frac{\nu^\alpha \cos\left[\nu\,(\pi/2 + |\arg(z)|)\right]}{\sin\pi\nu},$$

$$|\mathrm{Im}\,\nu| \to \infty, \quad |\arg(\nu)| \to \pi/2, \quad \alpha = |\mathrm{Re}\,\nu| - 1/2,$$

$$\mathrm{P}^{-|n|}_{\nu-1/2}(\cos\theta) = \frac{\sqrt{2}\,\Gamma(\nu - |n| + 1/2)}{\sqrt{\pi}\sin\theta\,\Gamma(\nu+1)} \cos\left(\nu\theta - |n|\frac{\pi}{2} + \frac{\pi}{4}\right)\left[1 + \mathrm{O}\left(\frac{1}{\nu}\right)\right],$$

$|n|$ is bounded, $\delta < \theta < \pi - \delta$, $\delta > 0$ small, $|\nu| \gg 1$, $|\arg(\nu)| < \pi$. The estimates for $\mathrm{K}_\nu(z)$ and $\mathrm{P}^{-|n|}_{\nu-1/2}(\cos\theta)$ can easily be found in [5.24].

Taking into account the representation (5.11), it is natural to look for the scattered field $U(kr, \omega, \omega_0)$ in an analogous form [5.29]

$$U(kr, \omega, \omega_0) = \frac{4}{\mathrm{i}\sqrt{2\pi}} \int\limits_{\mathrm{i}\mathbb{R}} \nu \sin \pi\nu \, u_\nu(\omega, \omega_0) \frac{\mathrm{K}_\nu(-ikr)}{\sqrt{-ikr}} \, d\nu. \tag{5.14}$$

It is worth commenting briefly on convergence of the integral in (5.14). The calculations described as follows are valid provided the Kontorovich–Lebedev integral converges rapidly on ν and uniformly on ω. The latter can be achieved by taking the wavenumber k complex. For the real values of k the same formulas are obtained by use of the analytic continuation with respect to k. The arguments of the analytic continuation as well as the use of other integral representations extending those of the KL type are systematically exploited in this and following chapters.

5.2.2 Formulation of the problem for the spectral function u_ν

To determine the spectral function $u_\nu(\omega, \omega_0)$ in (5.14) and to demonstrate that the representation delivers the desired solution it is important to formulate the conditions on the spectral function. The following ad hoc conditions, however, are very natural. Indeed, we should substitute the representation to the equation and the boundary conditions and, after some reductions, will arrive at the problem for $u_\nu(\omega, \omega_0)$. In this way, some necessary properties of $u_\nu(\omega, \omega_0)$ are required. Anticipating the result, we assume the following properties of the spectral function.

Let $u_\nu(\omega, \omega_0)$ be smooth on $\omega = (\theta, \varphi)$, $(\theta, \varphi) \in S^2 \setminus \overline{\Sigma}$, have a continuous derivative on σ for any ν belonging to the strip $\Pi_\delta = \{\nu : |\mathrm{Re}\,\nu| < 1 + \delta\}$, $\delta > 0$ is small. Besides, u_ν is even and regular in Π_δ with the same assumed for the derivative $\left.\frac{\partial u_\nu}{\partial N_\omega}\right|_\sigma$, where N_ω is the normal directed to the interior of σ and belonging to the corresponding tangent plane to S^2.

Now we formulate the problem for u_ν. Let $u_\nu(\omega, \omega_0)$ solve the equation

$$\left[\Delta_\omega + \left(\nu^2 - 1/4\right)\right] u_\nu(\omega, \omega_0) = 0 \tag{5.15}$$

and fulfill the boundary condition

$$\left.\frac{\partial \hat{U}_{\nu+1}}{\partial N_\omega}\right|_\sigma - \left.\frac{\partial \hat{U}_{\nu-1}}{\partial N_\omega}\right|_\sigma = -2\nu\eta \left.\hat{U}_\nu\right|_\sigma , \tag{5.16}$$

$\hat{U}_\nu = u_\nu + u^{\mathrm{i}}_\nu$, Δ_ω is the Laplace–Beltrami operator on S^2,

$$\Delta_\omega = (\sin\theta)^{-1} \, \partial_\theta \, (\sin\theta \, \partial_\theta) + (\sin\theta)^{-2} \, \partial^2_\varphi .$$

For convergence of the integral (5.14) $u_\nu(\omega, \omega_0)$ must exponentially vanish for any $\omega \in S^2 \setminus \Sigma$, or more exact, we assume that in Π_δ

$$|u_\nu(\omega, \omega_0)| \leq \text{const}\, \frac{|\nu|^\kappa}{|\cos[(\pi + \epsilon)\nu]|}, \quad \text{as } \omega \in M, \quad \kappa \geq 0, \tag{5.17}$$

$\epsilon > 0$ is some small number. As $\hat{\theta}'(\omega, \omega_0) \leq \pi$ (i.e., $\omega \in N$) we consider that

$$|u_\nu(\omega, \omega_0)| \leq \text{const}\, \frac{|\nu|^{\kappa_1}}{|\cos[\nu\hat{\theta}'(\omega, \omega_0)]|}, \quad \kappa_1 \geq -1/2. \tag{5.18}$$

It is remarkable that, provided one could specify the spectral function as a solution of the problem (5.15) and (5.16) in the prescribed class of functions, the Kontorovich–Lebedev integral representation of the wave field would satisfy the formulated diffraction problem.

More precisely, we can assert the following.

Let the spectral function $u_\nu(\omega, \omega_0)$ solve the problem (5.15) and (5.16) in the prescribed class of functions. Then, provided

$$\hat{\theta}'(\omega, \omega_0) > \pi/2 + |\arg(-ik)|, \quad |\arg(-ik)| \leq \pi/2, \tag{5.19}$$

the integral representation (5.14) is a classical solution of the Helmholtz equation (5.2) satisfying the boundary condition (5.3), Meixner's condition (5.5) and, as $kr \to \infty$, in the oasis ($\omega \in M$) has the asymptotics

$$U(kr, \omega, \omega_0) = D(\omega, \omega_0)\, \frac{e^{ikr}}{-ikr}\left[1 + O\left(\frac{1}{kr}\right)\right], \tag{5.20}$$

where

$$D(\omega, \omega_0) = \frac{2}{i} \int_{i\mathbb{R}} \nu \sin \pi\nu\, u_\nu(\omega, \omega_0)\, d\nu \tag{5.21}$$

is the scattering diagram (the diffraction coefficient).[2]

We note that, in view of the condition (5.19), the integral (5.14) converges exponentially, then by the direct substitution of (5.14) to (5.2) taking into account the Bessel equation for $K_\nu(z)$,

$$\left[\frac{\partial^2}{\partial z^2} + \frac{1}{z}\frac{\partial}{\partial z} + \left(1 + \frac{\nu^2}{z^2}\right)\right] K_\nu(z) = 0,$$

and (5.15), we show (see also [5.26]) that $U(kr, \omega, \omega_0)$ satisfies the Helmholtz equation.

We verify the condition (5.3), exploiting the representation (5.14) and taking into account the correlations

$$\frac{K_\nu(z)}{z} = \frac{K_{\nu+1}(z) - K_{\nu-1}(z)}{2\nu}, \quad \frac{\partial}{\partial n} = \frac{1}{r}\frac{\partial}{\partial N_\omega},$$

[2]Formula (5.21) is also known as Smyshlyaev's formula for the diffraction coefficient valid in the oasis.

and thus have

$$\int_{i\mathbb{R}} \frac{\nu \sin \pi \nu}{\sqrt{-ikr}} \left[\left.\frac{\partial \hat{U}_\nu}{\partial N_\omega}\right|_\sigma (-ik) \frac{\mathrm{K}_{\nu+1}(z) - \mathrm{K}_{\nu-1}(z)}{2\nu} - ik\eta \hat{U}_\nu\big|_\sigma \mathrm{K}_\nu(-ikr) \right] \mathrm{d}\nu$$

$$= \frac{-ik}{\sqrt{-ikr}} \left[\int_{i\mathbb{R}+1} \frac{\sin \pi \nu}{2} \left.\frac{\partial \hat{U}_{\nu+1}}{\partial N_\omega}\right|_\sigma \mathrm{K}_\nu(-ikr)\,\mathrm{d}\nu \right.$$

$$\left. - \int_{i\mathbb{R}-1} \frac{\sin \pi \nu}{2} \left.\frac{\partial \hat{U}_{\nu-1}}{\partial N_\omega}\right|_\sigma \mathrm{K}_\nu(-ikr)\,\mathrm{d}\nu + \int_{i\mathbb{R}} \frac{\sin \pi \nu}{2} 2\eta\,\nu\, \hat{U}_\nu\big|_\sigma \mathrm{K}_\nu(-ikr)\,\mathrm{d}\nu \right]$$

$$= \frac{-ik}{\sqrt{-ikr}} \int_{i\mathbb{R}} \frac{\sin \pi \nu}{2} \left[\left.\frac{\partial \hat{U}_{\nu+1}}{\partial N_\omega}\right|_\sigma - \left.\frac{\partial \hat{U}_{\nu-1}}{\partial N_\omega}\right|_\sigma + 2\eta\,\nu\, \hat{U}_\nu\big|_\sigma \right] \mathrm{K}_\nu(-ikr)\,\mathrm{d}\nu = 0.$$

When deforming the integration contour, we used the regularity of $\left.\frac{\partial \hat{U}_\nu}{\partial N_\omega}\right|_\sigma$ in the strip Π_σ and the condition (5.16).

To prove Meixner's condition we reduce (5.14) by means of

$$\mathrm{K}_\nu(z) = \frac{\pi\,[\mathrm{I}_{-\nu}(z) - \mathrm{I}_\nu(z)]}{2 \sin \pi \nu},$$

$$\mathrm{I}_\nu(z) = \exp(-\mathrm{i}\nu\pi/2)\,\mathrm{J}_\nu(\mathrm{i}z), \quad -\pi < \arg(z) \le \pi/2,$$

taking into account the regularity of $u_\nu(\omega, \omega_0)$ in a vicinity of the imaginary axis,

$$U(kr, \omega, \omega_0) = \frac{-\sqrt{2\pi}}{\mathrm{i}\sqrt{-ikr}} \int_{i\mathrm{R}} \nu\,[\mathrm{I}_\nu(-ikr) - \mathrm{I}_{-\nu}(-ikr)]\, u_\nu(\omega, \omega_0)\,\mathrm{d}\nu$$

$$= \frac{2\mathrm{i}\sqrt{2\pi}}{\sqrt{-ikr}} \int_{i\mathbb{R}} \nu\,\mathrm{I}_\nu(-ikr)\, u_\nu(\omega, \omega_0)\,\mathrm{d}\nu$$

$$= \frac{2\mathrm{i}\sqrt{2\pi}}{\sqrt{-ikr}} \int_{\infty e^{-\mathrm{i}\phi}+a}^{\infty e^{\mathrm{i}\phi}+a} \nu\, \mathrm{e}^{-\mathrm{i}\nu\pi/2}\, \mathrm{J}_\nu(kr)\, u_\nu(\omega, \omega_0)\,\mathrm{d}\nu = O\left((kr)^{-1/2+a}\right),$$

where $a > 0$ is small, $\phi = \pi/2 - \sigma$, $\sigma > 0$ is small, $|kr| \to 0$. The estimates justifying the previous reductions are considered next.

The formulas for the wave field for the real wavenumbers k are of the principal importance in applications. It follows from before that the representation (5.14) admits reductions and analytic continuation onto the real k without any limitations on $\hat{\theta}'(\omega, \omega_0)$

$$U(kr, \omega, \omega_0) = 4\mathrm{i}\sqrt{\frac{\pi}{2}} \int_{\infty e^{-\mathrm{i}\phi}}^{\infty e^{\mathrm{i}\phi}} \nu\, \mathrm{e}^{-\mathrm{i}\pi\nu/2} u_\nu(\omega, \omega_0) \frac{\mathrm{J}_\nu(kr)}{\sqrt{-ikr}} \mathrm{d}\nu, \tag{5.22}$$

where $\phi \in (0, \pi/2)$, the ends of integration in the latter integral can be conducted parallel to the real axis comprising the singularities of $u_\nu(\omega, \omega_0)$ with respect to ν, because they are located in some strip including the real axis. It is reminded that the integral above is called the Watson–Bessel integral, see Section 1.4.5.

Indeed, provided the absolute value of $\nu = |\nu|\exp(i\phi)$, $(|\phi| \le \pi/2)$ is sufficiently large, then $J_\nu(kr) \sim (kr/2)^\nu/\Gamma(\nu+1)$, the estimate is valid

$$|\exp(-i\pi\nu)u_\nu(\omega,\omega_0)\,J_\nu(kr)| \le C\,\exp[-|\nu|\log|\nu|\cos\phi$$

$$-|\nu|(\sin\phi(\arg k - \pi/2 - \phi) + |\sin\phi|\hat{\theta}'(\omega,\omega_0) - \cos\phi[1+\log(|k|r/2)])],$$

where $|kr|$ is arbitrarily fixed. For sufficiently large $|\nu|$, $\log|\nu| > 1+\log(|k|r/2)$ we deformed the integration contour in (5.22) into the contour C_ϕ $(\infty e^{-i\phi}, \infty e^{i\phi})$, where $\phi \in [0, \pi/2)$, thus obtaining the exponentially convergent integral (5.22) for positive k and $\omega \in S^2 \setminus \Sigma$.

The far-field representation in the oasis, $\omega \in M$ (see (5.20) and (5.21)) follows from (5.14) taking into account the asymptotics

$$K_\nu(z) = \sqrt{\frac{\pi}{2}}\frac{e^{-z}}{\sqrt{z}}\left[1 + O_\alpha\left(\frac{1}{z}\right)\right], \qquad \alpha = \frac{\nu}{z},\ z \to \infty.$$

The far field in the domain supplementary to the oasis $(\omega \in N)$ can also be computed; however, it requires development of a new technique, which will be discussed in the second part of the present chapter (see also [5.29]).

It is obvious that the problem [(5.15) and (5.16)] plays a key role for the study of the wave field U. The problem [(5.15) and (5.16)] is nonstandard because it is nonlocal with respect to the spectral variable ν in view of the boundary condition linking $u_\nu|_\sigma$ and its normal derivatives shifted on ν.

5.3 The boundary value problem for the spectral function $u_\nu(\omega, \omega_0)$

Instead of the problem (5.15) and (5.16) for u_ν it is convenient to study the equivalent one

$$\left[\Delta_\omega + \left(\nu^2 - 1/4\right)\right]u_\nu(\omega,\omega_0) = 0 \tag{5.23}$$

$$\left.\frac{\partial\hat{u}_\nu}{\partial N_\omega}\right|_\sigma = \eta\,A\,\hat{u}_\nu|_\sigma\,, \quad \hat{u}_\nu = u_\nu + u^{i}_\nu\,, \tag{5.24}$$

where

$$A\,\hat{u}_\nu|_\sigma := \frac{1}{2i}\int_{i\mathbb{R}} \frac{t\,\sin(\pi t)\,\hat{u}_\nu|_\sigma\,dt}{\cos(\pi\nu)+\cos(\pi t)}.$$

The equivalence of the boundary conditions (5.16) and (5.24) is easily verified with the aid of the technique discussed in Chapter 1. Namely, let $s(\nu)$ be regular in Π_σ and solve the equations

$$s(\nu \pm 1) - s(-\nu \pm 1) = \mp 2i\,H(\nu),$$

$(s(\nu)$ and $H(\nu)$ vanish exponentially as $\nu \to i\infty$, $s(\nu) = s(-\nu))$ then

$$s(\nu) = \frac{1}{4}\int_{i\mathbb{R}} H(\tau)\left[\frac{\sin(\pi\tau/2)}{\cos(\pi\tau/2)-\sin(\pi\nu/2)} + \frac{\sin(\pi\tau/2)}{\cos(\pi\tau/2)+\sin(\pi\nu/2)}\right]d\tau.$$

The equivalence follows from this simple observation.

Provided solution of the problem (5.23) and (5.24) exists, it is unique in the described class of functions. Note that proof of the uniqueness may be of interest because of the nonstandard boundary condition. Let $\operatorname{Re}\eta \geq 0$ and $\nu \in \mathrm{i}\mathbb{R}$, then solution of the homogeneous problem (5.23) and (5.24) ($u^i_\nu \equiv 0$) is trivial.

Indeed, multiply the equation (5.15) for U by $\overline{U}$ and integrate over $S^2 \setminus \Sigma$. Integrating by parts, one has

$$\int_\sigma \frac{\partial U}{\partial N}\overline{U}\mathrm{d}l - \int_{S^2\setminus\Sigma} |\nabla_\omega U|^2 \mathrm{d}S + (\nu^2 - 1/4) \int_{S^2\setminus\Sigma} |U|^2 \mathrm{d}S = 0, \tag{5.25}$$

where $\mathrm{d}S = \sin\theta\,\mathrm{d}\theta\,\mathrm{d}\varphi$, $\mathrm{d}l$ is the differential of the arc length of σ, and ν is purely imaginary. Exploit the homogeneous boundary condition; then from (5.25) one has

$$\int_{S^2\setminus\Sigma} \left[|\nabla_\omega U|^2 - (\nu^2 - 1/4)|U|^2\right] \mathrm{d}S - \frac{\eta}{2\mathrm{i}} \int_{\mathrm{i}\mathbb{R}} \frac{t \sin\pi t \int_\sigma U(t)\bar{U}(\nu)\mathrm{d}l}{\cos\pi t + \cos\pi\nu} \mathrm{d}t = 0.$$

Multiply the latter equality by $(\mathrm{i}\nu/2)\sin\pi\nu$, integrate over ν, and compute the imaginary part

$$\begin{aligned} &\int_{\mathrm{i}\mathbb{R}} \mathrm{d}\nu(\mathrm{i}\nu/2)\sin\pi\nu \left\{ \int_{S^2\setminus\Sigma} \left[|\nabla_\omega U|^2 - (\nu^2 - 1/4)|U|^2\right] \mathrm{d}S \right\} \\ &+ \operatorname{Re}(\eta) \int_{\mathrm{i}\mathbb{R}}\int_{\mathrm{i}\mathbb{R}} (\mathrm{i}/2)^2 \frac{t\sin(\pi t)\nu\sin(\pi\nu)\int_\sigma U(t)\overline{U}(\nu)\mathrm{d}l}{\cos\pi t + \cos\pi\nu} \mathrm{d}t\mathrm{d}\nu = 0. \end{aligned} \tag{5.26}$$

The first summand in (5.26) is nonnegative and $\operatorname{Re}\eta \geq 0$, so it is sufficient to demonstrate that the second summand, denoted by E_a, is also nonnegative. Introduce the integration variables $p = \cos\pi t$, $q = \cos\pi\nu$, exploit that

$$\begin{aligned} E_a &= \frac{1}{\pi^4}\int_\sigma\int_1^a\int_1^a \operatorname{arccosh}(p)U(\operatorname{arccosh}(p)/\pi)\operatorname{arccosh}(q)\bar{U}(\operatorname{arccosh}(q)/\pi)\frac{\mathrm{d}p\mathrm{d}q}{p+q}\mathrm{d}l \\ &= \frac{1}{\pi^4}\int_\sigma\int_1^a\int_1^a (h(p)h(q) + m(p)m(q))/(p+q)\mathrm{d}p\mathrm{d}q\mathrm{d}l \\ &= \frac{1}{\pi^4}\int_0^\infty\int_\sigma \left[\int_1^a h(p)\mathrm{e}^{-sp}\mathrm{d}p\right] \times \left[\int_1^a h(q)\mathrm{e}^{-sq}\mathrm{d}q\right] \mathrm{d}l\mathrm{d}s \\ &+ \frac{1}{\pi^4}\int_0^\infty\int_\sigma \left[\int_1^a m(p)\mathrm{e}^{-sp}\mathrm{d}p\right] \times \left[\int_1^a m(q)\mathrm{e}^{-sq}\mathrm{d}q\right] \mathrm{d}l\mathrm{d}s \geq 0, \end{aligned} \tag{5.27}$$

where $h(x)$ and $m(x)$ are real and imaginary parts of the function $\operatorname{arccosh}(x)\,U(\operatorname{arccosh}(x)/\pi)$, $a > 1$.

Denoting by E the limit of E_a as $a \to \infty$, we conclude that E is nonnegative and, therefore, $U = 0$.

We turn to the construction of the solution for the circular cone by means of the series expansion and to the derivation of the integral equation for the Fourier coefficients.

5.3.1 Separation of the angular variables for the circular cone

Let $\theta=\theta_1$ be the equation of the conical surface in the spherical coordinates attributed to the Cartesian ones. We look for the solution of the problem (5.23) and (5.24) in the form

$$u_\nu(\omega,\omega_0) = \sum_{n=-\infty}^{+\infty} \mathrm{i}^n\, \mathrm{e}^{-\mathrm{i}n\varphi}\, R(\nu,n)\, \frac{\mathrm{P}^{-|n|}_{\nu-1/2}(\cos\theta)}{\mathrm{d}_{\theta_1}\mathrm{P}^{-|n|}_{\nu-1/2}(\cos\theta_1)}, \tag{5.28}$$

where $\theta=\theta_1$ is also the equation of σ on the sphere S^2. From the summation formula for the Legendre functions we have the representation

$$\begin{aligned} u^{\mathrm{i}}_\nu(\omega,\omega_0) &= -\frac{\mathrm{P}_{\nu-1/2}(-\cos\hat\theta(\omega,\omega_0))}{4\cos(\pi\nu)} \\ &= \sum_{-\infty}^{+\infty} \mathrm{i}^n\, \mathrm{e}^{-\mathrm{i}n\varphi}\, R_i(\nu,n)\, \frac{\mathrm{P}^{-|n|}_{\nu-1/2}(-\cos\theta)}{\mathrm{d}_{\theta_1}\mathrm{P}^{-|n|}_{\nu-1/2}(-\cos\theta_1)}, \end{aligned} \tag{5.29}$$

$0<\theta_0<\theta$, $\cos\hat\theta(\omega,\omega_0)=\cos\theta\cos\theta_1+\sin\theta\sin\theta_1\sin(\varphi-\varphi_0)$, $\varphi_0=0$,

$$\begin{aligned} R_{\mathrm{i}}(\nu,n) &= \frac{\mathrm{i}^n}{-4\cos(\pi\nu)}\,\frac{\Gamma(\nu+|n|+1/2)}{\Gamma(\nu-|n|+1/2)} \\ &\quad\times \mathrm{d}_{\theta_1}\,\mathrm{P}^{-|n|}_{\nu-1/2}(-\cos\theta_1)\,\mathrm{P}^{-|n|}_{\nu-1/2}(\cos\theta_0). \end{aligned}$$

The series (5.28) solves the equation (5.23). Substituting (5.28) and (5.29) into the boundary condition, we obtain the equation for $R(\nu,n)$ ($\nu\in\mathrm{i}\mathbb{R}$)

$$R(\nu,n) = \frac{1}{2}\eta\int_{\mathrm{i}\mathbb{R}} \frac{w(t,n)\sin(\pi t)\,R(t,n)}{\cos(\pi t)+\cos(\pi\nu)}\,\mathrm{d}t + S_{\mathrm{i}}(\nu,n), \tag{5.30}$$

$$S_{\mathrm{i}}(\nu,n) = -R_{\mathrm{i}}(\nu,n) + \frac{\eta}{2}\int_{\mathrm{i}\mathbb{R}} \frac{w_{\mathrm{i}}(t,n)\sin(\pi t)\,R_{\mathrm{i}}(t,n)}{\cos(\pi t)+\cos(\pi\nu)}\,\mathrm{d}t,$$

$$w(\nu,n) = -\mathrm{i}\nu\,\frac{\mathrm{P}^{-|n|}_{\nu-1/2}(\cos\theta_1)}{\mathrm{d}_{\theta_1}\mathrm{P}^{-|n|}_{\nu-1/2}(\cos\theta_1)},\qquad w_{\mathrm{i}}(\nu,n) = -\mathrm{i}\nu\,\frac{\mathrm{P}^{-|n|}_{\nu-1/2}(-\cos\theta_1)}{\mathrm{d}_{\theta_1}\mathrm{P}^{-|n|}_{\nu-1/2}(-\cos\theta_1)}.$$

With the aid of

$$\mathrm{P}^{-|n|}_{\nu-1/2}(\cos\theta) = \sqrt{\frac{2}{\pi}}\,\frac{(\sin\theta)^{-|n|}}{\Gamma(|n|+1/2)}\int_0^\theta \frac{\cos(\nu t)\,\mathrm{d}t}{(\cos t-\cos\theta)^{-|n|+1/2}},\qquad n\in\mathbb{Z}$$

one can show that $w(\nu,n)>0$, if $\nu\in[0,+\mathrm{i}\infty)$. Besides,

$$w(\nu,n) = 1+\mathrm{O}_n(1/\nu),\qquad \nu\to\mathrm{i}\infty, \tag{5.31}$$

$|n|$ is fixed. Note that the asymptotics (5.31) is nonuniform with respect to $|n|$.

Remark: It is easily verified that $R(\nu, n)$ solves the functional equation

$$R(\nu + 1, n) - R(\nu - 1, n) = -2\mathrm{i}\,[\eta\, w(\nu, n)R(\nu, n) - G_{\mathrm{i}}(\nu, n)]\,, \tag{5.32}$$

n is a parameter and

$$G_{\mathrm{i}}(\nu, n) = (\mathrm{i}/2)\,[R_{\mathrm{i}}(\nu + 1, n) - R_{\mathrm{i}}(\nu - 1, n)] - \eta\, w_{\mathrm{i}}(\nu, n)R_{\mathrm{i}}(\nu, n).$$

5.3.2 Study of the integral equation for $R(\nu,n)$

The function $R(\nu, n)$ is even in ν so that from (5.30) we have

$$R(\nu, n) = \eta \int_0^{\mathrm{i}\infty} \frac{w(t, n)\, R(t, n)\, \sin \pi t}{\cos(\pi t) + \cos(\pi \nu)}\, \mathrm{d}t + S_{\mathrm{i}}(\nu, n)\,. \tag{5.33}$$

We reduce equation (5.33) introducing the new variable $x = 1/\cos(\pi\nu)$ and the new unknown function

$$\mathbf{r}(x, n) = R(\nu, n) \cos \pi\nu|_{x=1/\cos(\pi\nu)}\,.$$

Thus, we have

$$\mathbf{r}(x, n) + \frac{\eta}{\pi} \int_0^1 \frac{W(y, n)\,\mathbf{r}(y, n)\mathrm{d}y}{x + y} = \mathbf{r}_{\mathrm{i}}(x, n), \tag{5.34}$$

where

$$\mathbf{r}_{\mathrm{i}}(y, n) = S_{\mathrm{i}}(\nu, n) \cos \pi\nu|_{x=1/\cos(\pi\nu)}\,,$$
$$W(x, n) = w(\nu, n)|_{x=1/\cos(\pi\nu)}\,.$$

First, consider the case when $|n|$ is arbitrarily fixed, that is, $|n|$ is not a large parameter. Exploit the expression for R_{i} and the asymptotics of the Legendre function as $\nu \to \mathrm{i}\infty$; then

$$|R_{\mathrm{i}}(\nu, n)| \le \text{const}\, |\exp[\mathrm{i}\nu(\theta_1 - \theta_0)]|\,,$$
$$|\mathbf{r}_{\mathrm{i}}(x, n)| \le \text{const}\, |\log(2/x)|\, x^{(\theta_1-\theta_0)/\pi-1}, \quad x \in (0, 1]. \tag{5.35}$$

The estimate (5.35) shows that equation (5.34) is naturally studied in $L_p(0, 1)$ as $1 < p < p_* = \pi/[\pi - (\theta_1 - \theta_0)]$.

It is remarkable that, provided $\mathrm{Re}\,\eta > 0$ and n is fixed, the operator on the left-hand side of (5.34) is represented as a sum $D + K$, where D is a boundedly invertible operator[3] (the Dixon operator, e.g., [5.27])

$$(D\,\mathbf{r})(x) := \mathbf{r}(x) - \lambda \int_0^1 \frac{\mathbf{r}(y)\,\mathrm{d}y}{x + y}, \quad \lambda = -\eta/\pi, \tag{5.36}$$

[3]The equation and its splitting was considered in [5.27].

and the operator K,

$$(K\,\mathbf{r})(x) := \frac{\eta}{\pi} \int_0^1 \frac{W(y,n) - 1}{x+y}\, \mathbf{r}(y,n)\, dy \tag{5.37}$$

is compact. The latter is verified exploiting that, as $\nu \to i\infty$ and n is fixed, one has

$$w(\nu, n) = 1 + O_n(1/\nu), \quad W(x,n) = 1 + O_n(1/\log(1/x)).$$

The operator in equation (5.34) is then Fredholm and, moreover, the equation is uniquely solvable.

Note that together with equation (5.34) it is useful to consider the "adjoint" equation for $\rho(x,n) = W(x,n)\mathbf{r}(x,n)$

$$\rho(x,n) + \frac{\eta W(x,n)}{\pi} \int_0^1 \frac{\rho(y,n)}{x+y}\, dy = \rho_i(y,n), \tag{5.38}$$

with $\rho_i(x,n) = W(x,n)\mathbf{r}_i(x,n)$. Equations (5.38) and (5.34) are simultaneously Fredholm. It can be demonstrated that equation (5.38) is equivalent to an integral equation of the second kind with a self-adjoint integral operator [5.42] in which η plays the role of the spectral parameter.

We discuss briefly the case $|n| \gg 1$. The operator of the equation is then also boundedly invertible, which follows from the estimates

$$|w(\nu,n)| \leq \frac{|\nu|}{|n|}\, |\psi(\zeta)|, \tag{5.39}$$

where $\psi(\zeta)$ is bounded as $\nu \in [0, |n|/A]$, $|n| \gg 1$, A is a constant, which leads to

$$\lim_{|n|\to\infty} W(x,n) = 0$$

uniformly with respect to $x \in [\varepsilon, 1]$ for any small positive ε.

Remark: The estimate (5.39) is deduced by means of the asymptotics[4]

$$P_{\nu-1/2}^{-|n|}(\cos\theta) = \frac{(\sin\theta)^{|n|}}{\Gamma(|n|+1)}$$

$$\times \left\{ \frac{\left[\left(\zeta\cos\theta + \sqrt{1-\zeta^2\sin^2\theta}\right)/\left(1+\zeta^2\right)\right]^{\zeta}}{\cos\theta + \sqrt{1-\zeta^2\sin^2\theta}} \right\}^{|n|} \left[1 + O_\zeta\left(\frac{1}{|n|}\right)\right],$$

where $|n| \gg 1$, $\zeta = (\nu - 1/2)/n$, ζ outside a vicinity of the point $\pm 1/\sin\theta$.

The series for the spectral function u_ν converges uniformly with respect to the angular variables and vanishes exponentially as $\nu \to i\infty$ because $\cos\pi\nu\, u_\nu|_{x=1/\cos\pi\nu}$ belongs to $L_p(0,1)$. The functions u_ν and $\partial_{N_\omega} u_\nu$ are regular in the strip Π_δ.

[4] V. M. Babich, private communication.

As is obvious from the series representation of u_ν, it is a meromorphic function on the complex plane of the variable ν. The singularities of $R(\nu, n)$ and of u_ν in (5.28) are located in some strip containing the real axis.

5.4 Diffraction coefficient in the oasis *M* for a narrow cone

In numerous applications it is of interest to have an explicit but possibly approximate expression for the diffraction coefficient of the spherical wave from the vertex of the cone. Usually these results are connected with the possibility of the asymptotic solution of the problem at hand.

Our goal in the present section is to determine $u_\nu(\omega, \omega_0)$ for a narrow impedance cone and compute the leading term and the first correction of the scattering diagram (diffraction coefficient) (5.21)

$$D(\omega, \omega_0) = \frac{2}{\mathrm{i}} \int\limits_{\mathrm{i}\mathbb{R}} \nu \sin \pi\nu \, u_\nu(\omega, \omega_0)\, \mathrm{d}\nu$$

in the oasis, $\hat{\theta}'(\omega, \omega_0) > \pi$. Recall that in the oasis there are no other waves in the scattered field except the spherical one propagating from the vertex.

Provided the cone is narrow, the domain Σ is small and its diameter has the order $\beta \sim l_\sigma = \mathrm{mes}(\sigma)$, $\beta \ll 1$, β is a small parameter defined below. Let O belong to the domain Σ. We conduct the tangent plane R^2 to S^2 at the point O. Let x_1, x_2, x_3 be the coordinate system with the origin at O: $(x_1, x_2) \in R^2$, the axis x_3 is directed at the vertex T of the cone, κ_β is the projection of σ onto R^2. We assume that the domain with the boundary κ_β is small, that is, the points $(x_1, x_2) \in \kappa_\beta$ are the images of the points (X_1, X_2) of a fixed curve κ as a result of the corresponding similarity transform,

$$\kappa_\beta = \{(x_1, x_2) : x_i = \beta X_i,\ i = 1, 2;\ (X_1, X_2) \in \kappa\},\ \ 0 < \beta \ll 1,$$

β is the coefficient of the similarity transform.

It is useful to mention that the approach [5.5] for the Neumann condition is based on the asymptotic analysis developed for general elliptic problems in domain with small holes.

We define the so-called external and internal domains: the internal one is the subdomain of $S^2 \setminus \Sigma$, which is projected onto the circle

$$\rho = \sqrt{x_1^2 + x_2^2} < C_2 \beta^{\alpha_2},$$

and the external domain consists of the points of $S^2 \setminus \Sigma$, which is one-to-one projected onto the exterior of the circle

$$\rho = \sqrt{x_1^2 + x_2^2} < C_1 \beta^{\alpha_1},\ \ 0 < \alpha_2 < \alpha_1 < 1,\ \ C_1 > 0,\ C_2 > 0$$

and belongs to the lower hemisphere containing Σ. We also attach the upper hemishpere, which does not contain Σ, to the external domain. The internal and external domains overlap having a common part that is a circular ring.

Recall that the function u_ν satisfies the equation (5.23) and the boundary condition (5.24), u_ν^{i} is known.

In the internal domain, instead of the coordinates (x_1, x_2), we exploit the "extended" coordinates $X_i = x_i/\beta,\ i = 1$. We write the equation and the boundary condition in the coordinates (x_1, x_2) and then in the extended ones (X_1, X_2). We have

$$dl^2|_{S^2} = (dx_1^2 + dx_2^2 + dx_3^2)|_{S^2} = \sum_{i,j=1}^{2} h_{ij} dx_i dx_j ,$$

$$h_{ij} = \frac{x_i x_j}{1 - x_1^2 - x_2^2} + \delta_{ij}$$

and $x_3 = 1 - \sqrt{1 - x_1^2 - x_2^2}$ is the equation of the sphere in the coordinates x_1, x_2, x_3,

$$\Delta_\omega = \frac{1}{\sqrt{h}} \sum_{i,j=1}^{2} \frac{\partial}{\partial x_i} \left(h^{ij} \sqrt{h} \frac{\partial}{\partial x_j} \right), \quad h = \det\{h_{ij}\},$$

(h^{ij}) is the inverse matrix to (h_{ij}). We use the traditional agreement on summation over the repeating indices.

In the extended coordinates X_i we have

$$\left[\frac{1}{\beta^2 \sqrt{H}} \frac{\partial}{\partial X_i} \left(H^{ij} \sqrt{H} \frac{\partial}{\partial X_j} \right) + (\nu^2 - 1/4) \right] u_\nu = 0, \tag{5.40}$$

$$H = \det\{I + O(\beta^2)\}, \quad \sqrt{H} H^{ij} = \delta_{ij} + O(\beta^2)$$

and

$$\left[(H^{ij}/\beta) \cos(n, X_j) \frac{\partial}{\partial X_i} - \sin\zeta A \right] (u_\nu + u_\nu^{\mathrm{i}}) \Big|_\kappa = 0, \tag{5.41}$$

where n is the internal normal to $\kappa \subset R^2$ and I is the unit matrix. Introduce notation $A_\zeta = \sin\zeta A$.

To compute the asymptotics of u_ν in the internal and external domains we introduce the internal

$$u_\nu = \sum_{j=1}^{\infty} V_j(X_1, X_2) \beta^j \tag{5.42}$$

(V_j may depend on $\log\beta$ rationally) and external

$$u_\nu = \beta B_1 g_\nu^0 + \beta^2 \left(B_2 g_\nu^0 + B_{2j} \frac{\partial g_\nu^0}{\partial x_j^0} \right) + \cdots \tag{5.43}$$

expansions with

$$g_\nu^0(\omega, O) = -\frac{\mathrm{P}_{\nu-1/2}(-\cos\hat{\theta}(\omega, O))}{4\cos\pi\nu} .$$

In (5.42) V_j are the functions to be determined, in the expansion (5.43) $B_1, B_2, \ldots$ are the unknown coefficients (with possible rational dependence on $\log\beta$) which are specified by means of matching of (5.42) and (5.43) in their common domain of applicability.

Remark: The expansion (5.43) is an analog of the multipole expansion with the function $g^0_\nu(\omega, O)$ being a fundamental solution of the Laplace–Beltrami operator on the unit sphere.

5.4.1 Problems for the leading terms and for the first corrections

It is easily verified that V_1 and V_2 fulfill the Laplace equation

$$\Delta V_1 = 0, \tag{5.44}$$

$$\Delta V_2 = 0, \quad \Delta = \partial^2_{X_1} + \partial^2_{X_2}\,. \tag{5.45}$$

We deduce the boundary conditions and conditions at infinity as $R = \sqrt{X_1^2 + X_2^2} \to \infty$. Exploit the expansion

$$u^{\rm i}_\nu = (u^{\rm i}_\nu)^0 + \beta D^{\rm inc}_j X_j + \frac{1}{2}\beta^2 D_{ij} X_i X_j \ldots$$

with

$$(u^{\rm i}_\nu)^0 = -\frac{\mathrm{P}_{\nu-1/2}(-\cos\hat{\theta}(\omega_0,\omega))}{4\cos\pi\nu}\bigg|_{\omega=O},$$

$$D^{\rm inc}_j = \frac{\partial}{\partial x_j}\left[-\frac{\mathrm{P}_{\nu-1/2}(-\cos\hat{\theta}(\omega_0,\omega))}{4\cos\pi\nu}\right]\bigg|_{\omega=O},$$

$$D_{ij} = \frac{\partial^2}{\partial x_j \partial x_i}\left[-\frac{\mathrm{P}_{\nu-1/2}(-\cos\hat{\theta}(\omega_0,\omega))}{4\cos\pi\nu}\right]\bigg|_{\omega=O}$$

and from the boundary condition (5.24) thus obtain

$$\cos(n, x_i)\left[\delta_{ij} + \mathrm{O}(\beta^2)\right]\frac{\partial}{\partial X_j}\left[\beta(V_1 + D^{\rm inc}_l X_l) + \beta^2(V_2 + \frac{1}{2}D_{ij}X_iX_j) + \cdots\right]$$

$$= \beta A_\varsigma (u^{\rm i}_\nu)^0 + \beta^2 A_\varsigma (D^{\rm inc}_l X_l + V_1) + \cdots.$$

on κ. Equating the terms of $\mathrm{O}(\beta)$ and of $\mathrm{O}(\beta^2)$ correspondingly, we arrive at the boundary conditions for V_1 and V_2

$$\frac{\partial V_1}{\partial n}\bigg|_\kappa = -D^{\rm inc}_j \frac{\partial X_j}{\partial n}\bigg|_\kappa + A_\varsigma (u^{\rm i}_\nu)^0, \tag{5.46}$$

$$\frac{\partial V_2}{\partial n}\bigg|_\kappa = -D_{ij}\, X_i \frac{\partial X_j}{\partial n}\bigg|_\kappa + A_\varsigma (D^{\rm inc}_j X_j + V_1)|_\kappa, \tag{5.47}$$

where n is the internal normal to κ. The problems (5.44), (5.46) and (5.45), (5.47) should be supplemented by the corresponding conditions at infinity which are derived by means of matching of the internal and external expansions in the intermediate zone. We use the formulas

$$\cos\hat{\theta}(\omega, O) = \sqrt{1-\rho^2} = 1 - \rho^2/2 + O(\rho^4), \quad \rho = \sqrt{x_1^2 + x_2^2},$$

$$g_\nu^0(-\cos\hat{\theta}(\omega, O)) = \frac{1}{2\pi}\log\rho + \frac{G(\nu)}{2\pi} + O(\rho^2\log\rho),$$

$$G(\nu) = -\log 2 + C + \psi(\nu + 1/2) + (\pi/2)\cot\pi(\nu - 1/2),$$

where $C = \log\gamma$ is the Euler constant, $\psi(t)$ is the logarithmic derivative of the gamma function.

By (5.43) we find

$$u_\nu(\omega, \omega_0) = \beta B_1\left[\frac{1}{2\pi}\log(\beta R) + \frac{1}{2\pi}G(\nu) + O(\beta^2 R^2\log(\beta R))\right]$$

$$+\beta^2 B_2\left[\frac{1}{2\pi}\log(\beta R) + \frac{1}{2\pi}G(\nu) + O(\beta^2 R^2\log(\beta R))\right]$$

$$+\beta^2 B_{2j}\frac{\partial}{\partial x_j^0}\left[\frac{1}{2\pi}\log(\beta R) + \frac{1}{2\pi}G(\nu) + O(\beta^2 R^2\log(\beta R))\right] + \cdots,$$

where $R = \rho/\beta \to \infty$, $\rho \to 0$. Now the conditions for V_1 and V_2 at infinity, which ensure matching of the expansions (5.42) and (5.43), read

$$V_1(X_1, X_2) = \frac{B_1}{2\pi}\left[\log(\beta R) + G(\nu)\right] - \frac{\beta B_{2j}}{2\pi}\frac{\partial}{\partial x_j}\log(R) + O(R^{-2}), \tag{5.48}$$

$$V_2(X_1, X_2) = \frac{B_2}{2\pi}\left[\log(\beta R) + G(\nu)\right] + O(R^{-1}) \tag{5.49}$$

as $R \to \infty$, $\partial/\partial x_j = -\partial/\partial x_j^0$.

One can write the solvability conditions of the problem (5.44), (5.46), and (5.48)[5]

$$\int\limits_{X_1^2+X_2^2=r_1^2}\frac{\partial V_1}{\partial R}dl = \int\limits_{K}\frac{\partial V_1}{\partial n}dl$$

and thus obtain ($r_1 \to \infty$)

$$B_1 = -A_\zeta(u_\nu^{\mathrm{i}})^0 l_K, \quad l_K = \int\limits_K \mathrm{d}l. \tag{5.50}$$

[5]Integral of the normal derivative of a harmonic function does not depend on the closed integration contour.

It is obvious from (5.47) that to construct V_2 the expression for $V_1|_\kappa$ should be determined. The unknown V_1 is a solution of the problem (5.44), (5.46), and (5.48) which is uniquely solvable. Provided $V_1|_\kappa$ is known, the coefficient B_2 is determined from the solvability condition

$$\int\limits_{X_1^2+X_2^2=r_1^2} \frac{\partial V_2}{\partial n} \mathrm{d}l = \int\limits_\kappa D_{ij} X_i \frac{\partial X_j}{\partial n} \mathrm{d}l - \int\limits_\kappa A_\zeta (D_j^{\mathrm{inc}} X_j + V_1|_\kappa) \mathrm{d}l.$$

Let $r_1 \to \infty$, taking into account (5.49), we find

$$B_2 = -(D_{11} + D_{22}) S_\kappa - A_\zeta \left(D_j^{\mathrm{inc}} \int\limits_\kappa X_j \mathrm{d}l + \int\limits_\kappa V_1|_\kappa \mathrm{d}l \right). \tag{5.51}$$

As in the paper [5.5], to compute the leading term on the right-hand side of (5.51) we used the Green theorem, S_κ is the square of the domain bounded by the curve κ.

We account for the choice of the origin O for the coordinates (X_1, X_2) to ensure

$$D_j^{\mathrm{inc}} \int\limits_\kappa X_j \mathrm{d}l = 0. \tag{5.52}$$

Indeed, the latter integral can be written as

$$\int\limits_\kappa \boldsymbol{d} \cdot \boldsymbol{R} \, \mathrm{d}l = 0, \quad \boldsymbol{d} = (D_1^{\mathrm{inc}}, D_2^{\mathrm{inc}}).$$

Exploit the following simple observation: for any polygon with the vertices $A_1 \ldots A_n A_1$ there exists a point P such that the vectorial equality holds

$$\boldsymbol{PA_1} + \cdots \boldsymbol{PA_n} = \boldsymbol{0}.$$

We can consider a sequence of polygons $A_1 \ldots A_k A_1$, which in the limit $k \to \infty$ tends to the curve κ, so that the limiting point P_* of the sequence $\{P_k\}$ exists. Then we take $O = P_*$ and show that

$$\int\limits_\kappa \boldsymbol{R} \, \mathrm{d}l = \boldsymbol{0},$$

which leads to the desired equality (5.52).

As a result, the expression for B_2 takes the form

$$B_2 = -(D_{11} + D_{22}) S_\kappa - A_\zeta \int\limits_\kappa V_1|_\kappa \mathrm{d}l. \tag{5.53}$$

5.4.2 Calculation of V_1 and B_{2j}

Introduce auxiliary functions v_j, $j = 1, 2$ as solutions of the problems

$$\Delta v_j = 0, \quad \partial v_j / \partial n|_\kappa = -\partial X_j / \partial n|_\kappa, \quad v_j = \mathrm{O}(1/R), \quad R \to \infty.$$

Then the asymptotics of v_j has the form

$$v_j = d_{jk}\frac{\partial \log R}{\partial X_k} + \mathrm{O}\left(\frac{1}{R^2}\right), \quad R \to \infty,$$

where by definition d_{jk} is the polarization tensor, which is an integral characteristic of a domain with the boundary κ. Comparing the asymptotics (5.48) with those for v_j as $R \to \infty$, one has

$$-\frac{B_{2j}}{2\pi}\frac{\partial \log R}{\partial X_j} = D_i^{\mathrm{inc}} d_{ik}\frac{\partial \log R}{\partial X_k} = \mathrm{O}\left(\frac{1}{R}\right)$$

and

$$B_{2j} = -2\pi\, D_i^{\mathrm{inc}} d_{ij}. \tag{5.54}$$

The unknown constants in the asymptotics (5.48) are completely derived from (5.54) and (5.50).

Let κ be the circumference of a circle of unit radius. Then solution of the problem for V_1 can be written as (see, e.g., [5.48])

$$V_1(z) = -\frac{1}{\pi}\int\limits_{|\xi|=1}\left[u_1^+(\xi) - \frac{1}{4\pi}\int\limits_{|t|=1} u_1^+(t)\,\mathrm{d}l_t\right]\log|z-\xi|\,\mathrm{d}l_\xi + C_1(\nu), \tag{5.55}$$

$z = X_1 + \mathrm{i}X_2$, where $C_1(\nu)$ is found from the conditions (5.48) at infinity

$$C_1(\nu) = -\frac{l_\kappa A_\zeta (u_\nu^{\mathrm{i}})^0}{2\pi}\left[\log\beta + G(\nu)\right],$$

$$u_1^+ := \left.\frac{\partial V_1}{\partial n}\right|_\kappa = -D_j^{\mathrm{inc}}\left.\frac{\partial X_j}{\partial n}\right|_\kappa + A_\zeta (u_\nu^{\mathrm{i}})^0.$$

It is worth noting that the solution of (5.55) is unique, provided the asymptotics (5.48) is fixed. Solution $V_1(z)$ of the problem (5.44), (5.46), and (5.48) for an arbitrary boundary κ can be derived in terms of the conformal mapping $\xi = h(z)$ of the exterior domain to κ onto the exterior of the unit disk. If $V_1(z)$ ($z = X_1 + \mathrm{i}X_2$) is a harmonic function in the exterior of κ, then $\tilde{V}_1(\xi) = V_1(g(\xi))$ is a harmonic function outside the unit disk $|\xi| = 1$ and $g = g(\xi)$ is the inverse function to $h = h(z)$. Note that, provided $\partial V_1/\partial n|_\kappa$ is given, then

$$\left.\frac{\partial \tilde{V}_1}{\partial \tilde{n}}\right|_{|\xi|=1} = \left.\frac{\partial V_1}{\partial n}(g(\xi))|g'(\xi)|\right|_{|\xi|=1},$$

where $\tilde{n}$ is the normal to the circumference $|\xi| = 1$. We exploit formula (5.55) and obtain

$$V_1(z) = \frac{D_j^{\mathrm{inc}}}{\pi}\int\limits_\kappa \frac{\partial X_j}{\partial n}(\tau)\log|h(z)-h(\tau)|\,\mathrm{d}l_\tau \tag{5.56}$$

$$-\frac{A_\zeta(u_\nu^{\mathrm{i}})^0}{\pi}\int\limits_\kappa \log|h(z)-h(\tau)|\,\mathrm{d}l_\tau + \frac{l_\kappa A_\zeta(u_\nu^{\mathrm{i}})^0}{2\pi}\log|h(z)| + \frac{l_\kappa A_\zeta(u_\nu^{\mathrm{i}})^0}{2\pi}W + C_1(\nu),$$

where we took into account that $dl_\tau = |g'(\xi)| dl_\xi$ and W is the Wiener measure, $W = -\log r_0$, r_0 is the conformal radius of the exterior domain with respect to κ, $r_0 = \lim_{z\to\infty} z/h(z)$,

$$\int_{|z|=1} \log|z-\xi|\, d\, l_z = 2\pi \log|\xi|, \quad |\xi| \ge 1.$$

The constant C_1 is specified from (5.48).

In accordance with (5.53) one should find $\int_\kappa V_1|_\kappa(\tau)\, d\, l_\tau$. Then from (5.56) one has ($|h(t)| = 1,\ t \in \kappa$)

$$\int_\kappa V_1|_\kappa(\tau)\, dl_\tau = \frac{D_j^{\text{inc}}}{\pi} \int_\kappa \left.\frac{\partial X_j}{\partial n}\right|_\kappa (\tau) \left[\int_\kappa \log|h(\tau) - h(t)| dl_t\right] dl_\tau \tag{5.57}$$
$$- \frac{Z_\kappa}{\pi} A_\zeta (u_\nu^{\text{i}})^0 + \frac{l_\kappa^2 A_\zeta (u_\nu^{\text{i}})^0}{2\pi} W + C_1(\nu) l_\kappa,$$

where

$$Z_\kappa = \int_\kappa \int_\kappa \log|h(\tau) - h(t)| dl_t dl_\tau.$$

After some reduction (Section 5.11.1) one can show that the first summand on the right-hand side of (5.57) is equal to zero. Exploiting (5.57) then from (5.53), we obtain

$$B_2 = -(D_{11} + D_{22}) S_\kappa + \frac{Z_\kappa}{\pi} A_\zeta \left(A_\zeta (u_\nu^{\text{i}})^0\right)$$
$$+ \frac{l_\kappa^2}{2\pi} A_\zeta \left([\log\beta - W + G(\nu)] A_\zeta (u_t^{\text{i}})^0\right). \tag{5.58}$$

5.4.3 Basic formula for the diffraction coefficient of the spherical wave from the vertex of a narrow cone

We collect expressions (5.50), (5.54), and (5.58) for B_1, B_{2j}, and B_2 correspondingly and substitute (5.43) into the expression for the scattering diagram. Compute the integrals (see Section 5.11.2 and [5.28]), in particular, for the leading term

$$\frac{\pi}{4} \int_{i\mathbb{R}} \int_{i\mathbb{R}} d\nu dt \frac{\nu \tan(\pi\nu) t \tan(\pi t)\, \mathrm{P}_{\nu-1/2}(\cos\theta)\, \mathrm{P}_{t-1/2}(\cos\theta_0)}{\cos(\pi t) + \cos(\pi\nu)}$$
$$= -\frac{1}{(\cos\theta + \cos\theta_0)^2},$$

then arrive at the desired expression for the scattering diagram [5.31]

$$D(\omega, \omega_0) = -\frac{\beta l_\kappa}{4\pi} \frac{\sin(\zeta)}{(\cos\theta + \cos\theta_0)^2} - \frac{\beta^2 S_\kappa}{2\pi} \frac{1 + \cos\theta\cos\theta_0}{(\cos\theta + \cos\theta_0)^3}$$
$$-\beta^2 d_{ij} e_i e_j^0 \frac{\sin\theta\sin\theta_0}{(\cos\theta + \cos\theta_0)^3} - \frac{\beta^2 Z_\kappa \sin^2(\zeta)}{2\pi^2 (\cos\theta + \cos\theta_0)^3} \tag{5.59}$$

$$-\frac{\beta^2 l_\kappa^2 \sin^2(\zeta)[\log(\beta/2) - W + C]}{4\pi^2(\cos\theta + \cos\theta_0)^3} + \frac{\beta^2 l_\kappa^2 \sin^2(\zeta)}{2\pi} P_\psi(\theta, \theta_0) + \mathrm{O}(\beta^3 \log^2 \beta),$$

where

$$\theta + \theta_0 < \pi,$$

with

$$\theta_0 = \pi - \hat{\theta}(\omega_0, O), \quad \theta = \pi - \hat{\theta}(\omega, O),$$

which is equivalent to

$$\hat{\theta}(\omega_0, O) + \hat{\theta}(\omega, O) > \pi.$$

In formula (5.59) $e_j,\ j = 1, 2$ (correspondingly $e_j^0,\ j = 1, 2$) are the components of the unit vector $\boldsymbol{e}$ ($\boldsymbol{e}^0$) in the coordinates (x_1, x_2), which is tangent to the arc $\omega O (\omega_0 O)$ of the big circle connecting $\omega(\omega_0)$ and O on the unit sphere. They direct from $\omega(\omega_0)$ to O.

We have also used the notation

$$P_\psi(\theta, \theta_0) = \frac{\mathrm{i}}{32} \int\limits_{\mathrm{i}\mathbb{R}} \int\limits_{\mathrm{i}\mathbb{R}} \int\limits_{\mathrm{i}\mathbb{R}} \mathrm{d}\nu\, \mathrm{d}\tau\, \mathrm{d}t$$

$$\times \frac{\nu \tan(\pi\nu) t \sin(\pi t) \tau \tan(\pi\tau) \psi(t + 1/2)\, \mathrm{P}_{\nu-1/2}(\cos\theta)\, \mathrm{P}_{\tau-1/2}(\cos\theta_0)}{[\cos(\pi t) + \cos(\pi\nu)][\cos(\pi t) + \cos(\pi\tau)]}.$$

Remark: In practice it is also desirable to have the diffraction coefficient $F(\theta, \theta_0, r_0)$ for the illumination by a point source located at $\boldsymbol{r}_0$. Provided the source is not close to the vertex of the cone ($|kr_0| \gg 1$) one has

$$F(\theta, \theta_0, r_0) = \frac{\mathrm{e}^{\mathrm{i}kr_0}}{4\pi r_0} D(\theta, \theta_0) \left[1 + \mathrm{O}\left(\frac{1}{kr_0}\right)\right].$$

For the function $P_\psi(\theta, \theta_0)$ one can deduce a simplified expression; however, the multiple integral $P_\psi(\theta, \theta_0)$ was not computed. In the expression for the diagram the impedance $\eta := \sin\zeta$ is assumed to be bounded; however, it is easily verified that the expression (5.59) is reduced to the known result [5.5] as $\sin\zeta \to 0$, that is, for the Neumann boundary condition (under this condition one speaks of an acoustically hard cone).

Another important formal simplification is for the right-circular impedance cone. In this case, one has $l_\kappa = 2\pi$, $S_\kappa = \pi$, $Z_\kappa = 0$, $d_{ij} = \delta_{ij}$, where δ_{ij} is the Kronecker symbol.

For the so-called pseudoelliptic cone we have that the curve κ is an ellipse with the semi-axes a and b. The corresponding conformal mapping is

$$h(z) = \frac{z + \sqrt{z^2 + b^2 - a^2}}{a + b}, \quad h(\infty) = \infty.$$

The formulas for the integral characteristics l_κ, S_κ, W, and so forth are then written explicitly; for example, for the Wiener measure one has $W = -\log[(a + b)/2]$.

In the leading approximation (with respect to β) the diagram is given by a simple formula

$$D(\omega, \omega_0) \sim -\frac{l_\sigma}{4\pi}\frac{\sin\zeta}{(\cos\theta + \cos\theta_0)^2}, \quad \omega \in M.$$

It depends on the simplest integral characteristic of the scatterer, on the length $l_\sigma = \int_\sigma dl = \beta l_\kappa + O(\beta^2)$ of the contour σ.

5.5 Numerical calculation of the diffraction coefficient in the oasis M

Recall that the diffraction coefficient $D(\omega, \omega_0)$ in the oasis M is given by (5.21), whereas the spectral function $u_\nu(\omega, \omega_0)$ is expanded in the Fourier series (5.28) whose unknown coefficients $R(\nu, n)$, $n = -\infty, \ldots, +\infty$ solve the respective integral equations (5.30). Hence, the very first step in calculating the diffraction coefficient (aka scattering diagram) $D(\omega, \omega_0)$ lies in solving the integral equations (5.30), in the second step the expression (5.21) has to be evaluated; that is, an integration along the imaginary axis in the complex ν-plane needs to be carried out.

5.5.1 Numerical aspects

Owing to the evenness of the spectral functions $u_\nu(\omega, \omega_0)$ and $u_\nu^{\mathrm{i}}(\omega, \omega_0)$ in ν, and hence also of their Fourier coefficients $R(\nu, n)$ and $R_{\mathrm{i}}(\nu, n)$, the integration in the integral equations for $R_u(\nu, n)$ needs to be extended over only the upper half of the imaginary axis of the complex ν-plane. This allows us to confine our attention to (5.33).

To facilitate the numerical solution of (5.33), we follow Section 2.7 and take into account the asymptotic behavior of $R(\nu, n)$ from scratch. To this end we introduce new unknowns $\widetilde{R}(\nu, |n|)$ via

$$R(\nu, n) = \mathrm{i}^n \exp\left[\mathrm{i}\nu(\theta_1 - \theta_0)\right] \widetilde{R}(\nu, |n|).$$

Furthermore, the semi-infinite interval of (5.33) is converted into a finite one by means of

$$\nu = \frac{\mathrm{i}p}{\theta_1 - \theta_0} \ln\frac{1+\xi}{1-\xi}, \quad \xi \in [0, 1], \quad p \gg 1,$$

where the multiplication factor p offers an additional degree of freedom in mapping the finite domain $\xi \in [0, 1]$ onto the positive imaginary axis of the ν-plane.

In this way, the integral equations (5.33) are reduced to

$$\widetilde{R}(\nu, |n|) - \mathrm{e}^{-\mathrm{i}\nu(\theta_1-\theta_0)} \int_0^1 \frac{K(\xi')}{\cos(\pi t) + \cos(\pi\nu)} \widetilde{R}(t, |n|)\mathrm{d}\xi' = \widetilde{S}_{\mathrm{i}}(\nu, |n|), \tag{5.60}$$

with

$$K(\xi') = \frac{i2\eta p}{\theta_1 - \theta_0} e^{it(\theta_1 - \theta_0)} w(t, n) \sin(\pi t) \frac{1}{1 - (\xi')^2},$$

$$S_i(\nu, n) = i^n \exp[i\nu(\theta_1 - \theta_0)] \widetilde{S}_i(\nu, |n|),$$

$$t = \frac{ip}{\theta_1 - \theta_0} \ln \frac{1 + \xi'}{1 - \xi'}.$$

We then apply a quadrature method to the integral equations for $\widetilde{R}(\nu, |n|)$ (5.60), that is, evaluating the integrals contained in (5.60) on use of a Gauss-Legendre scheme of N abscissae ξ_j and weights w_j, $j = 1, \ldots, N$, enforcing the fulfillment of the approximations to (5.60) precisely at these N abscissae and lastly resolving the resultant matrix equation (see [5.49] for details).

After having solved the integral equations (5.60), $\widetilde{R}(\nu, |n|)$, and hence, the Fourier coefficients of the spectrum $u_\nu(\omega, \omega_0)$, are obtained with the aid of the integral extrapolations that turn out from (5.60)

$$\widetilde{R}(\nu, |n|) = e^{-i\nu(\theta_1 - \theta_0)} \sum_{j=1}^{N} \frac{w_j K(\xi_j) \widetilde{R}(t(\xi_j), |n|)}{\cos(\pi t(\xi_j)) + \cos(\pi \nu)} + \widetilde{S}_i(\nu, |n|).$$

The diffraction coefficient $D(\omega, \omega_0)$ (5.21) now reads

$$D(\omega, \omega_0) = -\frac{2}{\theta_s(\varphi_0) - \theta} \int_0^\infty \tau \left(1 - e^{-2\pi\tau}\right) \widetilde{u}_\nu(\omega, \omega_0) e^{-T} dT, \tag{5.61}$$

with $\nu = i\tau$, $T = [\theta_s(\varphi_0) - \theta]\,\tau$, and the smallest singular angle $\theta_s(\varphi_0) = 2\theta_1 - \theta_0 - \pi$. $\widetilde{u}_\nu(\omega, \omega_0)$ is related to the spectrum $u_\nu(\omega, \omega_0)$ according to

$$u_\nu(\omega, \omega_0) = \exp[i\nu(2\theta_1 - \theta_0 - \theta)]\, \widetilde{u}_\nu(\omega, \omega_0)$$

and hence, grows at most algebraically as $\operatorname{Im} \nu \to +\infty$. The property of (5.61) renders a quadrature scheme of the Gauss-Laguerre type (see [5.1]) the natural choice for executing the integration contained in (5.61).

The accuracy of the previously described numerical procedure depends upon four parameters at our disposal: the number of abscissae of the Gauss-Legendre scheme N; the number of abscissae of the Gauss-Laguerre scheme M; the number of terms in the Fourier series $2N_F + 1$; and the multiplication factor p. As a result of a detailed internal convergence study, it is found that the following set of parameters

$$N = 100, \quad M = 20, \quad N_F = 50, \quad p = 20$$

suffices for engineering purposes. Interested readers are referred to [5.32,5.43] for numerical details, including internal convergence studies, in similar problems.

5.5.2 A perturbation series for $|\eta| \gg 1$

To verify the aforementioned procedure, especially in the case of nonaxial incidence, let us derive for large amplitude of the normalized impedance a perturbation series solution to the functional difference equation (5.32) [5.50]. Obviously, a perturbation series can be constructed for $|\eta| \ll 1$ in a similar fashion.

Under this circumstance, the following series for $R(\nu, n)$ seems feasible

$$R(\nu, n) = R^{(0)}(\nu, n) + \eta^{-1} R^{(1)}(\nu, n) + \eta^{-2} R^{(2)}(\nu, n) + \cdots . \tag{5.62}$$

Inserting the previous Ansatz into the second-order functional difference equation for $R(\nu, n)$ (5.32) and equating the coefficients of like powers of η we get

$$\begin{aligned} R^{(0)}(\nu, n) &= -\frac{w_{\mathrm{i}}(\nu, n)}{w(\nu, n)} R_{\mathrm{i}}(\nu, n), \\ R^{(1)}(\nu, n) &= \frac{\mathrm{i}}{2w(\nu, n)} \Big[R^{(0)}(\nu + 1, n) - R^{(0)}(\nu - 1, n) \\ &\quad + R_{\mathrm{i}}(\nu + 1, n) - R_{\mathrm{i}}(\nu - 1, n) \Big], \\ R^{(2)}(\nu, n) &= \frac{\mathrm{i}}{2w(\nu, n)} \left[R^{(1)}(\nu + 1, n) - R^{(1)}(\nu - 1, n) \right]. \end{aligned}$$

As indicated by the boundary condition (5.3), the zeroth-order term $R^{(0)}(\nu, n)$ represents the exact solution of (5.32) for an acoustically soft cone, whereas the first-order and higher-order terms refine the approximate solution to the second-order functional difference equation (5.32) for large but finite $|\eta|$.

5.5.3 Examples

The first example in this subsection concerns the axisymmetric case, that is, the plane wave is incident along the axis of the cone at $\theta_0 = 0$. Figure 5.2 displays the amplitude of the diffraction coefficient $|D(\omega, \omega_0)|$ as a function of both the normalized surface impedance η and the co-latitude θ. The diffracted field, which is, in line with (5.20), proportional to $D(\omega, \omega_0)$, increases monotonically with θ for $\theta < \theta_s$. The angle $\theta_s = 2\theta_1 - \pi$ stands for the singular direction at

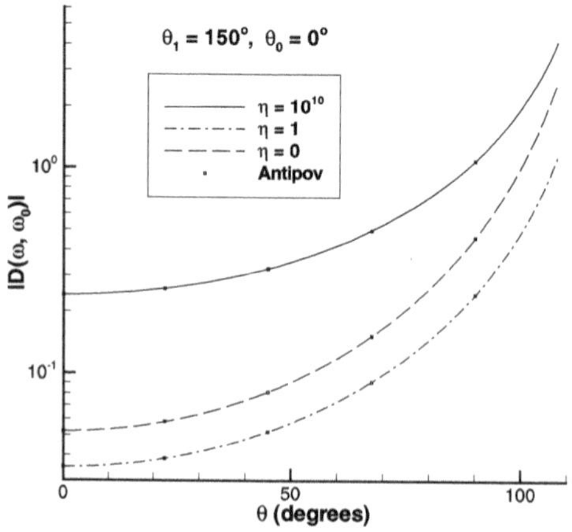

Figure 5.2 Diffraction by a right-circular impedance cone at axial incidence. Comparison with the results of Antipov [5.3].

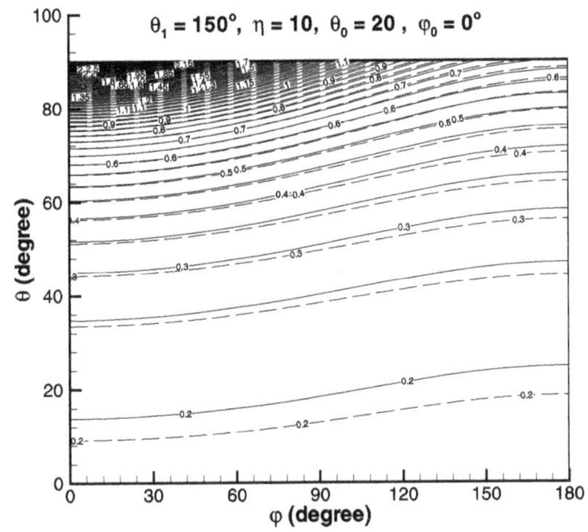

Figure 5.3 Contours of the amplitude of the diffraction coefficient for a right-circular impedance cone at nonaxial incidence (solid line: integral equations (5.30); broken line: the first two terms of the perturbation series (5.62)).

which the scattered field grows without limit, and hence (5.21) is no longer applicable. As to be expounded later in Section 5.10, in a neighborhood of the singular direction, the far field is described by the parabolic cylinder-function. Shown in Fig. 5.2 are also the data obtained by Antipov in [5.3], which corroborate our results.

The increase of the amplitude of the diffraction coefficient toward the singular direction, as made conspicuous by Fig. 5.2, hints at the physical character of the diffracted waves. The field at large distances changes its nature when approaching the region where nonspherical wave ingredients exist, and the "sources" of the diffracted waves in the oasis can be regarded as distributed on the shadow boundaries of the reflected waves that define the singular directions. These sources, as implied by Meixner's condition (5.4), are not located at the cone's tip, even if (5.21) applies in the oasis. This seeming contradiction can be resolved by noting that the diffracted waves are generated by the sources via the so-called transverse diffusion of the energy along the spherical wavefronts from the shadow boundaries into the oasis [5.51]. Thus, the "shining tip" is, in the words of Sommerfeld, merely "an optical delusion" [5.2].

At nonaxial incidence, the diffraction coefficient $D(\omega, \omega_0)$ depends in addition upon the azimuthal angle φ (Fig. 5.3). This holds also true for the singular direction $\theta_s(\varphi)$ with its smallest value occuring at $\varphi = \varphi_0$ (see Secton 5.5.1), and its maximum value appearing at $\varphi = \varphi_0 \pm \pi$: $\theta_s(\varphi_0 \pm \pi) = 2\theta_1 + \theta_0 - \pi$. This fact explains the tilted contours shown in Fig. 5.3, which also contains, for comparison purposes, data obtained on use of the first two terms of the perturbation series for $|\eta| \gg 1$ (5.62).

5.6 Sommerfeld–Malyuzhinets transform and analytic continuation

In the previous chapters we demonstrated that the Sommerfeld integral is a very convenient and natural tool to compute the far-field asymptotics. In the following sections we can efficiently

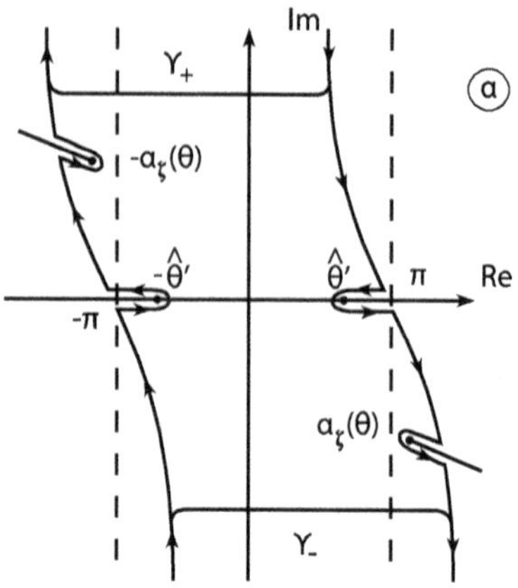

Figure 5.4 Deformation of the Sommerfeld contour

exploit it for the problem at hand, more exactly, for the derivation of the far field. To study the wave field scattered by a conical surface at infinity we use the Sommerfeld representation of the Macdonald function

$$\mathrm{K}_\nu(-ikr) = \frac{1}{4\sin\pi\nu}\int_\gamma \mathrm{e}^{-ikr\cos\alpha}\,\frac{\mathrm{e}^{\mathrm{i}\nu\alpha}-\mathrm{e}^{-\mathrm{i}\nu\alpha}}{2}\,\mathrm{d}\alpha, \tag{5.63}$$

where $\gamma = \gamma_+ \cup \gamma_-$ is the standard double-loop Sommerfeld contour (see Chapter 1 and Fig. 5.4). Let $\delta_0 > 0$ be small, implying that we deformed γ in such a way that it belongs to the strip $|\mathrm{Re}\,\alpha| < |\arg(-ik)| + \pi/2 + \delta_0$, substitute the representation (5.63) into (5.14) and change the order of integration. Thus, we have

$$U(r,\omega,\omega_0) = \frac{1}{\sqrt{-ikr}}\frac{1}{2\mathrm{i}\pi}\int_\gamma \mathrm{e}^{-ikr\cos\alpha}\,\Phi(\alpha,\omega,\omega_0)\,\mathrm{d}\alpha, \tag{5.64}$$

then integrating by parts

$$U(r,\omega,\omega_0) = \frac{\sqrt{-ikr}}{2\mathrm{i}\pi}\int_\gamma \mathrm{e}^{-ikr\cos\alpha}\,\sin\alpha\,\widetilde{\Phi}(\alpha,\omega,\omega_0)\,\mathrm{d}\alpha, \tag{5.65}$$

whereas

$$\Phi(\alpha,\omega,\omega_0) = \frac{\partial}{\partial\alpha}\widetilde{\Phi}(\alpha,\omega,\omega_0),$$

$$\Phi(\alpha,\omega,\omega_0) = \sqrt{2\pi}\int_{\mathrm{i}\mathbb{R}} \nu\, u_\nu(\omega,\omega_0)(\mathrm{e}^{\mathrm{i}\nu\alpha}-\mathrm{e}^{-\mathrm{i}\nu\alpha})/2\,\mathrm{d}\nu, \tag{5.66}$$

$$\widetilde{\Phi}(\alpha,\omega,\omega_0) = \sqrt{2\pi}\int_{\mathrm{i}\mathbb{R}} u_\nu(\omega,\omega_0)(\mathrm{e}^{\mathrm{i}\nu\alpha}+\mathrm{e}^{-\mathrm{i}\nu\alpha})/(2\mathrm{i})\,\mathrm{d}\nu \tag{5.67}$$

are the functions that we call the Sommerfeld transformants. $\Phi(\alpha, \omega, \omega_0)$ and $\widetilde{\Phi}(\alpha, \omega, \omega_0)$ are correspondingly odd and even with respect to α. Recall that $u_\nu(\omega, \omega_0)$ is a solution of the boundary-value problem for the Laplace–Beltrami operator on the unit sphere with the hole Σ.

Next we study the analytic properties of the Sommerfeld transformants $\widetilde{\Phi}(\alpha, \omega, \omega_0)$ and $\Phi(\alpha, \omega, \omega_0)$. In (5.64) and (5.65) the wavenumber can be taken purely real and positive, which is most important for application. It is remarkable that the representations of the wave field (5.64) and (5.65) are well adapted to the calculations of the far-field asymptotics and converge in weaker restrictions with respect to the wavenumber k in comparison with those for the Kontorovich–Lebedev integral representation (5.14).

5.6.1 Analytic properties of $\widetilde{\Phi}(\alpha, \omega, \omega_0)$ and $\Phi(\alpha, \omega, \omega_0)$

It is obvious that $\widetilde{\Phi}(\alpha, \omega, \omega_0)$ and $u_\nu(\omega, \omega_0)$ are connected by the Fourier transform, which means that the analytic properties of the first function on α are specified by the asymptotic behavior of the second one as $\nu \to i\infty$. The same is valid for $\Phi(\alpha, \omega, \omega_0)$. The following properties are verified in a standard manner.

The functions $\widetilde{\Phi}(\alpha, \omega, \omega_0)$ and $\Phi(\alpha, \omega, \omega_0)$ admit the estimates on α

$$
\begin{aligned}
|\Phi(\alpha, \omega, \omega_0)| &\leq q \exp(-a|\mathrm{Im}\,\alpha|), \quad a > 1, \\
|\widetilde{\Phi}(\alpha, \omega, \omega_0)| &\leq q \exp(-a|\mathrm{Im}\,\alpha|), \ |\mathrm{Im}\,\alpha| \to \infty,
\end{aligned}
\tag{5.68}
$$

where a is specified by the width of the regularity strip for $u_\nu(\omega, \omega_0)$ on ν in a vicinity of the imaginary axis of the ν-plane and q denotes a positive constant.

We now turn to the domain of regularity of $\widetilde{\Phi}(\alpha, \omega, \omega_0)$ and $\Phi(\alpha, \omega, \omega_0)$ and to analytic continuation to a broader domain. Consider the strip on the complex α-plane

$$
\Pi'_\omega = \{\alpha : \ |\mathrm{Im}\,\alpha| < \hat{\theta}'(\omega, \omega_0)\}.
$$

Provided the spectral function u_ν admits the estimate (5.17) in oasis M and the estimate (5.18) in its exterior N, exploiting the properties of the Fourier integral, one has that $\widetilde{\Phi}(\alpha, \omega, \omega_0)$ and $\Phi(\alpha, \omega, \omega_0)$ are regular in the strip Π'_ω as $\hat{\theta}' \leq \pi$ and in the strip $\Pi_{\pi+\varepsilon} = \{\alpha : \mathrm{Re}\,\alpha| < \pi + \varepsilon\}$, ($\varepsilon > 0$ small) as $\hat{\theta}' > \pi$, that is, in the strip being wider than π. In the latter case an observation point is located in the oasis.

In fact, the strip of regularity is specified by the location of singularities of $\widetilde{\Phi}(\alpha, \omega, \omega_0)$ and $\Phi(\alpha, \omega, \omega_0)$, which is discussed next. In particular, if an observation point is in the oasis, the strip $\Pi_{\pi+\varepsilon}$ does not contain singularities.

Consider the domain $D_C = \{\alpha \in \mathcal{C} : \ \mathrm{Im}\,\alpha < -C\}$ for some positive C.[6] Denote $D_C^\star$ the domain on the complex α-plane symmetric with respect to the origin. It can be verified that the function $\Phi(\alpha, \omega, \omega_0)$ ($\widetilde{\Phi}(\alpha, \omega, \omega_0)$) admits an analytic continuation as a regular function into the domain $D = \Pi'_\omega \cup D_C \cup D_C^\star$. The details of verification of the latter fact can be found, for example, in [5.43] and in Chapter 6.

As a result, we can assert that all singularities of $\Phi(\alpha, \omega, \omega_0)$ ($\widetilde{\Phi}(\alpha, \omega, \omega_0)$) with respect to α are in the domain $\mathcal{C} \setminus D$. The study of the singularities becomes a crucial task because

---

[6]For this constant the estimate from below could be given.

in the process of the asymptotic evaluation of the Sommerfeld integrals for the solution the Sommerfeld contour is deformed into the steepest-descent paths. In the process of such deformation the singularities of the Sommerfeld transformants can be captured. The corresponding contributions are responsible for the reflected or surface waves, whereas the saddle points $\pm\pi$ give rise to the spherical wave from the vertex of the cone.

5.6.2 Problems for the Sommerfeld transformants

Making use of equation (5.15) and of the Fourier transform, one can show that $\Phi(\alpha, \omega, \omega_0)$ satisfies the equation

$$\left(\Delta_\omega - \partial_\alpha^2 - 1/4\right) \Phi(\alpha, \omega, \omega_0) = 0 \tag{5.69}$$

and analogously

$$\left(\Delta_\omega - \partial_\alpha^2 - 1/4\right) \widetilde{\Phi}(\alpha, \omega, \omega_0) = 0. \tag{5.70}$$

It is instructive to verify the previous equations directly. Indeed, we have

$$\frac{r^2}{2\pi \mathrm{i}} \int\limits_\gamma \frac{\mathrm{e}^{-ikr\cos\alpha}}{\sqrt{-ikr}} \left[\frac{1}{r^2}\Delta_\omega \Phi(\alpha, \omega, \omega_0) - \frac{1}{r^2}\left(\partial_\alpha^2 + \frac{1}{4}\right)\Phi(\alpha, \omega, \omega_0)\right] \mathrm{d}\alpha$$

$$= \frac{r^2}{2\pi \mathrm{i}} \int\limits_\gamma \left[\frac{\mathrm{e}^{-ikr\cos\alpha}}{\sqrt{-ikr}}\frac{1}{r^2}\Delta_\omega \Phi(\alpha, \omega, \omega_0) - \frac{\Phi(\alpha, \omega, \omega_0)}{r^2}\left(\partial_\alpha^2 + \frac{1}{4}\right)\frac{\mathrm{e}^{-ikr\cos\alpha}}{\sqrt{-ikr}}\right] \mathrm{d}\alpha$$

$$= \frac{r^2}{2\pi \mathrm{i}} \int\limits_\gamma \left[\left(\frac{\partial^2}{\partial r^2} + \frac{2}{r}\frac{\partial}{\partial r} + k^2\right)\frac{\mathrm{e}^{-ikr\cos\alpha}}{\sqrt{-ikr}}\Phi(\alpha, \omega, \omega_0)\right.$$

$$\left. + \frac{\Delta_\omega \Phi(\alpha, \omega, \omega_0)}{r^2}\frac{\mathrm{e}^{-ikr\cos\alpha}}{\sqrt{-ikr}}\right] \mathrm{d}\alpha$$

$$= r^2\left(\Delta + k^2\right)\frac{1}{2\pi \mathrm{i}}\int\limits_\gamma \frac{\mathrm{e}^{-ikr\cos\alpha}}{\sqrt{-ikr}}\Phi(\alpha, \omega, \omega_0)\,\mathrm{d}\alpha = r^2\left(\Delta + k^2\right) U(kr, \omega, \omega_0) = 0,$$

where we integrated by parts and exploited the identity

$$\left(\frac{\partial^2}{\partial r^2} + \frac{2}{r}\frac{\partial}{\partial r} + k^2\right)\frac{\mathrm{e}^{-ikr\cos\alpha}}{\sqrt{-ikr}} = -\frac{1}{r^2}\left(\partial_\alpha^2 + \frac{1}{4}\right)\frac{\mathrm{e}^{-ikr\cos\alpha}}{\sqrt{-ikr}}.$$

Remark that the Sommerfeld transformant corresponding to u_ν^{i} can be explicitly computed

$$\widetilde{\Phi}_{\mathrm{i}}(\alpha, \omega, \omega_0) = -2^{-1}\sqrt{\pi}\left[\cos\alpha - \cos\hat{\theta}(\omega, \omega_0)\right]^{-1/2}, \tag{5.71}$$

where the square root is positive as $-\hat{\theta}(\omega, \omega_0) < \alpha < \hat{\theta}(\omega, \omega_0)$ and the branch cuts are conducted from $\pm\hat{\theta}(\omega, \omega_0)$ to $\pm\infty$ correspondingly. This expression can be obtained from the explicit formula for u_ν^{i} by use of the Fourier transform taking into accout the formula 7.216 from [5.24].

Let us turn to the boundary conditions

$$
\begin{aligned}
& r^{-1} \frac{\partial\left(U+U^{\mathrm{i}}\right)}{\partial N_\omega}\bigg|_C \\
&= \frac{1}{\sqrt{-ikr}} \frac{1}{2i\pi} \int_\gamma \mathrm{e}^{-ikr\cos\alpha}(-ik\sin\alpha) \frac{\partial}{\partial N_\omega}\left(\widetilde{\Phi}+\widetilde{\Phi}_{\mathrm{i}}\right)(\alpha,\omega,\omega_0)\bigg|_\sigma \mathrm{d}\alpha \qquad (5.72)\\
&= ik\eta\left(U+U^{\mathrm{i}}\right)\big|_C = \frac{ik\eta}{\sqrt{-ikr}} \frac{1}{2i\pi} \int_\gamma \mathrm{e}^{-ikr\cos\alpha} \frac{\partial}{\partial\alpha}\left(\widetilde{\Phi}+\widetilde{\Phi}_{\mathrm{i}}\right)(\alpha,\omega,\omega_0)\bigg|_\sigma \mathrm{d}\alpha,
\end{aligned}
$$

then, exploiting the Malyuzhinets theorem (see Chapter 1), obtain

$$
\sin\alpha \frac{\partial\left(\widetilde{\Phi}+\widetilde{\Phi}_{\mathrm{i}}\right)}{\partial N_\omega}(\alpha,\omega,\omega_0)\bigg|_\sigma = -\eta \frac{\partial\left(\widetilde{\Phi}+\widetilde{\Phi}_{\mathrm{i}}\right)}{\partial\alpha}(\alpha,\omega,\omega_0)\bigg|_\sigma, \tag{5.73}
$$

which is followed by

$$
\frac{\partial(\Phi+\Phi_{\mathrm{i}})}{\partial N_\omega}(\alpha,\omega,\omega_0)\bigg|_\sigma = -\eta\frac{\partial}{\partial\alpha}\frac{(\Phi+\Phi_{\mathrm{i}})(\alpha,\omega,\omega_0)}{\sin\alpha}\bigg|_\sigma, \tag{5.74}
$$

where

$$
\Phi_{\mathrm{i}}(\alpha,\omega,\omega_0) = \frac{\partial}{\partial\alpha}\widetilde{\Phi}_{\mathrm{i}}(\alpha,\omega,\omega_0).
$$

5.6.3 The singularity corresponding to the wave reflected from the conical surface

It is worth commenting on the solutions $\Phi(\alpha,\omega,\omega_0)$ and $\widetilde{\Phi}(\alpha,\omega,\omega_0)$ from the point of view of the ray method for real α. Equations (5.69) and (5.70) are of hyperbolic type for real α. The real singularities of solutions propagate along the rays on the unit sphere, which follows from the analysis below. We assume that the local behavior of the Sommerfeld transformant $\widetilde{\Phi}$ near the singularities $\alpha = \pm\hat{\theta}'(\omega,\omega_0)$ is as follows:[7]

$$
\begin{aligned}
\widetilde{\Phi}(\alpha,\omega,\omega_0) &= \frac{A^0(\omega,\omega_0)}{\sqrt{\cos\alpha-\cos\hat{\theta}'(\omega,\omega_0)}} \\
&+ A^1(\omega,\omega_0)\sqrt{\cos\alpha-\cos\hat{\theta}'(\omega,\omega_0)} + \cdots, \quad \omega\in N,
\end{aligned} \tag{5.75}
$$

where the dots stand for smoother terms at the singularity points and include the regular components. The branches of the square roots are specified by the chosen branch cuts shown in Fig. 5.4 and by the conditions $\sqrt{\cos\alpha-\cos\hat{\theta}'(\omega,\omega_0)} > 0$ as $-\hat{\theta}' < \alpha < \hat{\theta}'$. It is worth mentioning that, beside the real singularities giving rise to the wave reflected from the conical surface, some complex singularities may contribute to the asymptotics as $\operatorname{Im}\zeta < 0$ and are related to the surface waves (Fig. 5.4).

---

[7]Propagation of singularities of such a type is also considered in [5.19, 5.20] for a perfect cone.

We substitute the expansion into equation (5.69) and equate the summands of the same orders of singularities zero. In the leading order we obtain the eikonal equation

$$\left[\nabla_\omega \hat{\theta}'(\omega, \omega_0)\right]^2 = 1. \tag{5.76}$$

This equation is solved on the unit sphere and the solutions should fulfill the boundary conditions on σ

$$\hat{\theta}'(\omega, \omega_0)|_\sigma = \hat{\theta}(\omega, \omega_0)|_\sigma, \quad \nabla_\omega \hat{\theta}'(\omega, \omega_0) \cdot s|_\sigma = \nabla_\omega \hat{\theta}(\omega, \omega_0) \cdot s|_\sigma, \tag{5.77}$$

where s is the tangent to σ vector on the unit sphere.

In the next approximation we arrive at the “transport” equation

$$2\nabla_\omega \hat{\theta}' \cdot \nabla_\omega A^0(\omega, \omega_0) + (\Delta_\omega \hat{\theta}' - \cot \hat{\theta}')A^0(\omega, \omega_0) = 0, \tag{5.78}$$

where ∇_ω is the gradient operator on the unit sphere S^2, $\Delta_\omega = \nabla_\omega \cdot \nabla_\omega$.

The solution of the problem (5.76) and (5.77) for the eikonal equation is found by use of the traditional ray method [5.8]. The geodesics (rays) are the arcs of the big circle on the unit sphere which originate at the source ω_0 then reflect at the boundary σ in agreement with the geometrical optics then arrive at the point ω, whereas the corresponding rays, reflected from the cone, are shown in Fig. 5.5. In accordance with the Fermat principle the solution is given by $\hat{\theta}'(\omega, \omega_0) = \min_{s\in\sigma} \left(\hat{\theta}(\omega, s) + \hat{\theta}(s, \omega_0)\right)$.

The solutions of the equation (5.78) are also determined in the ray coordinates. Thus, we have

$$2\mathrm{d}A^0/\mathrm{d}\hat{\theta}' + (\Delta_\omega \hat{\theta}' - \cot \hat{\theta}')A^0 = 0 \tag{5.79}$$

and

$$A^0 = A^0|_\sigma \sqrt{\frac{J_\mathrm{r}}{\sin \hat{\theta}'}}\Bigg|_\sigma \sqrt{\frac{\sin \hat{\theta}'}{J_\mathrm{r}}},$$

where J_r is the geometrical spreading (divergence) of the reflected geodesics on the unit sphere. The explicit formula for J_r can be extracted from the work [5.7].

The unknown values $A^0|_\sigma$ should be determined from the boundary condition. We find $A^0|_\sigma = -(\sqrt{\pi}/2)R$, where R is the reflection coefficient and is found in an explicit form

$$R = \frac{\sin \hat{\theta}|_\sigma \sin \alpha_\mathrm{i} - \sin \zeta}{\sin \hat{\theta}|_\sigma \sin \alpha_\mathrm{i} + \sin \zeta},$$

where $\sin \alpha_\mathrm{i} = - \left.\partial\hat{\theta}/\partial\theta\right|_\sigma = -\nabla_\omega \hat{\theta}|_\sigma \cdot \boldsymbol{e}_\theta$.

In the leading approximation, we have thus obtained the local expression

$$\widetilde{\Phi}(\alpha, \omega, \omega_0) = -\frac{\sqrt{\pi}}{2} \sqrt{\frac{J_\mathrm{r}}{\sin \hat{\theta}'}}\Bigg|_\sigma \sqrt{\frac{\sin \hat{\theta}'}{J_\mathrm{r}}} \frac{R}{\sqrt{\cos \alpha - \cos \hat{\theta}'(\omega, \omega_0)}} + \ldots, \quad \omega \in N. \tag{5.80}$$

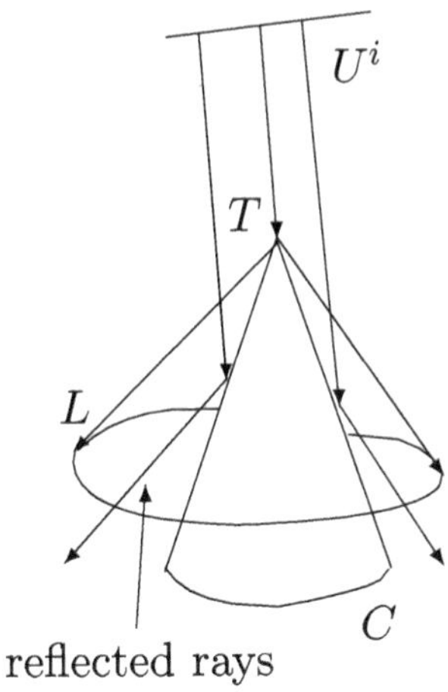

Figure 5.5 Scattering from a completely illuminated cone.

5.7 The reflected wave

The far-field asymptotics of the Sommerfeld integrals are found in a traditional manner. The Sommerfeld double-loop contour is transformed into the steepest-descent paths (Fig. 5.4) attributed to the saddle points $\pm\pi$. In this case, the singularities of the integrands, that is, of the Sommerfeld transformants, are captured and give rise to the corresponding components in the far-field asymptotics. In particular, the branch cuts from the points $\pm\hat{\theta}'$ for $\Phi(\alpha, \omega, \omega_0)$ are responsible for the reflected wave. We assume that the ends of the branch cuts are not close to the saddle points then their contributions can be separated in the asymptotics.

We substitute the local expression (5.80) into the Sommerfeld integral

$$U_{\mathrm{r}}(kr, \theta, \varphi)$$
$$= \frac{\sqrt{-\mathrm{i}kr}}{\pi \mathrm{i}} \int_{\gamma_{\mathrm{r}}} \mathrm{e}^{-\mathrm{i}kr\cos\alpha} \frac{A^0(\omega, \omega_0)\sin\alpha}{\sqrt{\cos\alpha - \cos\hat{\theta}'(\omega, \omega_0)}}\, \mathrm{d}\alpha \left[1 + \mathrm{O}\left(\frac{1}{kr\cos\frac{1}{2}\hat{\theta}'}\right)\right], \tag{5.81}$$

where the integration is performed over the contours γ_{r} in a small vicinity of the branch point $\hat{\theta}'$ along the sides of the corresponding branch cut (Fig. 5.4). We evaluate the previous integrals and demonstrate that asymptotically as $kr \to \infty$ we obtain the expressions for the reflected wave. To this end, we introduce the new variable of integration in (5.81) $p = \cos\alpha - \cos\hat{\theta}'(\omega, \omega_0)$, assuming that $\hat{\theta}'(\omega, \omega_0)$ is not close to the saddle point π. After some traditional calculations we obtain

$$U_{\mathrm{r}}(kr, \theta, \varphi)$$
$$= -\frac{2A^0(\omega, \omega_0)}{\pi} \mathrm{e}^{-\mathrm{i}kr\cos\hat{\theta}'(\omega,\omega_0)} \int_0^{c_1 kr} \frac{\mathrm{e}^{-\tau}}{\sqrt{\tau}}\, \mathrm{d}\tau \left[1 + \mathrm{O}\left(\frac{1}{kr\cos\frac{1}{2}\hat{\theta}'}\right)\right] \tag{5.82}$$

and, computing the integral in (5.82) asymptotically as $kr \to \infty$, we have

$$U_{\mathrm{r}}(kr, \theta, \varphi) = R\sqrt{\frac{\sin\hat{\theta}'\ J_{\mathrm{r}}|_{\sigma}}{\sin\hat{\theta}'|_{\sigma}\ J_{\mathrm{r}}}}\ \mathrm{e}^{-ikr\cos\hat{\theta}'(\omega,\omega_0)}\left[1 + \mathrm{O}\left(\frac{1}{kr\cos\frac{1}{2}\hat{\theta}'}\right)\right] \tag{5.83}$$

in the domain illuminated by the reflected rays. If the branch points are in close vicinities of the saddle points, the simple derivations exploited before in (5.81) must be modified accordingly.

Remark: The expression for the reflected wave (5.83) could be determined using the traditional ray method without solving the problem for the spectral function u_ν, that is, from the local considerations of geometrical optics.

5.8 Scattering diagram of the spherical wave from the vertex

Provided $\omega \in N$ (i.e., in the exterior of the oasis), the branch cuts (Fig. 5.4) pass through the saddle points. The integral over close vicinities of the saddle points $\pm\pi$ is asymptotically evaluated and enables us to write the expression for the spherical wave in the asymptotics (see also [5.29]):

$$U_{\mathrm{sph}}(kr, \omega, \omega_0) = D(\omega, \omega_0)\frac{\mathrm{e}^{ikr}}{-ikr}\left[1 + \mathrm{O}\left(\frac{1}{kr}\right)\right], \tag{5.84}$$

$$D(\omega, \omega_0) = -\sqrt{\frac{2}{\pi}}\lim_{\varepsilon\to 0+}\frac{\Phi(\pi + \mathrm{i}\varepsilon, \omega, \omega_0) + \Phi(\pi - \mathrm{i}\varepsilon, \omega, \omega_0)}{2}. \tag{5.85}$$

It is useful to mention that the formulas for $D(\omega, \omega_0)$ based on the Abel-Poisson summation [5.10] are not applicable in our case.

We have to develop analytic continuation for the formula (5.66) for the Sommerfeld transformant Φ because the integral in (5.66) diverges as $\omega \in N$. The analytic continuation of the formula for $\Phi(\alpha, \omega, \omega_0)$ given by

$$\Phi(\alpha, \omega, \omega_0) = \pm\frac{\sqrt{2\pi}}{\mathrm{i}}\int_{\mathrm{i}\mathbb{R}} \mathrm{i}\nu\, u_\nu(\omega, \omega_0)\mathrm{e}^{\pm\mathrm{i}\nu\alpha}\,\mathrm{d}\nu \tag{5.86}$$

from the domain $|\mathrm{Re}\,\alpha| < \hat{\theta}'(\omega, \omega_0)$ of definition of (5.86) onto a broader strip is performed by use of the formulas of the Fourier transform (1.67) and by use of the Fourier transform of the convolution (1.68).

Comparing with $u_\nu(\omega, \omega_0)$, we observe that the auxiliary function

$$u_\nu^{\mathrm{r}}(\omega, \omega_0) = -\frac{\mathrm{P}_{\nu-1/2}\left(-\cos\hat{\theta}'(\omega, \omega_0)\right)}{4\cos(\pi\nu)}, \tag{5.87}$$

up to a constant factor, has the same asymptotic behavior $C_{\pm}\mathrm{e}^{\pm\mathrm{i}\nu\hat{\theta}'(\omega,\omega_0)}/\sqrt{\nu}$ as $\nu \to \pm\mathrm{i}\infty$. Assuming that $|\mathrm{Re}\,\alpha| < \hat{\theta}'(\omega,\omega_0)$ (i.e., α belongs to Π'_{ω}), one has

$$\begin{aligned}
\Phi(\alpha,\omega,\omega_0) &= \frac{\sqrt{2\pi}}{\mathrm{i}}\frac{\mathrm{d}}{\mathrm{d}\alpha}\int\limits_{\mathrm{i}\mathbb{R}} u_\nu(\omega,\omega_0)\,\mathrm{e}^{\mathrm{i}\nu\alpha}\,\mathrm{d}\nu \\
&= \frac{\sqrt{2\pi}}{\mathrm{i}}\frac{\mathrm{d}}{\mathrm{d}\alpha}\int\limits_{\mathrm{i}\mathbb{R}} \left(\nu^2-a^2\right) u^{\mathrm{r}}_\nu(\omega,\omega_0)\frac{u_\nu(\omega,\omega_0)/u^{\mathrm{r}}_\nu(\omega,\omega_0)}{\nu^2-a^2}\mathrm{e}^{\mathrm{i}\nu\alpha}\mathrm{d}\nu \\
&= -\frac{\sqrt{2\pi}}{\mathrm{i}}\frac{\mathrm{d}}{\mathrm{d}\alpha}\left(\frac{\mathrm{d}^2}{\mathrm{d}\alpha^2}+a^2\right)\int\limits_{\mathrm{i}\mathbb{R}} u^{\mathrm{r}}_\nu(\omega,\omega_0)\frac{u_\nu(\omega,\omega_0)/u^{\mathrm{r}}_\nu(\omega,\omega_0)}{\nu^2-a^2}\mathrm{e}^{\mathrm{i}\nu\alpha}\mathrm{d}\nu \\
&= -\frac{1}{\mathrm{i}\sqrt{2\pi}}\frac{\mathrm{d}}{\mathrm{d}\alpha}\left(\frac{\mathrm{d}^2}{\mathrm{d}\alpha^2}+a^2\right)\int\limits_{\mathrm{i}\mathbb{R}} P_u(\alpha-\tau,\omega,\omega_0)\,r_u(\tau,\omega,\omega_0)\,\mathrm{d}\tau,
\end{aligned} \tag{5.88}$$

where

$$\begin{aligned}
P_u(\alpha,\omega,\omega_0) &= \frac{1}{\mathrm{i}}\int\limits_{\mathrm{i}\mathbb{R}} u^{\mathrm{r}}_\nu(\omega,\omega_0)\,\mathrm{e}^{\mathrm{i}\nu\alpha}\mathrm{d}\nu, \\
r_u(\alpha,\omega,\omega_0) &= \frac{1}{\mathrm{i}}\int\limits_{\mathrm{i}\mathbb{R}} \frac{u_\nu(\omega,\omega_0)/u^{\mathrm{r}}_\nu(\omega,\omega_0)}{\nu^2-a^2}\,\mathrm{e}^{\mathrm{i}\nu\alpha}\mathrm{d}\nu
\end{aligned} \tag{5.89}$$

and $a > 0$ is some constant. Note that the integrals in (5.89) converge absolutely and uniformly with respect to $\alpha \in \mathrm{i}\mathbb{R}$ and $\omega \in N$.

Using (7.216) from [5.24], one can obtain

$$\begin{aligned}
P_u(\alpha,\omega,\omega_0) &= -\frac{1}{2}\int\limits_0^\infty \frac{\mathrm{P}_{\mathrm{i}\tau-1/2}\left(\cos[\pi-\hat{\theta}']\right)}{\cos(\mathrm{i}\pi\tau)}\cos(\mathrm{i}\alpha\tau)\,\mathrm{d}\tau \\
&= -\left(2\sqrt{2}\right)^{-1}\left(\cos\alpha-\cos\hat{\theta}'\right)^{-1/2}, \quad \mathrm{i}\alpha > 0.
\end{aligned} \tag{5.90}$$

Now we use argumentation of the analytic continuation—first, onto the strip $|\mathrm{Re}\,\alpha| < \hat{\theta}'(\omega,\omega_0)$. To fix the regular branch of the right-hand side in (5.90) we conduct the branch cuts from $\pm\hat{\theta}'$ to $\pm\infty$ correspondingly and assume that $\sqrt{\cos\alpha-\cos\hat{\theta}'} > 0$ as $-\hat{\theta}' < \alpha < \hat{\theta}'$. The right-hand side of the formula (5.90) remains valid as $\alpha \in \Pi_{\pi+\delta}$ for some small positive δ outside the branch cuts if $\hat{\theta}' < \pi$; therefore, it gives also the analytic continuation of the left-hand side. The function $P_u(\alpha,\omega,\omega_0) = -(2\sqrt{2})^{-1}\left(\cos\alpha-\cos\hat{\theta}'\right)^{-1/2}$ is continuous on the sides of the branch cuts in the strip except the points $\pm\hat{\theta}'$.

Finally, introducing the notation

$$\mathcal{P}_u(\alpha,\omega,\omega_0) = -\frac{\mathrm{d}}{\mathrm{d}\alpha}\left(\frac{\mathrm{d}^2}{\mathrm{d}\alpha^2}+a^2\right)P_u(\alpha,\omega,\omega_0)$$
$$= \left(\frac{\mathrm{d}^2}{\mathrm{d}\alpha^2}+a^2\right)\frac{1}{4\sqrt{2}}\frac{\sin\alpha}{\left(\cos\alpha-\cos\hat{\theta}'\right)^{3/2}},$$

from (5.88) and (5.85) one has the formula

$$\Phi(\alpha,\omega,\omega_0) = \frac{1}{\mathrm{i}\sqrt{2\pi}}\int\limits_{\mathrm{i}\mathbb{R}} \mathcal{P}_u(\alpha-\tau,\omega,\omega_0)\, r_u(\tau,\omega,\omega_0)\,\mathrm{d}\tau \tag{5.91}$$

and for the scattering diagram

$$D(\omega,\omega_0) = \frac{\mathrm{i}}{2\pi}\int\limits_{\mathrm{i}\mathbb{R}} [\mathcal{P}_u(\pi+\mathrm{i}0-\tau,\omega,\omega_0)+\mathcal{P}_u(\pi-\mathrm{i}0-\tau,\omega,\omega_0)]\; r_u(\tau,\omega,\omega_0)\,\mathrm{d}\tau. \tag{5.92}$$

The numerical efficiency of the previous formula is to be additionally studied.

5.9 Surface wave at axial incidence

In the present section we assume that the exciting wave is axially symmetric incident. This leads to some formal simplifications in the formulas for the surface wave.

5.9.1 Ray solution for the surface wave

An acoustic surface wave is sought in the form of the ray expansion with complex eikonal τ[8]

$$U^{\mathrm{sw}}(kr,\theta) = \mathrm{e}^{\mathrm{i}k\tau(r,\theta)}\sum_{j=0}^{\infty}\frac{V_j(k,r,\theta)}{(-\mathrm{i}k)^{j+1/2}}. \tag{5.93}$$

The expression (5.93) satisfies asymptotically ($kr\to\infty$) the Helmholtz equation

$$(\Delta+k^2)\,U^{\mathrm{sw}} = 0, \tag{5.94}$$
$$\Delta = \frac{1}{r^2}\frac{\partial}{\partial r}r^2\frac{\partial}{\partial r}+\frac{1}{r^2}\Delta_\omega,$$

($\omega=(\theta,\varphi)$, $\omega\in S^2$), and the boundary condition on the conical surface C

$$\left(\frac{1}{r}\frac{\partial}{\partial\theta}-\mathrm{i}k\sin\zeta\right)U^{\mathrm{sw}}\bigg|_{\theta=\theta_1} = 0. \tag{5.95}$$

[8]It can be demonstrated that, due to the axial symmetry of the boundary, the wave field depends only on kr and θ. The surface wave with a general dependence on φ can be treated analogously by means of the Fourier series on φ.

The wave field does not depend on the azimuthal angle φ, the conical surface C is given by the equation $\theta = \theta_1$. Formal substitution of the series (5.93) into the Helmholtz equation (5.94) leads to the known eikonal and transport equations.

Solving the eikonal equation and leading transport equation by means of separation of variables, one has (see the third term in (5.7))

$$U^{\mathrm{sw}}(kr,\theta) = \mathrm{const}\,\frac{\mathrm{e}^{ikr\cos(\theta_1-\theta-\zeta)}}{\sqrt{-ikr}}$$

$$\times\sqrt{\frac{\sin\theta_1}{\sin\theta}}\,[kr\sin(\theta_1-\theta-\zeta)]^{-\mu}\left[1+\mathrm{O}\left(\frac{1}{kr}\right)\right], \tag{5.96}$$

where the exponent μ is purely imaginary and is determined equating the leading terms in the boundary condition,

$$\mu = (1/2)\tan\zeta\,\cot\theta_1.$$

The constant in (5.96) specifies the amplitude of the surface wave. The expression (5.96) is in fact a surface wave propagating from the vertex of the cone with the velocity $c_{\mathrm{sw}} = c/\cosh(|\zeta|)$, where c is the wave velocity in an acoustic medium. This wave vanishes exponentially as $kr \to \infty$ beyond the boundary and is of $\mathrm{O}(1/(kr)^{1/2+\mu})$ on C.

The excitation coefficient of the surface wave cannot be found from the local considerations and requires study of the whole diffraction problem at hand.

Remark: The surface wave propagating from infinity to the vertex can be determined in the same fashion

$$U^{\mathrm{r,sw}}(kr,\theta) = \mathrm{const}\,\frac{\mathrm{e}^{-ikr\cos(\theta_1-\theta+\zeta)}}{\sqrt{-ikr}}$$

$$\times\sqrt{\frac{\sin\theta_1}{\sin\theta}}\,[kr\sin(\theta_1-\theta+\pi+\zeta)]^{-\mu_0}\left[1+\mathrm{O}\left(\frac{1}{kr}\right)\right], \tag{5.97}$$

where $\mu_0 = -\mu = -(1/2)\tan\zeta\,\cot\theta_1$.

5.9.2 Singularities of the Sommerfeld transformants corresponding to the surface wave

We exploit the saddle-point technique, applied to the Sommerfeld integrals to evaluate the far field. As we have agreed upon, one has to deform the integration contour γ into the steepest-descent paths going through the saddle points $\pm\pi$ (see Fig. 5.4). These are the points of regularity provided $\zeta \neq 0$. In the process of such deformation some complex singularities of the Sommerfeld transformant can be also captured.

As was mentioned for real α the equation for $\Phi(\alpha,\theta)$ is of hyperbolic type. However, we actually need to describe also complex singularities of $\Phi(\alpha,\theta)$ in the strip $|\mathrm{Re}\,\alpha| < 3\pi/2$. Existence of such singularities can be anticipated from the asymptotic behavior of $R(\nu)$ and of its Fourier transform, also implying the experience already gained in [5.29].

The Sommerfeld transformant has two additional complex singularities at the points $\alpha = \pm\alpha_\zeta(\theta)$,

$$\alpha_\zeta(\theta) = \pi + \zeta + \theta_1 - \theta$$

located on the boundary of the strip Π as $\theta = \theta_1$. An additional analysis shows that these are branch points of a logarithmic singularity type $\exp\left[\lambda \log\left(\alpha - \alpha_\zeta\right)\right]$ then the local behavior of the transformant near the point $\alpha_\zeta(\theta)$ is sought in the form

$$\Phi(\alpha, \theta) \; = \; A(\alpha, \theta)[\alpha - \alpha_\zeta(\theta)]^\lambda \; + \; B(\alpha, \theta), \tag{5.98}$$

where

$$A(\alpha, \theta) = A_0(\theta) + A_1(\alpha)\left[\alpha - \alpha_\zeta(\theta)\right] + A_2(\alpha)\left[\alpha - \alpha_\zeta(\theta)\right]^2 + \cdots,$$

$$B(\alpha, \theta) = B_0(\theta) + B_1(\alpha)\left[\alpha - \alpha_\zeta(\theta)\right] + B_2(\alpha)\left[\alpha - \alpha_\zeta(\theta)\right]^2 + \cdots.$$

The complex constant λ is still unknown as well as the coefficients $A_j, B_j, \; j = 0, 1, 2, \ldots$ in the series. The branch of the analytic function in (5.98) will be fixed below.

We substitute the expansions (5.98) into the equation for $\Phi(\alpha, \theta)$ and equate the coefficients of the same powers $(\lambda + j)$ of $[\alpha - \alpha_\zeta(\theta)]^{\lambda + j}, \; j = -2, -1, 0, \ldots$ in the singular terms. In the leading approximations we obtain

$$2\lambda A_0'(\theta) + \lambda \cot\theta\, A_0(\theta) \; = \; 0, \tag{5.99}$$

$$2(\lambda + 1)A_1'(\theta) + (\lambda + 1)\cot\theta\, A_1(\theta) \; = \; A_0(\theta)/4 - A_0''(\theta) - \cot\theta\, A_0'(\theta).$$

The solution of (5.99) is[9]

$$A_0(\theta) \; = \; C_0\sqrt{\frac{\sin\theta_1}{\sin\theta}}\,.$$

The constant C_0 cannot be found from the local considerations. It is a functional of $R(\nu, n)|_{n=0} =: R(\nu)$, which is discussed in Section 5.11.3. The exponent λ is specified from the boundary condition (5.74) substituting the expansion and equating to zero the leading terms,

$$\lambda \; = \; -1 + \mu, \;\; \mu \; = \; (1/2)\tan\zeta\,\cot\theta_1. \tag{5.100}$$

Remark that μ is purely imaginary and for the leading singular term one has

$$\Phi(\alpha, \theta) \; = \; C_0\sqrt{\frac{\sin\theta_1}{\sin\theta}}\,[\alpha - \alpha_\zeta(\theta)]^{-1+\mu} \; + \cdots \tag{5.101}$$

in a vicinity of the singular point $\alpha_\zeta(\theta)$. Taking into account that $\Phi(\alpha, \theta)$ is odd, we also determine the local behavior at the point $-\alpha_\zeta(\theta)$. The Sommerfeld transformant $\widetilde{\Phi}(\alpha, \theta)$

---

[9]The higher-order terms are not considered because we are looking for the leading terms of the far-field asymptotics.

behaves locally as

$$\widetilde{\Phi}(\alpha,\theta) = \frac{C_0}{\mu}\sqrt{\frac{\sin\theta_1}{\sin\theta}}\,[\alpha-\alpha_\zeta(\theta)]^\mu + \cdots . \tag{5.102}$$

The branch cuts on the α-plane for the expressions (5.101) and (5.102) are conducted from the points $\pm\alpha_\zeta(\theta)$ to $\pm\infty$ correspondingly and are symmetric with respect to the origin (Fig. 5.4). The branch is fixed by a condition proposed in the next subsection.

5.9.3 Asymptotic evaluation of the surface wave

The integral representation (5.66) for $\Phi(\alpha,\theta)$ with

$$u_\nu(\omega,\omega_0) = R(\nu)\frac{\mathrm{P}_{\nu-1/2}(\cos\theta)}{\mathrm{d}_{\theta_1}\mathrm{P}_{\nu-1/2}(\cos\theta_1)}$$

(the series for u_ν contains only one summand) is valid in the strip Π'_ω of its regularity; otherwise an analytic continuation is required. Such a continuation can be performed in different ways.[10] In particular, the behavior of the Sommerfeld transformants near the singularities $\pm\alpha_\zeta(\theta)$ is of importance.

Now we turn to the surface wave. We evaluate asymptotically the integral over the contour l_ζ^+ comprising the branch cut from the point α_ζ[11]

$$U^{\mathrm{sw}}(kr,\theta) = \frac{1}{\pi\mathrm{i}}\int_{l_\zeta^+}\frac{\mathrm{e}^{-\mathrm{i}kr\cos\alpha}}{\sqrt{-\mathrm{i}kr}}\,\Phi(\alpha,\theta)\,\mathrm{d}\alpha. \tag{5.103}$$

In our conditions ($kr\to\infty$) it suffices to restrict ourselves to the local considerations, taking into account the expansion

$$\cos\alpha = \cos\alpha_\zeta(\theta) - \sin\alpha_\zeta(\theta)[\alpha-\alpha_\zeta(\theta)] + \mathrm{O}([\alpha-\alpha_\zeta(\theta)]^2),\quad \alpha-\alpha_\zeta(\theta) = \mathrm{O}(1/kr)$$

near the singular point and the local behavior (5.101) in the integrand, thus having

$$U^{\mathrm{sw}}(kr,\theta) = \frac{C_0}{\pi\mathrm{i}}\sqrt{\frac{\sin\theta_1}{\sin\theta}}\,\frac{\mathrm{e}^{-\mathrm{i}kr\cos\alpha_\zeta(\theta)}}{\sqrt{-\mathrm{i}kr}}$$

$$\times\int_{l_\zeta^+}\mathrm{e}^{\mathrm{i}kr\sin\alpha_\zeta(\theta)[\alpha-\alpha_\zeta(\theta)]}[\alpha-\alpha_\zeta(\theta)]^{-1+\mu}\,\mathrm{d}\alpha\,[1+\mathrm{O}(1/kr)]. \tag{5.104}$$

Now it is reasonable to introduce a new variable of integration

$$\tau = -\mathrm{i}kr\sin\alpha_\zeta(\theta)[\alpha-\alpha_\zeta(\theta)]$$

---

[10] We considered a way of such a continuation in formula (5.91).

[11] The contour l_ζ^+ is located in a small vicinity of the branch point, goes along the upper side of the branch cut, enveloping the point α_ζ, then passes along the lower side of the cut as shown in Fig. 5.4.

and choose the branch of $[\alpha - \alpha_\zeta(\theta)]^\mu$ or equivalently of τ^μ as follows:

$$a_0 < \arg\left[\alpha - \alpha_\zeta(\theta)\right] < 2\pi + a_0,$$

$$a_0 := -\arg\left[-ikr\sin\alpha_\zeta(\theta)\right] = \arg\left[\alpha - \alpha_\zeta(\theta)\right]\Big|_{l_\zeta^{+,\mathrm{up}}},$$

where $l_\zeta^{+,\mathrm{up}}$ is the upper side of the branch cut from $\alpha_\zeta(\theta)$ to infinity,

$$\tau^\mu = \exp(\mu\log|\tau| + \mathrm{i}\,\mu\arg\tau), \quad \arg\tau = a_0 + \arg\left[\alpha - \alpha_\zeta(\theta)\right].$$

We evaluate asymptotically the integral in (5.104)

$$\int_{l_\zeta^+} \mathrm{e}^{ikr\sin\alpha_\zeta(\theta)[\alpha-\alpha_\zeta(\theta)]}[\alpha - \alpha_\zeta(\theta)]^{-1+\mu}\,\mathrm{d}\alpha$$

$$= [-ikr\sin\alpha_\zeta(\theta)]^{-\mu}\left(\mathrm{e}^{2\pi\mathrm{i}\mu} - 1\right)\int_0^{Ckr} \mathrm{e}^{-\tau}\tau^{-1+\mu}\mathrm{d}\tau$$

$$= [-ikr\sin\alpha_\zeta(\theta)]^{-\mu}\left(\mathrm{e}^{2\pi\mathrm{i}\mu} - 1\right)\Gamma(\mu)[1 + \mathrm{o}(1)],$$

where $\Gamma(z)$ is the Euler gamma-function. Collecting the previous formulas, we obtain from (5.104)

$$U^{\mathrm{sw}}(kr,\theta) = \frac{C_0}{\pi\mathrm{i}}\frac{\left(\mathrm{e}^{2\pi\mathrm{i}\mu} - 1\right)\Gamma(\mu)}{[\mathrm{i}\sin(\theta_1 - \theta + \zeta)]^\mu}\sqrt{\frac{\sin\theta_1}{\sin\theta}}$$

$$\times\frac{\mathrm{e}^{ikr\cos[\theta_1-\theta+\zeta]}}{\sqrt{-ikr}}\,\mathrm{e}^{-\cot\theta_1\tan(\zeta)\log(kr)/2}\,[1 + \mathrm{O}(1/kr)], \tag{5.105}$$

where $\mu = (1/2)\cot\theta_1\tan\zeta$, ζ is purely imaginary. The derivation of the constant C_0 is not elementary and will be discussed in Section 5.11.3.

We notice that under the sufficient conditions

$$\theta - \theta_1 - \mathrm{Re}\,\zeta - \mathrm{gd}\,(\mathrm{Im}\,\zeta) > 0, \quad \mathrm{Im}\,\zeta < 0, \quad \mathrm{Re}\,\zeta = 0,$$

where $\mathrm{gd}(x)$ is the Gudermann function (see Chapter 1), the corresponding singularities are indeed captured, and the outgoing surface waves are really excited.

The expression for the surface wave (5.105) is elementary except the derivation of the constant C_0 (see Section 5.11.3) which requires calculation of a functional of the solution $R(\nu, 0)$ of the integral equation (5.33). As we have already mentioned, the expression (5.105) can be obtained by separating variables in the eikonal and transport equations or by directly applying the technique developed in [5.11–5.13].

It is also worth noting that the factor $\exp\left[-\mathrm{i}\,\Phi_g(kr)\right]$ in (5.105) with the geometrical phase

$$\Phi_g(kr) = -\cot\theta_1\tan(\mathrm{Im}\,\zeta)\log(kr)/2$$

depends weakly on the normalized distance kr as $\log(kr)$. The geometrical phase (sometimes also known as the Berry phase) is expressed by use of the mean curvature of the conical surface, which is in full agreement with the result in [5.13]. The geometrical (or topological) phase is encountered with in many research areas of quantum physics or in the theory of waves.

5.10 Uniform asymptotics of the far field and the parabolic cylinder functions

The (with respect to the angular variables) uniform asymptotic expression of the scattered space field W (reflected plus spherical waves) is sought in the form [5.6, 5.34]

$$
\begin{aligned}
W(kr,\omega,\omega_0) = {} & \frac{\exp[ikl(\boldsymbol{r})]}{k^{1/4}}\left\{\mathrm{D}_{-3/2}\left(\sqrt{k}\,\mathrm{e}^{-\mathrm{i}\pi/4}m(\boldsymbol{r})\right)\left[A_0(\boldsymbol{r})+\frac{A_1(\boldsymbol{r})}{ik}+\cdots\right]\right.\\
& \left.+\frac{\mathrm{e}^{\mathrm{i}\pi/4}}{\sqrt{k}}\mathrm{D}'_{-3/2}\left(\sqrt{k}\,\mathrm{e}^{-\mathrm{i}\pi/4}m(\boldsymbol{r})\right)\left[B_0(\boldsymbol{r})+\frac{B_1(\boldsymbol{r})}{ik}+\cdots\right]\right\}\\
& +\frac{\mathrm{e}^{ikr}}{k}\left[C_0(\boldsymbol{r})+\frac{C_1(\boldsymbol{r})}{ik}+\cdots\right],
\end{aligned}
\tag{5.106}
$$

with $(kr \gg 1)$

$$
l(\boldsymbol{r}) = r\sin^2[\hat{\theta}'(\omega,\omega_0)/2],\quad m(\boldsymbol{r}) = -2\sqrt{r}\cos[\hat{\theta}'(\omega,\omega_0)/2]
$$

and $\mathrm{D}_p(z)$ is the parabolic cylinder (Weber) function specified by the equality

$$
\mathrm{D}_p(z) = \frac{\mathrm{e}^{-z^2/4}}{\Gamma(-p)}\int_0^\infty \mathrm{e}^{-zx-x^2/2}x^{-p-1}\mathrm{d}x,\quad \mathrm{Re}\,p<0,
$$

having the asymptotics $(p=-3/2)$

$$
\begin{aligned}
\mathrm{D}_{-3/2}(z) &= \mathrm{e}^{-z^2/4}z^{-3/2}\left[1+\mathrm{O}(|z|^{-1})\right]\\
&\quad -\frac{\sqrt{2\pi}}{\Gamma(3/2)}\mathrm{e}^{-\mathrm{i}3\pi/2}\mathrm{e}^{z^2/4}z^{1/2}\left[1+\mathrm{O}(|z|^{-1})\right],\quad \arg z = 3\pi/4,\\
\mathrm{D}_{-3/2}(z) &= \mathrm{e}^{-z^2/4}z^{-3/2}\left[1+\mathrm{O}(|z|^{-1})\right],\quad \arg z = -\pi/4,\ |z|\to\infty.
\end{aligned}
\tag{5.107}
$$

The unknown coefficients A_j, B_j, C_j, $j=0,1,\ldots$ are smooth across the singular directions. The Ansatz (5.106) is substituted into the Helmholtz equation, and the coefficients of $k^{7/4}\mathrm{D}_{-3/2}$ and $k^{5/4}\mathrm{D}'_{-3/2}$ are equated to zero (see (5.2) in [5.6]). The corresponding equations are valid for arbitrary A_0, B_0 as long as $A_0^2+B_0^2 \neq 0$ holds because, in our case, $l+m^2/4=r$ and $l+m^2/4=-r\cos\hat{\theta}'(\omega,\omega_0)$ (equations (5.3) and (5.4) in [5.6] are then satisfied), $\left[\nabla(l\pm m^2/4)\right]^2=1$. Equating the term of $\mathrm{O}(kr)$ zero results in the transport equation for $C_0(\boldsymbol{r})$[12]

$$
2\nabla r\cdot\nabla C_0(\boldsymbol{r})+\Delta r\,C_0(\boldsymbol{r})=0
$$

[12] The terms of $\mathrm{O}((kr)^2)$ are identically zero because $(\nabla r)^2=1$.

and, therefore, $C_0(\boldsymbol{r}) = C(\omega, \omega_0)/r$, where $C(\omega, \omega_0)$ is a smooth function of ω to be determined from matching with the local asymptotics.

Equating the coefficients of $k^{3/4}\mathrm{D}_{-3/2}$ and $k^{1/4}\mathrm{D}'_{-3/2}$ zero, after tedious calculations (see also equations (5.5) and (5.6) in [5.6] which contain misprints) one has

$$\begin{aligned}
&2\nabla(l + m^2/4) \cdot \nabla \left(A_0\, m^{-3/2} - \frac{m^{-1/2}}{2} B_0 \right) \\
&+ \Delta(l + m^2/4) \left(A_0\, m^{-3/2} - \frac{m^{-1/2}}{2} B_0 \right) = 0, \\
&2\nabla(l - m^2/4) \cdot \nabla \left(A_0\, m^{1/2} + \frac{m^{3/2}}{2} B_0 \right) \\
&+ \Delta(l - m^2/4) \left(A_0\, m^{1/2} + \frac{m^{3/2}}{2} B_0 \right) = 0.
\end{aligned} \tag{5.108}$$

The first equation in (5.108) is easily integrated ($l + m^2/4 = r$)

$$A_0\, m^{-3/2} - \frac{m^{-1/2}}{2} B_0 = \frac{D_+(\omega, \omega_0)}{r}. \tag{5.109}$$

The second one with $l - m^2/4 = -r \cos \hat{\theta}'(\omega, \omega_0)$, after some calculations, leads to

$$A_0\, m^{1/2} + \frac{m^{3/2}}{2} B_0 = D_-(\omega, \omega_0) \tag{5.110}$$

with

$$D_-(\omega, \omega_0) = D_-(\omega|_\sigma, \omega_0) \sqrt{\frac{\sin \hat{\theta}'(\omega, \omega_0)}{\sin \hat{\theta}'(\omega|_\sigma, \omega_0)} \frac{J_{\mathrm{r}}(\omega|_\sigma, \omega_0)}{J_{\mathrm{r}}(\omega, \omega_0)}}, \tag{5.111}$$

the spreading $J_{\mathrm{r}}(\omega, \omega_0)$ coincides with that in (5.83). The coefficients A_0 and B_0 in (5.109) and (5.110) take the form

$$\begin{aligned}
A_0 &= \frac{D_+(\omega, \omega_0)}{r} m^{3/2}/2 + D_-(\omega, \omega_0) m^{-1/2}/2, \\
B_0 &= -\frac{D_+(\omega, \omega_0)}{r} m^{1/2} + D_-(\omega, \omega_0) m^{-3/2},
\end{aligned} \tag{5.112}$$

whereas

$$C_0(\boldsymbol{r}) = \frac{C(\omega, \omega_0)}{r}. \tag{5.113}$$

The unknown functions $D_+(\omega, \omega_0)$, $D_-(\omega|_\sigma, \omega_0)$, and $C(\omega, \omega_0)$ in (5.111)–(5.113) are determined from matching, in the leading approximation, of the uniform formula (5.106) with those nonuniform (5.83), (5.84), and (5.85) as $m \to -\infty$, $\omega \in N$. We introduce

$$A(\omega, \omega_0) = r^{1/4} A_0(\boldsymbol{r}), \quad B(\omega, \omega_0) = r^{3/4} B_0(\boldsymbol{r}) \tag{5.114}$$

and exploit (5.107) in the expression (5.106). Comparing the result, as $m \to -\infty$, with the local asymptotics (5.83), (5.84), and (5.85)

$$W(kr, \omega, \omega_0)$$

$$= \mathrm{e}^{-ikr\cos\hat{\theta}'(\omega,\omega_0)} R_\zeta \sqrt{\frac{\sin\hat{\theta}'(\omega,\omega_0)}{\sin\hat{\theta}'(\omega|_\sigma,\omega_0)} \frac{J_\mathrm{r}(\omega|_\sigma,\omega_0)}{J_\mathrm{r}(\omega,\omega_0)}} \left[1 + \mathrm{O}\left(\frac{1}{kr\cos^2\left(\frac{1}{2}\hat{\theta}'\right)}\right)\right]$$

$$+ \frac{\mathrm{e}^{ikr}}{-ikr} D(\omega,\omega_0)\left[1 + \mathrm{O}\left(\frac{1}{kr}\right)\right],$$

we arrive at the equalities

$$4\mathrm{e}^{-\mathrm{i}\pi/8} A(\omega,\omega_0)[\cos(\hat{\theta}'/2)]^{1/2} + 4\mathrm{e}^{-\mathrm{i}9\pi/8} B(\omega,\omega_0)[\cos(\hat{\theta}'/2)]^{3/2}$$

$$= R_\zeta \sqrt{\frac{\sin\hat{\theta}'(\omega,\omega_0)}{\sin\hat{\theta}'(\omega|_\sigma,\omega_0)} \frac{J(\omega|_\sigma,\omega_0)}{J(\omega,\omega_0)}}, \tag{5.115}$$

$$2^{-3/2}\mathrm{e}^{-\mathrm{i}9\pi/8} A(\omega,\omega_0)[\cos(\hat{\theta}'/2)]^{-3/2} + 2^{-3/2}\mathrm{e}^{-\mathrm{i}9\pi/8} B(\omega,\omega_0)[\cos(\hat{\theta}'/2)]^{-1/2}$$

$$+ C(\omega,\omega_0) = \mathrm{i}D(\omega,\omega_0).$$

In (5.115) we have two equations for three unknown functions A, B, and C. To get the third equation we study (5.85) for the diffraction coefficient $D(\omega,\omega_0)$ near the singular directions ω_s, $\hat{\theta}'(\omega_\mathrm{s},\omega_0) = \pi$. We have the representation outside the singular directions ($\omega \sim \omega_\mathrm{s}$)

$$D(\omega,\omega_0) = \frac{D_1(\omega,\omega_0)}{[\cos(\hat{\theta}'(\omega,\omega_0)/2)]^{3/2}} + D_2(\omega,\omega_0), \tag{5.116}$$

where $D_{1,2}(\omega,\omega_0)$ are regular functions of ω. Provided one could compute the value $D_2(\omega,\omega_0)$ in (5.116), from the second equation in (5.115) one has

$$C(\omega,\omega_0) = \mathrm{i}D_2(\omega,\omega_0) \tag{5.117}$$

which gives the desired third equation.

Equations (5.115) and (5.117) specify the coefficients A, B, and C and therefore A_0, B_0, and C_0 in the uniform asymptotics (5.106) in the leading approximation.

In practical use of the asymptotic expansion (5.106) and its numerical elaboration there are several important advances to be made. First of all, an efficient numerical elaboration for $D(\omega,\omega_0)$ as $\omega \in N$ should be developed. Second, the splitting (5.116) implies possibility to compute $D_1(\omega,\omega_0)$ and $D_2(\omega,\omega_0)$ having efficient formulas for $D(\omega,\omega_0)$. Assuming that $D_{1,2}(\omega,\omega_0)$ are analytic functions of $\cos\left[\hat{\theta}'(\omega,\omega_0)/2\right]$, that is,

$$D_{1,2}(\omega,\omega_0) = \sum_{s=0}^{\infty} D_{1,2}^{(s)}(\omega_0)\cos^s[\hat{\theta}'(\omega,\omega_0)/2],$$

from (5.116) recurrent formulas can easily be developed to determine $D_{1,2}^{(s)}(\omega_0)$, $s = 0, 1, \ldots$ in terms of $D(\omega,\omega_0)$.

5.11 Appendices

5.11.1 Appendix A

In the present section we verify the equality

$$\frac{D_j^{\rm inc}}{\pi}\int_{\kappa}\left.\frac{\partial X_j}{\partial n}\right|_{\kappa}(\tau)\left[\int_{\kappa}\log|h(\tau)-h(t)|\,\mathrm{d}l_t\right]\mathrm{d}l_\tau=0.$$

We exploit the following simple statement. Let v be a harmonic function outside κ and fulfill the condition

$$\left.\frac{\partial v}{\partial n}\right|_{\kappa}=\mathrm{const}$$

as well as that at infinity

$$v=C\log R+C_0+\mathrm{O}(1/R)\,.$$

Then its solution reads

$$v(z)=-\frac{\mathrm{const}}{\pi}\int_{\kappa}\log|h(z)-h(t)|\,\mathrm{d}l_t+\frac{\mathrm{Const}}{2\pi}l_\kappa\log|h(z)|+C_1,$$

where $h(z)$ is the conformal mapping described in Section 5.4. We use the Green theorem for two harmonic functions v and $D_j^{\rm inc}X_j$ in the exterior of κ and in the interior of the circumference S_R of radius R and thus have

$$\frac{D_j^{\rm inc}}{\pi}\left(\int_{\kappa}\left.\frac{\partial X_j}{\partial n}\right|_{\kappa}v|_\kappa\,\mathrm{d}l_\tau+\int_{S_R}\left.\frac{\partial X_j}{\partial R}\right|_{S_R}v|_{S_R}\,\mathrm{d}l_R\right)$$

$$=\frac{D_j^{\rm inc}}{\pi}\left(\int_{\kappa}X_j|_\kappa\left.\frac{\partial v}{\partial n}\right|_{\kappa}\mathrm{d}l_\tau+\int_{S_R}X_j|_{S_R}\left.\frac{\partial v}{\partial R}\right|_{S_R}\mathrm{d}l_R\right).$$

Exploiting the boundary condition and that at infinity for v, we find

$$\frac{D_j^{\rm inc}}{\pi}\int_{\kappa}\left.\frac{\partial X_j}{\partial n}\right|_{\kappa}v|_\kappa\,\mathrm{d}l_\tau=\frac{D_j^{\rm inc}}{\pi}\int_{\kappa}X_j|_\kappa\mathrm{const}\,\mathrm{d}l_\tau.$$

Now, to prove the desired equality remark that

$$v|_\kappa(\tau)=-\frac{\mathrm{const}}{\pi}\int_{\kappa}\log|h(\tau)-h(t)|\,\mathrm{d}l_t+C_1,$$

because $|h(z)|=1,\;\; z\in\kappa$, then use (5.52).

5.11.2 Appendix B. Reduction of integrals

Consider the integral

$$J(\theta,\theta_0)=\frac{\mathrm{i}}{32}\int\limits_{\mathrm{i}\mathbb{R}}\int\limits_{\mathrm{i}\mathbb{R}}\int\limits_{\mathrm{i}\mathbb{R}} \mathrm{d}\nu \mathrm{d}\tau \mathrm{d}t \frac{\nu\tan(\pi\nu)t\sin(\pi t)\tau\tan(\pi\tau)\,\mathrm{P}_{\nu-1/2}(\cos\theta)\,\mathrm{P}_{\tau-1/2}(\cos\theta_0)}{[\cos(\pi t)+\cos(\pi\nu)]\,[\cos(\pi t)+\cos(\pi\tau)]},$$

as $\theta+\theta_0<\pi$. The latter integral can be explicitly computed

$$J(\theta,\theta_0)=-\frac{1}{2\pi}\frac{1}{(\cos\theta+\cos\theta_0)^3}.$$

Indeed, we use the Fock representation for the conical functions $\mathrm{P}_{\nu-1/2}(\cosh v)$ (see [5.24]), and after simple transformations we find

$$J(\theta,\theta_0)=-\frac{1}{2\pi^2}\int\limits_0^\infty \mathrm{d}t\, t\sinh(\pi t)\left\{\int\limits_0^\infty\int\limits_0^\infty \frac{\mathrm{d}w_1\mathrm{d}w_2}{\sqrt{\cosh(w_1+v)-\cosh v}\sqrt{\cosh(w_2+v_0)-\cosh v_0}}\right.$$

$$\left.\times\int\limits_0^\infty\int\limits_0^\infty \frac{\mathrm{d}x\mathrm{d}y\; x\sinh[x(w_1+v)]\; y\sinh[y(w_2+v_0)]}{(\cosh(\pi t)+\cosh(\pi x))(\cosh(\pi t)+\cosh(\pi y))}\right\},$$

where $v=-\mathrm{i}\theta+0,\ v_0=-\mathrm{i}\theta_0+0$. Note that

$$\partial_v\cos[x(w_1+v)]=-x\sin[x(w_1+v)],$$
$$\partial_{v_0}\cos[x(w_2+v_0)]=-x\sin[x(w_2+v_0)],$$

then exploit (3.983.2) from [5.24] and write

$$J(\theta,\theta_0)$$

$$=-\frac{1}{2\pi^2}\int\limits_0^\infty \mathrm{d}t\frac{t}{\sinh(\pi t)}\left\{\int\limits_0^\infty\int\limits_0^\infty \frac{\mathrm{d}w_1\mathrm{d}w_2}{\sqrt{\cosh(w_1+v)-\cosh v}\sqrt{\cosh(w_2+v_0)-\cosh v_0}}\right.$$

$$\left.\times\frac{\partial}{\partial v}\frac{\sin[t(w_1+v)]}{\sinh(w_1+v)}\frac{\partial}{\partial v_0}\frac{\sin[t(w_2+v_0)]}{\sinh(w_2+v_0)}\right\}.$$

We have formally changed the order of integration, which is justified. Take into account the identity

$$t\frac{\partial}{\partial v}\frac{\sin[t(w_1+v)]}{\sinh(w_1+v)}=\frac{\partial}{\partial v}\left[\frac{1}{\sinh(w_1+v)}\right]$$

$$\times\frac{\partial}{\partial v}\{-\cos[t(w_1+v)]\}+\frac{1}{\sinh(w_1+v)}\frac{\partial^2}{\partial^2 v}\{-\cos[t(w_1+v)]\},$$

change the order of integration with respect to t, w_1, and w_2, and then exploit (3.981.5) from [5.24] for the internal integral on t. We arrive at the representation

$$J(\theta,\theta_0)=\frac{1}{4\pi^2}\int_0^\infty\int_0^\infty \frac{dw_1 dw_2}{\sqrt{\cosh(w_1+v)-\cosh v}\sqrt{\cosh(w_2+v_0)-\cosh v_0}}$$

$$\times\left\{\frac{1}{2}\frac{\partial}{\partial v}\left[\frac{1}{\sinh(w_1+v)}\right]\frac{\partial^2}{\partial v\partial v_0}\frac{1}{\cosh(w_1+v)+\cosh(w_2+v_0)}\right.$$

$$\left.+\frac{1}{2}\frac{\partial}{\partial v_0}\left[\frac{1}{\sinh(w_2+v_0)}\right]\frac{\partial^2}{\partial v_0\partial v}\frac{1}{\cosh(w_1+v)+\cosh(w_2+v_0)}\right\},$$

where the integral expression is reduced to a symmetrized form. After simple calculations we find

$$J(\theta,\theta_0)=\frac{1}{4\pi^2}\int_0^\infty\int_0^\infty \frac{dw_1 dw_2}{\sqrt{\cosh(w_1+v)-\cosh v}\sqrt{\cosh(w_2+v_0)-\cosh v_0}}$$

$$\times(-6)\frac{\sinh(w_2+v_0)\sinh(w_1+v)}{\cosh(w_1+v)+\cosh(w_2+v_0)}$$

$$=-\frac{6}{\pi^2}\int_0^\infty\int_0^\infty\frac{dp dq}{(p+q+\cosh v+\cosh v_0)^4}.$$

The double integral can be computed explicitly in the polar coordinates (r,φ) on the plane (p,q). As a result, we find the desired expression for $J(\theta,\theta_0)$.

In a similar manner, we arrive at the new representation for $P_\psi(\theta,\theta_0)$ in (5.59)

$$P_\psi(\theta,\theta_0)$$

$$=-\frac{1}{2\pi^2}\int_0^\infty dt\frac{t\psi_s(it+1/2)}{\sinh(\pi t)}\left\{\int_0^\infty\int_0^\infty\frac{dw_1 dw_2}{\sqrt{\cosh(w_1+v)-\cosh v}\sqrt{\cosh(w_2+v_0)-\cosh v_0}}\right.$$

$$\left.\times\frac{\partial}{\partial v}\frac{\sin[t(w_1+v)]}{\sinh(w_1+v)}\frac{\partial}{\partial v_0}\frac{\sin[t(w_2+v_0)]}{\sinh(w_2+v_0)}\right\},$$

where $\psi_s(t)=[\psi(t)+\psi(-t)]/2$.

5.11.3 Appendix C. Derivation of the constant C_0

To compute the constant C_0 we use (5.73), written as

$$\Phi(\alpha,\theta_1) = -\frac{\sin\alpha}{\sin\zeta}\left.\frac{\partial\widetilde{\Phi}(\alpha,\theta)}{\partial\theta}\right|_{\theta=\theta_1} - \frac{\psi_{\rm i}(\alpha)}{\sin\zeta} \tag{5.118}$$

$$= -\frac{\sin\alpha}{\sin\zeta}\sqrt{2\pi}\int_{{\rm i}\mathbb{R}} R(\nu)\,\frac{{\rm e}^{{\rm i}\,\nu\alpha}+{\rm e}^{-{\rm i}\,\nu\alpha}}{2{\rm i}}\,{\rm d}\nu - \frac{\psi_{\rm i}(\alpha)}{\sin\zeta}$$

with

$$\psi_{\rm i}(\alpha) = -\sin\alpha\left.\frac{\partial\widetilde{\Phi}(\alpha,\theta)}{\partial\theta}\right|_{\theta=\theta_1} - \Phi_{\rm i}(\alpha,\theta_1).$$

Our nearest goal is to perform an analytic continuation of the right-hand side in the expression

$$\left.\frac{\partial\widetilde{\Phi}(\alpha,\theta)}{\partial\theta}\right|_{\theta=\theta_1} = \sqrt{2\pi}\int_{{\rm i}\mathbb{R}} R(\nu)\,\frac{{\rm e}^{{\rm i}\,\nu\alpha}+{\rm e}^{-{\rm i}\,\nu\alpha}}{2{\rm i}}\,{\rm d}\nu$$

from the strip $|\mathrm{Re}\,\alpha| < \pi$ onto the wider strip $|\mathrm{Re}\,\alpha| < \pi + \delta_0$ for some positive δ_0 so that $\alpha_\zeta(\theta_1) = \pi + \zeta$ belongs to that strip. Such a continuation can be obtained by a procedure proposed in [5.27] and is given by the formula

$$\left.\frac{\partial\widetilde{\Phi}(\alpha,\theta)}{\partial\theta}\right|_{\theta=\theta_1} = \sqrt{2\pi}\,\frac{\sin\zeta\,[V(\alpha)+V_1(\alpha)+V_2(\alpha)] - T_{\rm i}(\alpha)}{\sin\alpha\, +\, \sin\zeta}, \tag{5.119}$$

with

$$V(\alpha) = \frac{1}{\rm i}\int_{{\rm i}\mathbb{R}} [w(\nu)+{\rm i}\tan(\nu b)]\; R(\nu)\,{\rm e}^{{\rm i}\nu\alpha}\,{\rm d}\nu\;,\;\; b > 2\pi,$$

$$V_1(\alpha) = \frac{1}{\rm i}\int_{{\rm i}\mathbb{R}} \left[\mathrm{sgn}({\rm i}\nu) - {\rm i}\tan(\nu b)\right]\; R(\nu)\,{\rm e}^{{\rm i}\nu\alpha}\,{\rm d}\nu,$$

$$V_2(\alpha) = \frac{1}{\rm i}\int_{{\rm i}\mathbb{R}} \left[1 - \mathrm{sgn}({\rm i}\nu)\right]\; R(\nu)\,{\rm e}^{{\rm i}\nu\alpha}\,{\rm d}\nu,$$

$$T_{\rm i}(\alpha) = \frac{1}{\rm i}\int_{{\rm i}\mathbb{R}} G_{\rm i}(\nu)\,{\rm e}^{{\rm i}\nu\alpha}{\rm d}\nu.$$

V_1, V_2, and $T_{\rm i}$ are regular at $\alpha_\zeta(\theta_1)$.

The formula for $V(\alpha)$ can be transformed as

$$V(\alpha) = \frac{1}{2\pi \mathrm{i}} \int\limits_{\mathrm{i}\mathbb{R}} H_b(\alpha - \tau)\, \psi(\tau)\, \mathrm{d}\tau,$$

where

$$H_b(\alpha) = \frac{1}{\mathrm{i}} \int\limits_{\mathrm{i}\mathbb{R}} [w(\tau) + \mathrm{i}\tan(b\nu)]\, \mathrm{e}^{\mathrm{i}\nu\alpha}\, \mathrm{d}\alpha\,, \qquad H_b(\alpha) = -H_b(-\alpha),$$

$H_b(\alpha)$ is regular in the upper half-plane of the variable α and continuous on the real axis except the points $2m\theta_1$, $2mb$, $m = \pm 1, \pm 2, \ldots$. The estimate is valid $|H_b(\alpha)| \leq C \exp[-\pi\, |\mathrm{Im}\,\alpha| / (4\theta_1)]$, $\mathrm{Im}\,\alpha \to \infty$.

Finally, we exploit the analytic continuation (5.119) in formula (5.118) and also take into account the local behavior (5.101) as $\theta = \theta_1$, thus obtaining

$$C_0 = -\sqrt{2\pi}\tan\zeta \lim_{\alpha \to \alpha_\zeta(\theta_1)} \left([\alpha - \alpha_\zeta(\theta)]^{-\mu} \left[(V(\alpha) + V_1(\alpha) + V_2(\alpha)) - T_{\mathrm{i}}(\alpha)/\sin\zeta \right] \right).$$

The previous procedure delivers an alternative way of analytic continuation also for the Sommerfeld transformants.

CHAPTER 6

Electromagnetic wave scattering by a circular impedance cone

This chapter is devoted to the study of electromagnetic scattering of a plane wave by a circular cone with impedance boundary conditions on its surface. The technique developed in the previous chapter is extended and applied to the electromagnetic diffraction problem with the aim to compute the far field.

By means of the Kontorovich–Lebedev integral representations for the Debye potentials and a "partial" separation of variables, the problem is reduced to coupled functional-difference equations for the relevant spectral functions. For a circular cone the functional-difference equations are then further reduced to integral equations, which are shown to be Fredholm-type equations. We then solve numerically the integral equations. The radar cross section in the domain which is free from both the reflected and the surface waves has been computed numerically. Certain useful further integral representations for the solution of "Watson–Bessel" and Sommerfeld types are developed, which give, as in acoustic case, a theoretical basis for subsequent calculation of the far-field (high frequency) asymptotics for the diffracted field. We also discuss the asymptotic expressions for the surface waves propagating from the conical vertex to infinity.

While the problem we are considering is physically sound, from the mathematical viewpoint it is particularly attractive as containing various earlier studied models, both with ideal and nonideal boundaries, as particular or limiting cases. At the same time one has to deal with some novel effects of coupling, which provides in total a good test ground for advancing appropriate technical tools, both analytical and numerical, as we are aiming in this chapter.

6.1 Formulation and reduction to the problem for the Debye potentials

Specifically, we consider a conical surface, in particular but not necessarily with a circular cross section and with an exterior Ω ($\theta < \theta_1$ for the circular cone's case) (see Fig. 6.1). We assume that a plane wave is incident from the exterior and illuminates completely[1] the conical surface S. Of interest is the scattered field in the exterior.

[1] In this case the diffraction phenomena can be analyzed in a simpler way although the case of arbitrary incidence can also be studied.

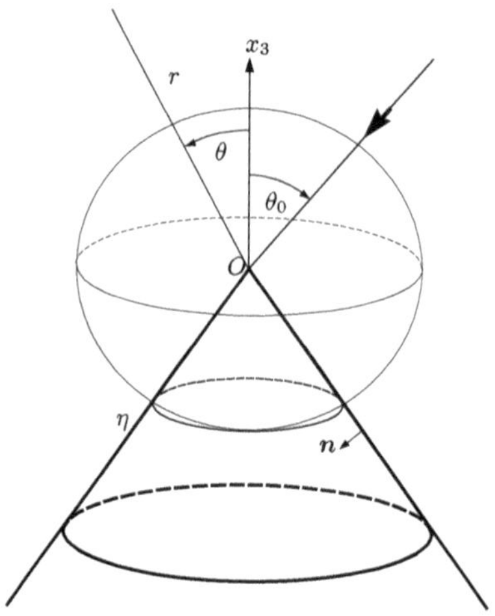

Figure 6.1 Diffraction by a right circular impedance cone.

The time-harmonic wave field (with suppressed factor e^{-ikct} where k and c are the wavenumber and the speed of light in the exterior domain Ω, respectively) is characterized by vector functions $\boldsymbol{E}(\boldsymbol{x})$ (electric field) and $\boldsymbol{H}(\boldsymbol{x})$ (magnetic field times Z_0, the intrinsic impedance of the exterior domain Ω), solving time-harmonic Maxwell's equations

$$\mathrm{i}k\,\boldsymbol{H} = \mathrm{curl}\,\boldsymbol{E}, \quad \mathrm{i}k\,\boldsymbol{E} = -\mathrm{curl}\,\boldsymbol{H}. \tag{6.1}$$

The boundary conditions for the total wave field read

$$\boldsymbol{E} - (\boldsymbol{E}\cdot\boldsymbol{n})\boldsymbol{n} = -\eta\,\boldsymbol{n}\times\boldsymbol{H}, \tag{6.2}$$

where $\boldsymbol{n}$ is the unit normal vector to the conical surface S (in the present problem we select $\boldsymbol{n}$ to point into the cone, that is, to the interior.)

Assume that the dimensionless parameter η (with respect to Z_0 normalized impedance) describing the electromagnetic properties of the conical surface is such that $\mathrm{Re}\,\eta \geq 0$ with $\mathrm{Re}\,\eta = 0$ corresponding to a nonabsorbing ("reactive") surface and $\mathrm{Re}\,\eta > 0$ relating to an energy-absorbing conducting surface. So we assume η to be generally complex with $\mathrm{Re}\,\eta \geq 0$, which is reasonable from physical and justified from mathematical points of view, as next.

The usual spherical coordinates (r, θ, φ) are related to the Cartesian coordinates (x_1, x_2, x_3) via

$$x_1 = r\,\sin\theta\,\cos\varphi, \quad x_2 = r\,\sin\theta\,\sin\varphi, \quad x_3 = r\,\cos\theta. \tag{6.3}$$

The origin O of both coordinate systems coincides with the cone's vertex and the x_3 axis points along the cone's axis outside the cone.

A normalized plane wave $\boldsymbol{E}^0$, $\boldsymbol{H}^0$ is incident from the exterior of the cone and is characterized by the angles of incidence θ_0 $(0 \leq \theta_0 < \theta_1)$, φ_0,[2] and the angle of polarization β:

$$\begin{aligned} \boldsymbol{E}^0 &= (\boldsymbol{e}_{\theta_0}\sin\beta + \boldsymbol{e}_{\varphi_0}\cos\beta)\,e^{-ikr\cos\widehat{\theta}(\omega,\omega_0)}, \\ \boldsymbol{H}^0 &= (\boldsymbol{e}_{\theta_0}\cos\beta - \boldsymbol{e}_{\varphi_0}\sin\beta)\,e^{-ikr\cos\widehat{\theta}(\omega,\omega_0)}. \end{aligned} \tag{6.4}$$

[2]Owing to the symmetry of the cone, the angle of incidence φ_0 can in principle be set to zero, that is, $\varphi_0 = 0$; however, we preserve it in what follows for generality.

Here $\boldsymbol{\omega}_0 = (\theta_0, \varphi_0)$ is the unit vector in the direction opposite to the direction of incidence or, equivalently, ω_0 is a point (the endpoint of vector $\boldsymbol{\omega}_0$) on the unit sphere S^2 centered at the vertex O corresponding to the direction from which the plane wave is incident.[3] Unit vectors $\boldsymbol{e}_{\theta_0}$ and $\boldsymbol{e}_{\varphi_0}$ are tangent to S^2 at ω_0 and point in the direction of increase of θ and φ, respectively. Similarly, the vector $\boldsymbol{\omega} = (\theta, \varphi) \in S^2$ corresponds to the direction of observation. As in the previous chapter the argument $\widehat{\theta}(\omega, \omega_0)$ in

$$\cos \widehat{\theta}(\omega, \omega_0) = \cos\theta \cos\theta_0 + \sin\theta \sin\theta_0 \cos(\varphi - \varphi_0)$$

is the geodesic distance on S^2 between the points ω and ω_0 (or, what is the same, the angle between the vectors $\boldsymbol{\omega}$ and $\boldsymbol{\omega}_0$). In the case of other types of sources rather than the incident plane wave (6.4), Maxwell's equations (6.1) may also have inhomogeneous terms on the right-hand sides.

The total (incident plus scattered) field has to satisfy Maxwell's equations (6.1) in the exterior domain and the boundary conditions (6.2) on the conical surface. Additionally, Meixner-type conditions of absence of sources at the cone's tip have to be imposed. Traditionally, we require for this the "finiteness of energy" near the cone's tip, that is, the convergence of the integral

$$\int_{V_\delta} \left(|\boldsymbol{E}|^2 + |\boldsymbol{H}|^2\right) \mathrm{d}V < \infty, \tag{6.5}$$

where V_δ is the part of a ball of a small positive radius δ centered at the cone's tip and belonging to Ω. It is also convenient to impose a "vanishing flux" version of Meixner's conditions:

$$\int_{S_\delta} \boldsymbol{e}_r \cdot \left(\boldsymbol{E} \times \overline{\boldsymbol{H}} + \overline{\boldsymbol{E}} \times \boldsymbol{H}\right) \mathrm{d}S \to 0, \quad \text{when } \delta \to 0, \tag{6.6}$$

where S_δ is the part within Ω of the sphere of a small radius δ centered at the tip, $\boldsymbol{e}_r$ is the unit vector in the radial direction pointing outside the sphere, and the bar denotes the complex conjugate.

Finally, the "scattered" field should satisfy appropriate outgoing radiation-type conditions when $r \to \infty$ (or some form of "limiting-absorption" conditions). In the case of "finite sources" rather than of an incident plane wave, traditional Sommerfeld-type[4] radiation conditions are required to hold [6.1], (I.20). We shall use their weaker integral form

$$\int_{S_R} |\boldsymbol{E} + \boldsymbol{e}_r \times \boldsymbol{H}|^2 \,\mathrm{d}S \to 0, \quad R \to \infty, \tag{6.7}$$

$$\int_{S_R} |\boldsymbol{H} - \boldsymbol{e}_r \times \boldsymbol{E}|^2 \,\mathrm{d}S \to 0, \quad R \to \infty, \tag{6.8}$$

where S_R is the part, within Ω, of the sphere of a large radius R centered at the tip.[5]

---

[3]For the axisymmetric incidence one has $\theta_0 = 0$.

[4]In the electromagnetic theory these are often called Silver-Müller radiation conditions.

[5]For a discussion on the outgoing surface waves propagating at infinity as $\operatorname{Re}\eta = 0$ see Section 6.9.

Then the problem can be shown to be well posed and in particular the uniqueness of its solution can be established, initially for absorbing boundaries (i.e., $\operatorname{Re}\eta > 0$, k real positive). Namely, the following uniqueness statement holds.

Proposition 6.1. *Let* $\operatorname{Re}\eta > 0$, $\operatorname{Im}k = 0$, $\operatorname{Re}k > 0$, *and let a classical solution satisfy the homogeneous problem (6.1), (6.2), (6.5), (6.6), (6.7), and (6.8). Then it is identically zero.*

The proof is standard and details could be found in the work [6.2], Appendix A. The plane-wave incidence can be "defined" as the limit of a point source problem when the source moves away to infinity (see [6.3–6.7]). Proposition 6.1 ensures, in particular, that the classical solution of the electromagnetic diffraction problem for the point-source illumination is unique. Then, after the mentioned limiting procedure, the solution for the plane-wave incidence is also expected to be unique. From the mathematical viewpoint the limiting procedure itself requires, of course, a justification. Having the latter at hand, one can ensure that the scattering problem for the illumination by a plane wave for real positive k is well posed. In some specific cases the uniqueness for the case of the plane-wave incidence can be also shown (see, e.g., [6.2]).

The solution for a range of values of η and k (including those nonabsorbing) as well as for other directions of incidence can be "defined" via analytic continuation, including for positive real k (i.e., by a "limiting absorption"-type procedure). The subsequent construction using the Kontorovich–Lebedev transform, further supplemented by "Bessel–Watson's integral"-type transformations of the Kontorovich–Lebedev integral, see also the previous chapter, ensures the existence of such an analytic continuation. Therefore, the above representations are valid for the real wavenumbers and give the unique solution understood in the sense of the limiting absorption.

6.1.1 The far-field pattern

One can see that the nature of the boundary condition (6.2) is such that, away from the conical tip, the (high frequency) incident wave partly reflects with appropriate reflection coefficient back into the exterior domain in accordance with the reflection law of Geometrical Optics (GO).

Apart from this, in a usual fashion, a wave with a spherical front diffracted by the vertex of the cone is generated and is of particular interest. Developing and implementing a method for evaluating the corresponding diffraction coefficients is one of the main goals of the present chapter.

A detailed study of several other issues, including the derivation of the other components in the far-field (high-frequency) asymptotics, will be considered in the second half of this chapter. This includes important problems of dealing with the effects of the waves scattered from the surface into Ω[6] as well as of the surface waves of the Rayleigh type initiated in the vicinity of the conical surface. An experience gained from studying the scalar problem of diffraction by an impedance cone indicates that the surface waves do exist; see Chapter 5 and [6.8]. Their possible type (electric or magnetic) is anticipated to be specified by the sign of the imaginary part of the surface impedance η. Another important issue in the diffraction by conical

[6]The reflected waves components are similarly treated as in the scalar acoustic case, so they are only briefly discussed herein.

surfaces is dealing with the effect of the singular directions, see [6.5] and Chapter 5, which correspond to the surface of termination of the rays reflected from the cone. In these directions, the diffraction coefficients become singular as typical for a special transitional boundary-layer asymptotic behavior. A uniform asymptotics of the reflected and scattered fields should be constructed in this case; see [6.9].

6.1.2 The Debye potentials

As for example in [6.4], [6.10], and [6.7], the present problem is reformulated in terms of the Debye potentials, u and v in the exterior domain:

$$\begin{aligned} \boldsymbol{E} &= \operatorname{curl}\operatorname{curl}\,(r\boldsymbol{e}_r u) + ik\operatorname{curl}\,(r\boldsymbol{e}_r v)\,, \\ \boldsymbol{H} &= \operatorname{curl}\operatorname{curl}\,(r\boldsymbol{e}_r v) - ik\operatorname{curl}\,(r\boldsymbol{e}_r u) \end{aligned} \tag{6.9}$$

in Ω.

The Debye potentials are required to satisfy the Helmholtz equation in Ω:

$$\Delta u + k^2 u = 0\,, \quad \Delta v + k^2 v = 0, \tag{6.10}$$

automatically ensuring thereby that the wave fields $(\boldsymbol{E}, \boldsymbol{H})$ satisfy Maxwell's equations (6.1) in Ω.

In this chapter, the following convention is adopted for the "scattered" potentials u^s and v^s:

$$u = u_0 + u^s, \quad v = v_0 + v^s. \tag{6.11}$$

The "incident" Debye potentials u_0 and v_0 in Ω are the same as in [6.7], or indeed as in the case of a perfectly conducting cone [6.4]. We give their expressions later in terms of the so-called spectral functions.

As usual, additional conditions have to be imposed to ensure that the "scattered" parts satisfy appropriate radiation (or limiting-absorption) conditions and Meixner's conditions.

Sufficient conditions for the wave field to satisfy the requirement of local integrability (6.5) are the estimates

$$|\boldsymbol{E}| \le Cr^m, \quad |\boldsymbol{H}| \le Cr^m, \; m > -3/2, \tag{6.12}$$

where C is a positive constant, for small enough r uniformly with respect to angular variables. Let us then assume that

$$|u| \le Cr^p, \quad |v| \le Cr^p, \tag{6.13}$$

uniformly with respect to (θ, φ), with (6.13) surviving formal differentiation in r as well as $\omega = (\theta, \varphi)$.[7] Then we have that, as $r \to 0$, for example,

$$E_r \sim \partial_r^2(ru) + k^2 ru \sim \; p(p+1)Cr^{p-1}$$

[7]More precisely, we assume that, for example, $u = U(\omega)r^p + o(r^p)$ with U smooth away from the interface l defined as follows and $\left|\partial_r^\alpha \partial_\omega^\beta u\right| \le C_{\alpha\beta} r^{l-\alpha}$ uniformly in M.

and conclude from (6.12) that (see also the discussion in [6.7])

$$\text{either } p > -1/2 \text{ or } p = -1. \tag{6.14}$$

Now we turn to the radiation conditions for the scattered potentials u^s and v^s.[8] For simplicity we assume that the incident plane wave illuminates completely the conical surface.

Consider a unit sphere centered at the vertex of the cone. The conical surface S divides the sphere into two parts, M and M_{i} corresponding to Ω and to the cone's interior, respectively, with the boundary l. We introduce the length of the geodesics (broken for the reflected wave, see also the previous chapter, [6.7], and [6.5] for definitions of $\widehat{\theta}'(\omega, \omega_0)$) on the unit sphere

$$\widehat{\theta}'(\omega, \omega_0) = \min_{\omega' \in l} \left(\widehat{\theta}(\omega', \omega_0) + \widehat{\theta}(\omega, \omega') \right), \tag{6.15}$$

where $\widehat{\theta}(\omega', \omega_0)$ is the (geodesic) distance between the points ω_0 and ω' on the unit sphere. The domain M can in turn be subdivided into two parts $M' = \{\theta, \varphi : \ \widehat{\theta}'(\omega, \omega_0) > \pi\}$ (oasis), the angular domain where only the spherical wave from the vertex propagates to infinity, and $M'' = \{\theta, \varphi : \ \widehat{\theta}'(\omega, \omega_0) < \pi\}$, the subdomain containing in addition the rays reflected from the cone's surface and the surface waves. The common boundary σ of M' and M'' is the curve of the so-called singular directions (corresponding to the directions where the reflected wave terminates; see [6.5] and Chapter 5).

In the subdomain Ω' of Ω corresponding to M', that is, to those directions satisfying[9]

$$\widehat{\theta}'(\omega, \omega_0) > \pi, \tag{6.16}$$

the conditions at infinity are required to have the form

$$u^{\text{s}}(r, \omega, \omega_0) = D_u(\omega, \omega_0) \frac{\mathrm{e}^{ikr}}{-ikr} \left[1 + \mathrm{O}\left(\frac{1}{kr} \right) \right], \quad kr \to \infty, \tag{6.17}$$

$$v^{\text{s}}(r, \omega, \omega_0) = D_v(\omega, \omega_0) \frac{\mathrm{e}^{ikr}}{-ikr} \left[1 + \mathrm{O}\left(\frac{1}{kr} \right) \right], \quad kr \to \infty, \tag{6.18}$$

where $D_{u,v}(\omega, \omega_0)$ are smooth functions of ω satisfying (6.16), away from the singular directions σ, where they can blow up.

In the subdomain Ω'' corresponding to M'', that is, to

$$\widehat{\theta}(\omega|_S, \omega_0) < \widehat{\theta}'(\omega, \omega_0) < \pi, \tag{6.19}$$

the potentials should, together with (6.17), also contain the rays reflected from the cone and the surface waves. The singular directions form the surface $\partial\Omega'$ in Ω corresponding to σ, that is,

$$\widehat{\theta}'(\omega, \omega_0) = \pi. \tag{6.20}$$

Finally, the potentials u^{s} and v^{s} have to satisfy appropriate boundary conditions on S to ensure that the conditions (6.2) are satisfied. Those are described next.

---

[8] By the "scattered" fields we understand "total less incident" fields in the exterior domain Ω.

[9] In some of our publications and in Chapter 5 this domain is also called oasis.

6.1.3 Boundary conditions for the Debye potentials

In the case of impedance boundary conditions [6.7], it is convenient to introduce an orthogonal right-handed vector basis $\boldsymbol{e}_r, \boldsymbol{n}, \boldsymbol{s}$, where $\boldsymbol{n}$ is the unit vector normal to S pointing into Ω_{i}, and $\boldsymbol{s}$ is the unit vector tangent to S and orthogonal to $\boldsymbol{e}_r$ (hence in the case of a circular cone $\boldsymbol{n} = \boldsymbol{e}_\theta$ and $\boldsymbol{s} = \boldsymbol{e}_\varphi$). Denoting by $\partial/\partial n$ and $\partial/\partial s$ the derivatives along $\boldsymbol{n}$ and $\boldsymbol{s}$, respectively, introduce also the "spherical" normal and tangent derivatives $\partial/\partial m$ and $\partial/\partial t$, respectively, by [10] $\partial/\partial m := r\partial/\partial n$, $\partial/\partial t := r\partial/\partial s$.

Then, as a result of substitution of (6.9) into (6.2), we obtain via routine manipulation

$$\eta\left[-\mathrm{i}k\,\partial_m u - \partial_t\left(r^{-1}\partial_r(r\,v)\right)\right] = \left(\partial_r^2(r\,u) + k^2 r\,u\right)\big|_S, \tag{6.21}$$

$$\eta\left[\partial_r^2(r\,v) + k^2 r\,v\right] = -\mathrm{i}k\,\partial_m v|_S + \partial_t\left(r^{-1}\partial_r(r\,u)\right)\big|_S. \tag{6.22}$$

We can assert the following.

Proposition 6.2. *Let the potentials u and v be classical solutions (i.e., from $C^2\left(\overline{\Omega}\setminus O\right)$) of (6.10), (6.21)–(6.22), and satisfy (6.13), (6.14), and (6.17), (6.18) in the domain (6.16). Then the electromagnetic wave field (6.9) is the classical solution of Maxwell's equations (6.1) satisfying (6.2), (6.12), and having the leading terms of the asymptotics*

$$\begin{Bmatrix} \boldsymbol{E} \\ \boldsymbol{H} \end{Bmatrix} \cong -2\pi\,\frac{\mathrm{e}^{\mathrm{i}kr}}{kr}\begin{Bmatrix} \boldsymbol{\mathcal{E}}(\omega) \\ \boldsymbol{\mathcal{H}}(\omega) \end{Bmatrix}, \qquad kr \to \infty, \tag{6.23}$$

in the subdomain Ω' of Ω corresponding to (6.16), with some vector functions $\boldsymbol{\mathcal{E}}(\omega)$ and $\boldsymbol{\mathcal{H}}(\omega)$ smooth in M'.

The Debye potentials u and v can be related to the so-called spectral functions g_u and g_v, for example, via the Kontorovich–Lebedev transform (see, e.g., [6.7]), as we review next.

6.2 Kontorovich–Lebedev (KL) integrals and spectral functions

6.2.1 KL integral representations

Following the pattern of [6.7] in the case of an impedance cone, we seek representations for the scattered parts of the Debye potentials u^{s} and v^{s} in the form of the Kontorovich–Lebedev (KL) integrals

$$(u^{\mathrm{s}}, v^{\mathrm{s}}) = \frac{-2}{k}\sqrt{\frac{2}{\pi}}\left[\int_{\mathrm{i}\mathbb{R}} \frac{\nu\sin\pi\nu}{\nu^2 - 1/4}\,\frac{\mathrm{K}_\nu(-\mathrm{i}kr)}{\sqrt{-\mathrm{i}kr}}\,g_{u,v}(\omega,\omega_0,\nu)\,\mathrm{d}\nu \right.$$
$$\left. + \mathrm{i}\pi\,\frac{\mathrm{K}_{1/2}(-\mathrm{i}kr)}{\sqrt{-\mathrm{i}kr}}\,g_{u,v}(\omega,\omega_0,1/2)\right]. \tag{6.24}$$

[10]Those would correspond to introducing the "spherical" coordinates $\boldsymbol{\omega} = (m,t)$ near the boundary l on the associated spherical domain M: m being the geodesic distance to l along the sphere and t being the arclength of the "projection" point on l. For the circular cone $\partial/\partial m = \partial/\partial\theta$, $\partial/\partial t = \sin^{-1}\theta\,\partial/\partial\varphi$.

It turns out that the integrals converge exponentially provided

$$\widehat{\theta}'(\omega, \omega_0) > \pi/2 + |\arg(-ik)|, \tag{6.25}$$

$$|\arg(-ik)| \le \pi/2, \tag{6.26}$$

$\omega = (\theta, \varphi)$, $\omega_0 = (\theta_0, \varphi_0)$, and provided the sought-for spectral functions $g_{u,v}$ satisfy the growth conditions to be imposed below.

In (6.24) $\mathrm{K}_\nu(z)$ is the modified Bessel function (the Macdonald function, see the previous chapter)

$$\mathrm{K}_\nu(z) = -(\mathrm{i}\pi/2)\exp(-\mathrm{i}\pi\nu/2)\,\mathrm{H}_\nu^{(2)}(-\mathrm{i}z),$$

where $\mathrm{H}_\nu^{(2)}$ is the Hankel function; in particular there is

$$\mathrm{K}_{1/2}(-\mathrm{i}kr) = \sqrt{\pi/(-2\mathrm{i}kr)}\,\exp(\mathrm{i}kr).$$

One of the aims is to reformulate the problem for the Debye potentials as a problem for the spectral functions, thereby separating in a sense the radial variable. To ensure the convergence of the integrals in (6.24) the wavenumber k has first to be taken to have a sufficiently large imaginary part and the ranges of ω may have to be restricted appropriately too. The condition (6.25) of convergence in Ω is the same as in the previous chapter. Varying the argument of k, the integrals can be made convergent everywhere (assuming ω_0 is such that the cone is fully illuminated). Then the solutions are expected to be analytically continued with respect to k, including in particular for real values of k.

Remark: An alternative is to seek the representation for u and v in terms of "Watson's" integral over a deformed contour C_ϕ; see next and the previous chapter, with the "standard" Bessel functions $\mathrm{J}_\nu(z)$ instead of the modified Bessel functions K_ν, with both approaches apparently equivalent. Both approaches have their advantages and limitations, and may supplement each other for various applications, such as for evaluating of the diffraction coefficients in various domains, as we discuss herein and in the previous chapter exploiting Watson–Bessel integrals.

For the potentials of the incident plane wave (6.4), we have (see, e.g., [6.7])

$$(u^0, v^0) = \frac{-2}{k}\sqrt{\frac{2}{\pi}}\left[\int_{\mathrm{i}\mathbb{R}} \frac{\nu\sin\pi\nu}{\nu^2 - 1/4}\,\frac{\mathrm{K}_\nu(-\mathrm{i}kr)}{\sqrt{-\mathrm{i}kr}}\, g^0_{u,v}(\omega, \omega_0, \nu)\,d\nu \right. \\ \left. + \mathrm{i}\pi\,\frac{\mathrm{K}_{1/2}(-\mathrm{i}kr)}{\sqrt{-\mathrm{i}kr}}\, g^0_{u,v}(\omega, \omega_0, 1/2)\right], \tag{6.27}$$

where

$$g^0_u(\omega, \omega_0, \nu) = -L_u\, u^{\mathrm{i}}_\nu(\omega, \omega_0), \quad g^0_v(\omega, \omega_0, \nu) = -L_v\, u^{\mathrm{i}}_\nu(\omega, \omega_0) \tag{6.28}$$

with

$$u^{\mathrm{i}}_{\nu}(\omega, \omega_0) = -\frac{\mathrm{P}_{\nu-1/2}(-\cos\widehat{\theta}(\omega, \omega_0))}{4\cos(\pi\nu)}, \tag{6.29}$$

$$L_u := \sin\beta\frac{\partial}{\partial\theta_0} + \cos\beta\frac{\partial}{\sin\theta_0\partial\varphi_0}, \tag{6.30}$$

$$L_v := \cos\beta\frac{\partial}{\partial\theta_0} - \sin\beta\frac{\partial}{\sin\theta_0\partial\varphi_0}. \tag{6.31}$$

The integrals in (6.27) converge under the conditions

$$\widehat{\theta}(\omega, \omega_0) > \pi/2 + |\arg(-ik)|, \quad |\arg(-ik)| \leq \pi/2. \tag{6.32}$$

As already noted, the alternative approach based on the Bessel functions is free from this limitation, and the corresponding ("Watson's") integrals will converge for real k. It also provides the aforementioned analytic continuation. On the other hand, the Sommerefeld-type representations are very efficient for study of the far-field asymptotics.

The spectral functions g_u and g_v are to be determined, which are meromorphic functions of the complex variable ν and even in ν, that is

$$g_{u,v}(\omega, \omega_0, -\nu) = g_{u,v}(\omega, \omega_0, \nu).$$

Our main aim in the present section is to formulate the problem for the spectral functions in an appropriate functional class, so it ensures that the potentials constructed by (6.24) correspond to the classical solution of the problem for the potentials as stated already.

6.2.2 Properties of the spectral functions

Guided by the precedent in diffraction by both perfectly conducting and acoustic impedance cones, the spectral functions are first required to satisfy the following growth conditions when $|\mathrm{Im}\,\nu| \to \infty$:

$$\left|g_{u,v}(\omega, \omega_0, \nu)\right| \leq C_0\frac{1}{|\cos(\pi\nu)|}, \quad \text{in } M', \tag{6.33}$$

$$\left|g_{u,v}(\omega, \omega_0, \nu)\right| \leq C_0\frac{|\nu|^{1/2}}{\left|\cos\left(\nu\widehat{\theta}'(\omega, \omega_0)\right)\right|}, \quad \text{in } M'', \tag{6.34}$$

with the same functions $\widehat{\theta}'$ and $\widehat{\theta}$ as before. In (6.33) and (6.34), the constant C_0 is independent of ω, ω_0, and ν (hence, in particular, it is uniform with respect to real part of ν). The spectral functions $g_{u,v}$ are also required to be regular in a vertical strip containing the imaginary axis, that is, in

$$\Pi_{3/2+m} = \{\nu : \ |\mathrm{Re}\,\nu| \leq 3/2 + m\}, \quad m > -3/2. \tag{6.35}$$

Another natural requirement, [6.7], is for the boundary values of the spectral functions $g_{u,v}\big|_S$, to be regular in the strip

$$\Pi_{1+\delta} = \{\nu : \ |\mathrm{Re}\,\nu| \leq 1 + \delta\}$$

with some small positive δ.

It can be demonstrated (see also discussion in [6.7], Remarks 5 and 6) that the previously given regularity properties imply the following regularity conditions for certain boundary-value combinations of the spectral functions:

$$\left(\nu+\frac{1}{2}\right)\left[g_u(\nu)-\eta\frac{\partial}{\partial t}\frac{g_v(\nu)}{\nu+1/2}\right] \text{ regular in } 0<-\operatorname{Re}\nu<1+\delta, \tag{6.36}$$

$$\left(\nu-\frac{1}{2}\right)\left[g_u(\nu)+\eta\frac{\partial}{\partial t}\frac{g_v(\nu)}{\nu-1/2}\right] \text{ regular in } 0<\operatorname{Re}\nu<1+\delta, \tag{6.37}$$

$$\left(\nu+\frac{1}{2}\right)\left[g_v(\nu)+\eta^{-1}\frac{\partial}{\partial t}\frac{g_u|_l(\nu)}{\nu+1/2}\right] \text{ regular in } 0<-\operatorname{Re}\nu<1+\delta, \tag{6.38}$$

$$\left(\nu-\frac{1}{2}\right)\left[g_v(\nu)-\eta^{-1}\frac{\partial}{\partial t}\frac{g_u|_l(\nu)}{\nu-1/2}\right] \text{ regular in } 0<\operatorname{Re}\nu<1+\delta. \tag{6.39}$$

Remark that the conditions above are also valid for g_u^0 and g_v^0 because the incident field is regular at the vertex; see [6.7]. We shall demonstrate in particular that imposing these conditions is precisely what is necessary for ensuring the boundary conditions for the potentials.

It is verified in a traditional manner (see, e.g., [6.11]) that, provided the spectral functions satisfy the equations

$$\left(\Delta_\omega+\nu^2-1/4\right)g_u=0,\quad \left(\Delta_\omega+\nu^2-1/4\right)g_v=0, \tag{6.40}$$

and, in particular, for $\nu=1/2$,

$$\Delta_\omega\, g_{u,v}(\omega,\omega_0,1/2)=0, \tag{6.41}$$

then u and v satisfy automatically the Helmholtz equation.

Further, in the case of an impedance cone [6.7] as well as in the case of a perfectly conducting cone [6.4] we require $g_{u,v}(\omega,\omega_0,1/2)$ to satisfy the Cauchy-Riemann-like equations

$$\begin{aligned}\frac{\partial(g_u+g_u^0)(\omega,\omega_0,1/2)}{\partial\theta}&=\frac{1}{\sin\theta}\frac{\partial(g_v+g_v^0)(\omega,\omega_0,1/2)}{\partial\varphi},\\ \frac{\partial(g_v+g_v^0)(\omega,\omega_0,1/2)}{\partial\theta}&=-\frac{1}{\sin\theta}\frac{\partial(g_u+g_u^0)(\omega,\omega_0,1/2)}{\partial\varphi}.\end{aligned} \tag{6.42}$$

These conditions ensure the absence of "sources" at the cone's tip to comply with Meixner's condition as well as of the incoming waves, other than the incident waves, to satisfy the radiation (limiting-absorption) conditions.

6.2.3 Boundary conditions for the spectral functions

Guided further by the analogy with the diffraction by an acoustic impedance cone we will argue that the spectral functions have also to satisfy the following boundary conditions. The latter are verified next by a direct substitution of the KL integral representations (6.24) into the conditions (6.21)–(6.22) for the Debye potentials and read

$$\frac{\mathrm{i}}{2}\left[\left(g_u+g_u^0\right)(\nu+1)-\left(g_u+g_u^0\right)(\nu-1)\right]\Big|_l-\eta\,\frac{\mathrm{i}\nu}{\nu^2-1/4}\,\frac{\partial(g_u+g_u^0)(\nu)}{\partial m}$$
$$=-\frac{\mathrm{i}\eta}{2}\frac{\partial}{\partial t}\left[\frac{(g_v+g_v^0)(\nu+1)}{\nu+1/2}+\frac{(g_v+g_v^0)(\nu-1)}{\nu-1/2}\right], \tag{6.43}$$

$$\frac{\mathrm{i}}{2}\left[(g_v+g_v^0)(\nu+1)-(g_v+g_v^0)(\nu-1)\right]-\eta^{-1}\,\frac{\mathrm{i}\nu}{\nu^2-1/4}\,\frac{\partial(g_v+g_v^0)(\nu)}{\partial m}$$
$$=\frac{\mathrm{i}\eta^{-1}}{2}\frac{\partial}{\partial t}\left[\frac{\left(g_u+g_u^0\right)(\nu+1)}{\nu+1/2}+\frac{\left(g_u+g_u^0\right)(\nu-1)}{\nu-1/2}\right]\Bigg|_l, \tag{6.44}$$

and (formally following from before)

$$\partial_m(g_u+g_u^0)(1/2)=\partial_t(g_v+g_v^0)(1/2),$$
$$\partial_m\left(g_v+g_v^0\right)(1/2)\Big|_l=-\,\partial_t\left(g_u+g_u^0\right)(1/2)\Big|_l, \tag{6.45}$$

The nonlocality in ν of the conditions (6.43)–(6.44) for the spectral functions is a consequence of a lack of separation of variables in the radial and the spherical components, in the related boundary conditions (6.21)–(6.22) for the Debye potentials u and v. For this reason, the separation of the (spherical) spectral functions from the (radial) modified Bessel (Macdonald) functions K_ν is achieved via employing appropriate functional relations for K_ν together with the above listed regularity properties of the spectral functions. This is what leads to the nonlocality in ν, as we clarify next.

We recall again that apart from the (nonlocal in ν) boundary conditions (6.43)–(6.44) the sought spectral functions have also to satisfy the regularity conditions (6.36)–(6.39). In fact, the latter imply some further regularity properties, which are used in what follows.

Now we are ready to formulate the main statement of this section.

Proposition 6.3. *Let the spectral functions $g_{u,v}(\omega,\omega_0,\nu)$ satisfy, in the classical sense, equations (6.40), (6.41) and conditions (6.43)–(6.44) in the class of meromorphic even functions of ν satisfying (6.33)–(6.39). Then the potentials (6.24) are classical solutions of the boundary-value problem for the potentials (including Meixner's condition (6.13), (6.14), and the conditions at infinity (6.17), (6.18)), provided the inequalities (6.16), (6.25)–(6.26) are valid.*

As mentioned before, verification of the equations for the potentials is standard, see [6.11]. Hence, we turn to the boundary and other conditions.

6.2.4 Verification of the boundary and other conditions

It is convenient to introduce the following notation

$$\begin{aligned} z &:= -ikr\,, \\ \Omega &:= \sqrt{z}\,u\,, \\ A'(u) &:= \partial_z A(u) = \sqrt{z}\,\left[\partial_r^2(ru) + k^2 ru\right] \\ &\quad = -ikz\left\{\Omega'' + \Omega'/z - \left[1 + 1/\left(4z^2\right)\right]\Omega\right\}, \\ B'(u) &:= \sqrt{z}\,\partial_t\,\left[r^{-1}\partial_r(ru)\right] = -ik\partial_t\left[\Omega/(2z) + \Omega'\right], \\ C'(\partial_m u) &:= \sqrt{z}\,\partial_m u. \end{aligned} \tag{6.46}$$

Then, using the modified Bessel equation and the functional relations for the Macdonald functions

$$\mathrm{K}_\nu''(z) + (1/z)\mathrm{K}_\nu'(z) - \left(1 + \nu^2/z^2\right)\mathrm{K}_\nu(z) = 0,$$

$$\mathrm{K}_\nu'(z) = -\frac{1}{2}\left[\mathrm{K}_{\nu-1}(z) + \mathrm{K}_{\nu+1}(z)\right], \quad \frac{\mathrm{K}_\nu(z)}{z} = \frac{\mathrm{K}_{\nu+1}(z) - \mathrm{K}_{\nu-1}(z)}{2\nu},$$

we obtain

$$\begin{aligned} A'(u_1^{\mathrm{s}}) &= A_0\,\frac{-ik}{2}\int\limits_{\mathrm{i}\mathbb{R}} 2\nu\sin\pi\nu\; g_{u_1}(\omega,\omega_0,\nu)\,\frac{\mathrm{K}_\nu(z)}{z}\,\mathrm{d}\nu \\ &= A_0\,\frac{-ik}{2}\int\limits_{\mathrm{i}\mathbb{R}} \sin\pi\nu\; g_u(\omega,\omega_0,\nu)\left[\mathrm{K}_{\nu+1}(z) - \mathrm{K}_{\nu-1}(z)\right]\mathrm{d}\nu, \end{aligned} \tag{6.47}$$

$$A_0 = -\frac{2}{k}\sqrt{\frac{2}{\pi}}.$$

In a similar manner,

$$\begin{aligned} B'(v) = A_0\,\frac{-ik}{2}\Bigg\{\partial_t\int\limits_{\mathrm{i}\mathbb{R}} \sin\pi\nu\; g_v(\omega,\omega_0,\nu)\left[-\frac{\mathrm{K}_{\nu+1}(z)}{\nu+1/2} - \frac{\mathrm{K}_{\nu-1}(z)}{\nu-1/2}\right]\mathrm{d}\nu \\ - 2\pi\mathrm{i}\,\mathrm{K}_{1/2}(z)\,\partial_t\,g_v(\omega,\omega_0,1/2)\Bigg\}, \end{aligned} \tag{6.48}$$

$$\begin{aligned} C'(\partial_m u) = -ik\,A_0\Bigg[\int\limits_{\mathrm{i}\mathbb{R}} \frac{\nu\sin\pi\nu}{\nu^2 - 1/4}\,\mathrm{K}_\nu(z)\,\frac{\partial\, g_u(\omega,\omega_0,\nu)}{\partial m}\,\mathrm{d}\nu \\ + \,\mathrm{i}\pi\,\mathrm{K}_{1/2}(z)\,\frac{\partial g_u}{\partial m}(\omega,\omega_0,1/2)\Bigg]. \end{aligned} \tag{6.49}$$

We then substitute the KL integrals (6.24) in the boundary condition (6.21) to obtain

$$\begin{aligned}
&-\eta \int_{\mathrm{i}\mathbb{R}} \frac{\nu \sin \pi\nu}{\nu^2 - 1/4} \frac{\partial}{\partial m} \left(g_u + g_u^0\right)(\nu)\Big|_l \mathrm{K}_\nu(z)\, \mathrm{d}\nu \\
&+ \int_{\mathrm{i}\mathbb{R}} \frac{\sin \pi\nu}{2} \left(g_u + g_u^0\right)(\nu)\big|_l \left[\mathrm{K}_{\nu+1}(z) - \mathrm{K}_{\nu-1}(z)\right] \mathrm{d}\nu \\
&-\eta \int_{\mathrm{i}\mathbb{R}} \frac{\sin \pi\nu}{2} \frac{\partial}{\partial t} (g_v + g_v^0)(\nu)\Big|_l \left[\frac{\mathrm{K}_{\nu+1}(z)}{\nu + 1/2} + \frac{\mathrm{K}_{\nu-1}(z)}{\nu - 1/2}\right] \mathrm{d}\nu \\
&- \mathrm{i}\pi\eta \left[\frac{\partial}{\partial m} (g_u + g_u^0)(1/2)\Big|_l + \frac{\partial}{\partial t} (g_v + g_v^0)(1/2)\Big|_l\right] \mathrm{K}_{1/2}(z) = 0.
\end{aligned} \tag{6.50}$$

Introducing the new variables of integration $\nu + 1 \to \nu$, $\nu - 1 \to \nu$ in the second and third integrals, we have

$$\begin{aligned}
&-\eta \int_{\mathrm{i}\mathbb{R}} \frac{\nu \sin \pi\nu}{\nu^2 - 1/4} \frac{\partial}{\partial m} (g_u + g_u^0)(\nu)\Big|_l \mathrm{K}_\nu(z)\, \mathrm{d}\nu \\
&+ \int_{\mathrm{i}\mathbb{R}+1} \frac{\sin \pi(\nu - 1)}{2} \left[\left(g_u + g_u^0\right)(\nu - 1)\big|_l - \eta \frac{\partial}{\partial t} \frac{(g_v + g_v^0)(\nu - 1)}{\nu - 1/2}\right] \mathrm{K}_\nu(z)\, \mathrm{d}\nu \\
&- \int_{\mathrm{i}\mathbb{R}-1} \frac{\sin \pi(\nu + 1)}{2} \left[\left(g_u + g_u^0\right)(\nu + 1)\big|_l + \eta \frac{\partial}{\partial t} \frac{(g_v + g_v^0)(\nu + 1)}{\nu + 1/2}\right] \mathrm{K}_\nu(z)\, \mathrm{d}\nu \\
&- \mathrm{i}\pi\eta \left[\frac{\partial}{\partial m} (g_u + g_u^0)(1/2) + \frac{\partial}{\partial t} (g_v + g_v^0)(1/2)\right] \mathrm{K}_{1/2}(z) = 0.
\end{aligned}$$

We then intend to deform the "shifted" integration contours $\mathrm{i}\mathbb{R} \pm 1$ back into the imaginary axis. The spectral functions $g_{u,v}^0$ are regular in the strip $|\mathrm{Re}\,\nu| < 1 + \delta$, for some small positive δ. When deforming the integration contours, the poles of the integrands are captured. Via the Cauchy residue theorem the poles at $\nu = \pm 1/2$ contribute to the nonintegral terms. However, there will be no contributions from the poles at points $\nu_0 \neq \pm 1/2$ since we required that (see (6.36) and (6.37))

$$\left(\tau - \frac{1}{2}\right) \left[g_u(\tau)|_l + \eta \frac{\partial}{\partial t} \frac{g_v(\tau)}{\tau - 1/2}\right] \tag{6.51}$$

be regular in the strip $0 < \mathrm{Re}\,\nu < 1 + \delta$ and, in the same manner,

$$\left(\tau + \frac{1}{2}\right) \left[g_u(\tau)|_l - \eta \frac{\partial}{\partial t} \frac{g_v(\tau)}{\tau + 1/2}\right] \tag{6.52}$$

be regular in the strip $0 < -\mathrm{Re}\,\nu < 1 + \delta$. Recall that the same conditions are valid for the incident wave spectral functions [6.7].

After the contour deformations the boundary condition (6.21) is now verified from (6.43) and (6.45):

$$
\begin{aligned}
&-\eta \int_{i\mathbb{R}} \frac{\nu \sin \pi \nu}{\nu^2 - 1/4} \frac{\partial}{\partial m} \left(g_u + g_u^0\right)(\nu)\, \mathrm{K}_\nu(z)\, d\nu \\
&- \int_{i\mathbb{R}} \frac{\sin \pi \nu}{2} \left[\left(g_u + g_u^0\right)(\nu - 1)\big|_l - \eta \frac{\partial}{\partial t} \frac{(g_v + g_v^0)(\nu - 1)}{\nu - 1/2} \right] \mathrm{K}_\nu(z)\, d\nu \\
&+ \int_{i\mathbb{R}} \frac{\sin \pi \nu}{2} \left[\left(g_u + g_u^0\right)(\nu + 1)\big|_l + \eta \frac{\partial}{\partial t} \frac{(g_v + g_v^0)(\nu + 1)}{\nu + 1/2} \right] \mathrm{K}_\nu(z)\, d\nu \\
&- i\pi \eta \left[\frac{\partial}{\partial m} \left(g_u + g_u^0\right)(1/2) - \frac{\partial}{\partial t} \left(g_v + g_v^0\right)(1/2) \right] \mathrm{K}_{1/2}(z) = 0 .
\end{aligned}
\tag{6.53}
$$

The other condition in (6.22) is demonstrated in the same manner, which is omitted here.

The estimates (6.13) (together with (6.14)) ensure the validity of Meixner's conditions (6.5) and (6.6) for the wave fields. These estimates are in turn verified by the use of the regularity properties of the spectral functions in the strip $\Pi_{m+3/2}, m > -3/2$, by reducing the KL integrals to those depending on the Bessel function and by appropriately deforming the integration contours (see also the previous chapter).

6.2.5 Diffraction coefficients

Now we turn to the behavior at infinity. The electromagnetic field components in the spherical coordinates can be written as

$$
\begin{aligned}
E_r &= k^2 r u + \partial_r^2 (r u), \quad E_\theta = r^{-1}\, \partial_r \partial_\theta (r u) + \frac{ik}{\sin\theta} \frac{\partial v}{\partial \varphi}, \\
E_\varphi &= r^{-1} \frac{1}{\sin\theta} \frac{\partial}{\partial r} \left(r \frac{\partial u}{\partial \varphi} \right) - ik \frac{\partial v}{\partial \theta}, \\
H_r &= k^2 r v + \partial_r^2 (r v), \quad H_\theta = r^{-1}\, \partial_r \partial_\theta (r v) - \frac{ik}{\sin\theta} \frac{\partial u}{\partial \varphi}, \\
H_\varphi &= r^{-1} \frac{1}{\sin\theta} \frac{\partial}{\partial r} \left(r \frac{\partial v}{\partial \varphi} \right) + ik \frac{\partial u}{\partial \theta} .
\end{aligned}
\tag{6.54}
$$

Their behavior at infinity in Ω' (see also (6.17) and (6.18)) is easily verified using the asymptotics of the modified Bessel functions in the integrands of the KL integrals provided the latter remain convergent, which is ensured in turn as long as the conditions (6.16) and (6.26) are met. [11] In this way, we further obtain the formulas for the diffraction coefficients in (6.23) as follows.

Indeed, we consider the subdomain of Ω where there are no waves reflected by the conical surface S (which corresponds to the spherical subdomain M', i.e., to oasis). For this

[11] Note that the study of the far field in the supplementary domain M'' is more delicate and is considered herein.

subdomain the modified Bessel functions in the representation (6.24) can be replaced by their asymptotics

$$\mathrm{K}_\nu(-ikr) = \sqrt{\frac{\pi}{2}}\frac{\mathrm{e}^{ikr}}{\sqrt{-ikr}}\left[1+\mathrm{O}\left(\frac{1}{kr}\right)\right], \quad kr \to \infty. \tag{6.55}$$

Since $\widehat{\theta}'(\omega, \omega_0) > \pi$, the integrals remain convergent, see (6.33), and one has

$$(u^s, v^s) = \frac{\mathrm{e}^{ikr}}{ikr}\frac{2}{k}\left[\int\limits_{i\mathbb{R}} \frac{\nu \sin \pi\nu}{\nu^2 - 1/4} g_{u,v}(\omega, \omega_0, \nu)\,\mathrm{d}\nu + \mathrm{i}\pi\, g_{u,v}(\omega, \omega_0, 1/2)\right]\left[1+\mathrm{O}\left(\frac{1}{kr}\right)\right]. \tag{6.56}$$

Substituting this into (6.9) we have, in the spherical coordinates (r, θ, φ), see (6.3) and (6.54)

$$E_\theta = -2\pi\frac{\mathrm{e}^{ikr}}{kr}\mathcal{E}_\theta, \quad E_\varphi = -2\pi\frac{\mathrm{e}^{ikr}}{kr}\mathcal{E}_\varphi, \quad E_r = \mathrm{O}\left(\frac{\mathrm{e}^{ikr}}{(kr)^2}\right),$$

$$\mathcal{E}_\theta = -\left\{\frac{\mathrm{i}}{\pi}\int\limits_{i\mathbb{R}} \frac{\nu \mathrm{e}^{-\mathrm{i}\pi\nu}}{\nu^2 - 1/4}\left[\frac{\partial g_u(\omega, \omega_0, \nu)}{\partial\theta} + \frac{1}{\sin\theta}\frac{\partial g_v(\omega, \omega_0, \nu)}{\partial\varphi}\right]\mathrm{d}\nu\right. \tag{6.57}$$

$$\left. + \mathrm{i}\left[\frac{\partial g_u(\omega, \omega_0, 1/2)}{\partial\theta} + \frac{1}{\sin\theta}\frac{\partial g_v(\omega, \omega_0, 1/2)}{\partial\varphi}\right]\right\}\left[1+\mathrm{O}\left(\frac{1}{kr}\right)\right],$$

$$\mathcal{E}_\varphi = -\left\{\frac{\mathrm{i}}{\pi}\int\limits_{i\mathbb{R}} \frac{\nu \mathrm{e}^{-\mathrm{i}\pi\nu}}{\nu^2 - 1/4}\left[\frac{1}{\sin\theta}\frac{\partial g_u(\omega, \omega_0, \nu)}{\partial\varphi} - \frac{\partial g_v(\omega, \omega_0, \nu)}{\partial\theta}\right]\mathrm{d}\nu\right.$$

$$\left. + \mathrm{i}\left[\frac{1}{\sin\theta}\frac{\partial g_u(\omega, \omega_0, 1/2)}{\partial\varphi} - \frac{\partial g_v(\omega, \omega_0, 1/2)}{\partial\theta}\right]\right\}\left[1+\mathrm{O}\left(\frac{1}{kr}\right)\right].$$

Introducing the notation

$$f_{u,v}(\omega, \omega_0) = -\frac{\mathrm{i}}{\pi}\int\limits_{i\mathbb{R}} \frac{\nu \mathrm{e}^{-\mathrm{i}\pi\nu}}{\nu^2 - 1/4} g_{u,v}(\omega, \omega_0, \nu)\,\mathrm{d}\nu - \mathrm{i}\, g_{u,v}(\omega, \omega_0, 1/2), \tag{6.58}$$

the electromagnetic diffraction coefficients are then determined by (6.23), where

$$\begin{aligned} \mathcal{E}(\omega) &:= \mathrm{grad}_\omega f_u(\omega, \omega_0) + \mathrm{grad}_\omega f_v \times \omega, \\ \mathcal{H}(\omega) &:= \mathrm{grad}_\omega f_v(\omega, \omega_0) - \mathrm{grad}_\omega f_u \times \omega, \end{aligned} \tag{6.59}$$

in the domain M'. Here the spherical gradient operator grad_ω is defined as $\mathrm{grad}_\omega = \boldsymbol{e}_\theta \partial_\theta + (1/\sin\theta)\boldsymbol{e}_\varphi \partial_\varphi$. Note that, as expected, the above expressions for the diffraction coefficients are formally indistinguishable from those in the case of perfectly conducting cones, with the "ideal" spectral functions g_D and g_N replaced by g_u and g_v, respectively [6.4].

6.3 Separation of angular variables and reduction to functional-difference (FD) equations

In the case of a circular cone one can try separating further the angular variables in the standard spherical coordinates (θ, φ), see (6.3), for the spectral functions. However, contrary to the case of a perfectly conducting cone, the corresponding reductions lead to a problem for functional difference (FD) equations of the second order, which then can be transformed to integral equations of the second kind.

In full analogy with the case of an acoustic impedance cone detailed in Chapter 5, we represent the spectral functions $g_{u,v}(\omega, \omega_0, \nu)$ in the form of a Fourier series ($\omega = (\theta, \varphi)$, $\omega_0 = (\theta_0, \varphi_0)$, $\theta = \theta_1$ is the cone's surface):

$$g_{u,v}(\omega, \omega_0, \nu) = \sum_{n=-\infty}^{+\infty} \mathrm{i}^n \mathrm{e}^{-in\varphi} R_{u,v}(\nu, n, \omega_0) \frac{\mathrm{P}_{\nu-1/2}^{-|n|}(\cos\theta)}{\mathrm{P}_{\nu-1/2}^{-|n|}(\cos\theta_1)}, \tag{6.60}$$

(see [6.7]), where the coefficients R_u and R_v are to be determined, and $\mathrm{P}_{\nu-1/2}^{-|n|}(x)$ is the associated Legendre function; see, for example, the previous chapter and [6.12, 6.13]. For the incident wave, the associated spectral functions $g_{u,v}^0$ are the same as in [6.7],

$$g_{u,v}^0(\omega, \omega_0, \nu) = \sum_{n=-\infty}^{+\infty} \mathrm{i}^n \mathrm{e}^{-in\varphi} R_{u,v}^0(\nu, n, \omega_0) \frac{\mathrm{P}_{\nu-1/2}^{-|n|}(-\cos\theta)}{\mathrm{P}_{\nu-1/2}^{-|n|}(-\cos\theta_1)}, \tag{6.61}$$

$$R_{u,v}^0(\nu, n, \omega_0) = \frac{\mathrm{i}^n}{-4\cos\pi\nu} \frac{\Gamma(\nu + |n| + 1/2)}{\Gamma(\nu - |n| + 1/2)} \mathrm{P}_{\nu-1/2}^{-|n|}(-\cos\theta_1)$$

$$\times(-L_{u,v}) \left[\mathrm{P}_{\nu-1/2}^{-|n|}(\cos\theta_0)\, \mathrm{e}^{in\varphi_0}\right],$$

where L_u and L_v are given by (6.30) and (6.31).

The functions (6.60) satisfy automatically the equations (6.40). Substituting these further into (6.43) results in

$$\begin{aligned}
&\frac{\mathrm{i}}{2}[R_u(\nu+1) - R_u(\nu-1)] = \eta\, w(\nu)\, R_u(\nu) \\
&\quad - \eta \frac{n}{2\sin\theta_1}\left[\frac{R_v(\nu+1)}{\nu+1/2} + \frac{R_v(\nu-1)}{\nu-1/2}\right] - G_u^0(\nu), \\
&\frac{\mathrm{i}}{2}[R_v(\nu+1) - R_v(\nu-1)] = \eta^{-1}\, w(\nu)\, R_v(\nu) \\
&\quad + \frac{n\eta^{-1}}{2\sin\theta_1}\left[\frac{R_u(\nu+1)}{\nu+1/2} + \frac{R_u(\nu-1)}{\nu-1/2}\right] - G_v^0(\nu),
\end{aligned} \tag{6.62}$$

Here $w(\nu) = \dfrac{\mathrm{i}\nu}{\nu^2 - 1/4} \dfrac{\partial_{\theta_1} \mathrm{P}_{\nu-1/2}^{-|n|}(\cos\theta_1)}{\mathrm{P}_{\nu-1/2}^{-|n|}(\cos\theta_1)}$,

$$G_u^0(\nu) := \frac{\mathrm{i}}{2}\left[R_u^0(\nu+1) - R_u^0(\nu-1)\right] - \eta\, w_{\mathrm{i}}(\nu)\, R_u^0(\nu)$$

$$+\frac{n\eta}{2\sin\theta_1}\left[\frac{R_v^0(\nu+1)}{\nu+1/2}+\frac{R_v^0(\nu-1)}{\nu-1/2}\right],$$

$$G_v^0(\nu) := \frac{\mathrm{i}}{2}\left[R_v^0(\nu+1)-R_v^0(\nu-1)\right]-\eta^{-1}\,w_{\mathrm{i}}(\nu)\,R_v^0(\nu)$$

$$-\frac{n\eta^{-1}}{2\sin\theta_1}\left[\frac{R_u^0(\nu+1)}{\nu+1/2}+\frac{R_u^0(\nu-1)}{\nu-1/2}\right],$$

$$w_{\mathrm{i}}(\nu) = \frac{\mathrm{i}\nu}{\nu^2-1/4}\,\frac{\partial_{\theta_1}\mathrm{P}_{\nu-1/2}^{-|n|}(-\cos\theta_1)}{\mathrm{P}_{\nu-1/2}^{-|n|}(-\cos\theta_1)}.$$

To simplify the notation, we also do not display henceforth the dependence on n and ω_0 in the arguments of $R_{u,v}$ and $R_{u,v}^0$.

The solution of the FD equations is sought in the class of meromorphic functions, which are regular in the strip $|\mathrm{Re}\,\nu| < 1+\delta$, vanish exponentially as $\nu \to \mathrm{i}\infty$, and are even, that is, $R_{u,v}(\nu) = R_{u,v}(-\nu)$. More accurate growth estimates for $R_{u,v}$ when $\nu \to \mathrm{i}\infty$ follow from the corresponding estimates for $g_{u,v}(\omega, \omega_0, \nu)\big|_l$, see also (6.33):

$$\left|R_{u,v}(\nu)\right| \le C_n\left|\nu^{-1/2}\mathrm{e}^{\mathrm{i}\nu(\theta_1-\theta_0)}\right|. \tag{6.63}$$

The system of the FD equations (6.62) for functions $R_{u,v}(\nu)$ with the above described regularity and growth conditions, should also be supplemented by an additional relation following from (6.45). The second equation in (6.45) leads to

$$R_u(1/2) = -\mathrm{i}\,(n/|n|)\,R_v(1/2). \tag{6.64}$$

The derived system of functional difference relations (6.62) contains two unknown functions R_u and R_v.

The functional difference equations can be put into a more compact matrix form, by introducing new functions $\rho_\pm(\nu) = R_u(\nu) \pm (\eta/\mathrm{i})R_v(\nu)$:

$$(\mathrm{i}/2)\left[a(\nu)\rho(\nu+1) - a(-\nu)\rho(\nu-1)\right] = W(\nu)\rho(\nu) - F^0(\nu). \tag{6.65}$$

The matrices $a(\nu)$ and $W(\nu)$ read

$$a(\nu) = \begin{bmatrix}1 & 0\\ 0 & 1\end{bmatrix} + \frac{n}{\sin\theta_1}\,\frac{1}{\nu+1/2}\begin{bmatrix}1 & 0\\ 0 & -1\end{bmatrix}, \tag{6.66}$$

$$W(\nu) = \frac{1}{2}\begin{bmatrix} w(\nu)\left(\eta+\eta^{-1}\right) & w(\nu)\left(\eta-\eta^{-1}\right)\\ w(\nu)\left(\eta-\eta^{-1}\right) & w(\nu)\left(\eta+\eta^{-1}\right)\end{bmatrix}; \tag{6.67}$$

the column vectors $\rho(\nu)$ and $F^0(\nu)$ are given by

$$\rho(\nu) = \begin{bmatrix}\rho_+(\nu)\\ \rho_-(\nu)\end{bmatrix}, \quad F^0(\nu) = \begin{bmatrix} G_u^0(\nu) - \mathrm{i}\eta\, G_v^0(\nu)\\ G_u^0(\nu) + \mathrm{i}\eta\, G_v^0(\nu)\end{bmatrix}, \tag{6.68}$$

which are even functions in ν. From (6.64) one gets

$$\rho_+(1/2) = \frac{1+\eta n/|n|}{1-\eta n/|n|}\rho_-(1/2). \tag{6.69}$$

In the special case $\eta = 1$, the matrix $W(\nu)$ becomes diagonal. Therefore, both the functional difference equations (6.65) and the respective integral equations to be given in the next section are decoupled.

6.4 Fredholm integral equations for the Fourier coefficients

6.4.1 Reduction to integral equations

Reduction of functional difference equations to integral equations is quite a traditional tool (see, e.g., [6.15, 6.14] and the previous chapters). To this end, first we introduce an auxiliary function $\Psi(\nu, \Lambda)$ [6.10] which is an even meromorphic solution of the equation in ν depending on a real parameter Λ.

$$\left(1 \pm \frac{\Lambda}{\nu \pm 1/2}\right)\Psi(\nu \pm 1, \Lambda) = \left(1 \pm \frac{\Lambda}{-\nu \pm 1/2}\right)\Psi(-\nu \pm 1, \Lambda). \tag{6.70}$$

This function is found explicitly:

$$\Psi(\nu, \Lambda) =$$

$$\left\{\prod_{(\pm)} \frac{\Gamma\left(\frac{1}{2}+\frac{1}{4}\left(\nu \pm 1 + \frac{r(\Lambda)}{2}\right)\right)\Gamma\left(\frac{1}{2}-\frac{1}{4}\left(\nu \pm 1 - \frac{r(\Lambda)}{2}\right)\right)}{\Gamma\left(\frac{1}{2}+\frac{1}{4}\left(\nu \pm 1 + \left|\frac{1}{2}+\Lambda\right|\right)\right)\Gamma\left(\frac{1}{2}-\frac{1}{4}\left(\nu \pm 1 - \left|\frac{1}{2}+\Lambda\right|\right)\right)}\right\}^{r(\Lambda)}, \tag{6.71}$$

$r(\Lambda) := \operatorname{sgn}(1/2 + \Lambda)$, see [6.10] and [6.7] for details of the derivation and the properties of $\Psi(\nu, \Lambda)$.

Omitting simple technical details, we display next only the final equations. Introduce new unknown functions by

$$\rho_\pm(\nu) = \Psi(\nu, \Lambda_\pm)\,\chi_\pm(\nu), \tag{6.72}$$

$\Lambda_\pm := \pm n/\sin\theta_1$. The system (6.66) is then converted into that for $\chi_\pm(\nu)$ with constant coefficients on the left-hand side. As in the previous chapter the corresponding difference operator on the left-hand side looks like

$$\chi_\pm(\nu+1) - \chi_\pm(\nu-1) = H_\pm(\nu)$$

and is "inverted." Then the resulting system of integral equations of the second kind reads, for imaginary ν:

$$\begin{bmatrix}\chi_+(\nu)\\ \chi_-(\nu)\end{bmatrix} = \frac{-\pi}{\cos\pi\nu}\begin{bmatrix}\dfrac{1-r(\Lambda_+)}{2}\rho_+(1/2)\,\mathrm{res}_{1/2}\,(1/\Psi(\nu,\Lambda_+))\\ \dfrac{1-r(\Lambda_-)}{2}\rho_-(1/2)\,\mathrm{res}_{1/2}\,(1/\Psi(\nu,\Lambda_-))\end{bmatrix}$$

$$+\frac{1}{2}\int\limits_{\mathrm{i}\mathbb{R}}\begin{bmatrix}\Psi(\tau,\Lambda_+)\Psi(\tau,\Lambda_-)W_{11}(\tau) & \Psi(\tau,\Lambda_-)\Psi(\tau,\Lambda_-)W_{12}(\tau)\\ \Psi(\tau,\Lambda_+)\Psi(\tau,\Lambda_+)W_{21}(\tau) & \Psi(\tau,\Lambda_-)\Psi(\tau,\Lambda_+)W_{22}(\tau)\end{bmatrix}\begin{bmatrix}\chi_+(\tau)\\ \chi_-(\tau)\end{bmatrix} \tag{6.73}$$

$$\times\frac{\sin\pi\tau\,\mathrm{d}\tau}{\cos\pi\tau+\cos\pi\nu}-\frac{1}{2}\int\limits_{\mathrm{i}\mathbb{R}}\begin{bmatrix}\Psi(\tau,\Lambda_-)\left(\psi_u^0(\tau)-\mathrm{i}\eta\,\psi_v^0(\tau)\right)\\ \Psi(\tau,\Lambda_+)\left(\psi_u^0(\tau)+\mathrm{i}\eta\,\psi_v^0(\tau)\right)\end{bmatrix}\frac{\sin\pi\tau\,\mathrm{d}\tau}{\cos\pi\tau+\cos\pi\nu},$$

where $W_{ij}(\nu)$ are the entries of the matrix $W(\nu)$, and

$$\mathrm{res}_{1/2}\,(1/\Psi(\nu,\Lambda)) = -4\pi\,(2\pi)^{1/2}$$

$$\times\left\{\prod_{(\pm)}\Gamma\left(\frac{1}{2}+\frac{1}{4}\left(\frac{1}{2}\pm 1+\left|\frac{1}{2}+\Lambda\right|\right)\right)\Gamma\left(\frac{1}{2}-\frac{1}{4}\left(\frac{1}{2}\pm 1-\left|\frac{1}{2}+\Lambda\right|\right)\right)\right\}^{-1}$$

The first term on the right-hand side of (6.73) is a solution of a simple homogeneous equation

$$\chi(\nu+1)\ -\ \chi(\nu-1)\ = 0$$

and has poles at $\nu=\pm 1/2$, whereas the product $[1-r(\Lambda)]\,\Psi(\nu,\Lambda)/\cos\pi\nu$ is regular in the strip $\Pi_{1+\delta}$, which was assumed. This term depends in (6.73) on the unknown constant $\rho_+(1/2)$ if $r(\Lambda_+)<0$ $(n<0)$, and on $\rho_-(1/2)$ if $r(\Lambda_-)<0$, $(n>0)$. Both of these constants should be known to determine the sought wave field. They are mutually related by (6.69).

It is worth commenting on (6.73). The correlations (6.73) may also serve for analytic continuation of $\chi_\pm$ from the imaginary axis of the complex ν-plane onto the whole strip $|\mathrm{Re}\,\nu|<1$ because the right-hand side is regular in this strip and provided $\chi_\pm$ are known on the imaginary axis.

Let $n>0, r(\Lambda_+)>0$. We need to derive one additional linear relation connecting $\rho_+(1/2)$ with $\rho_-(1/2)$. One has $(\Psi(1/2,\Lambda_-)=0)$

$$\rho_+(1/2)-\left\{\Psi(1/2,\Lambda_+)\frac{1}{2}\int\limits_{\mathrm{i}\mathbb{R}}\tan(\pi\tau)\left[W_{11}(\tau)\Psi(\tau,\Lambda_+)\Psi(\tau,\Lambda_-)\chi_+^2(\tau)\right.\right.$$

$$\left.\left.+\ W_{12}(\tau)\Psi(\tau,\Lambda_-)\Psi(\tau,\Lambda_-)\chi_-^2(\tau)\right]\mathrm{d}\tau\right\}\rho_-(1/2)$$

$$=\Psi(1/2,\Lambda_+)\left\{\frac{1}{2}\int\limits_{\mathrm{i}\mathbb{R}}\tan(\pi\tau)\left[W_{11}(\tau)\Psi(\tau,\Lambda_+)\Psi(\tau,\Lambda_-)\chi_+^1(\tau)\right.\right. \tag{6.74}$$

$$\left.+\ W_{12}(\tau)\Psi(\tau,\Lambda_-)\Psi(\tau,\Lambda_-)\chi_-^1(\tau)\right]\mathrm{d}\tau-\frac{1}{2}\int\limits_{\mathrm{i}\mathbb{R}}\Psi(\tau,\Lambda_-)\Big(\psi_u^0(\tau)$$

$$\left.-\,\mathrm{i}\,\eta\,\psi_v^0(\tau)\Big)\ \tan(\pi\tau)\,\mathrm{d}\tau\right\},$$

where $\chi^2 = \left[\chi^2_+, \chi^2_-\right]^{\mathrm{T}}$, $\chi^1 = \left[\chi^1_+, \chi^1_-\right]^{\mathrm{T}}$ are solutions of the equation (6.73) corresponding to inhomogeneous terms

$$S^2(\nu) = \left[0, -\pi\, \mathrm{res}_{1/2}\, (1/\Psi(\nu, \Lambda_-)) / \cos\pi\nu\right]^{\mathrm{T}}$$

for χ^2 and $S^1(\nu)$, coinciding with the last term on the right-hand side of (6.73), for χ^1. Actually, we should in total solve the matrix integral equations

$$(I - \mathbf{B})\,\chi^j = S^j, \quad j = 1, 2, 3 \tag{6.75}$$

for three different right-hand sides (as explained next), where $\mathbf{B}$ is the integral operator in the second summand on the right-hand side of (6.73). The corresponding solutions χ^1 and χ^2 are then substituted into (6.74).

In the case $n < 0$, ($\Lambda_- > 0$), the required additional (to (6.69)) linear relation takes the form

$$\begin{aligned}
&\rho_-(1/2) - \Big\{\Psi(1/2, \Lambda_-)\frac{1}{2}\int_{\mathrm{i}\mathbb{R}} \tan(\pi\tau)\Big[W_{21}(\tau)\Psi(\tau, \Lambda_+)\Psi(\tau, \Lambda_+)\chi^3_+(\tau)\\
&+W_{22}(\tau)\Psi(\tau, \Lambda_+)\Psi(\tau, \Lambda_-)\chi^3_-(\tau)\Big]\mathrm{d}\tau\Big\}\,\rho_+(1/2)\\
&= \Psi(1/2, \Lambda_-)\Big\{\frac{1}{2}\int_{\mathrm{i}\mathbb{R}} \tan(\pi\tau)\Big[W_{21}(\tau)\Psi(\tau, \Lambda_+)\Psi(\tau, \Lambda_+)\chi^1_+(\tau)\\
&+W_{22}(\tau)\Psi(\tau, \Lambda_+)\Psi(\tau, \Lambda_-)\chi^1_-(\tau)\Big]\mathrm{d}\tau - \frac{1}{2}\int_{\mathrm{i}\mathbb{R}} \Psi(\tau, \Lambda_+)\Big[\psi^0_u(\tau)\\
&+\mathrm{i}\eta\,\psi^0_v(\tau)\Big]\,\tan(\pi\tau)\,\mathrm{d}\tau\Big\},
\end{aligned} \tag{6.76}$$

$$S^3(\nu) = \left[-\pi\,\mathrm{res}_{1/2}\,(1/\Psi\,(\nu, \Lambda_+)) / \cos(\pi\nu), 0\right]^{\mathrm{T}}.$$

The solution of the integral equation (6.73) is then represented in the form

$$\chi(\nu) = \chi^1(\nu) + \rho_-(1/2)\chi^2(\nu), \quad n > 0, \tag{6.77}$$

or

$$\chi(\nu) = \chi^1(\nu) + \rho_+(1/2)\chi^3(\nu), \quad n < 0. \tag{6.78}$$

In the case $n = 0$ there is obviously $\chi(\nu) = \chi^1(\nu)$.

The next subsection deals with the Fredholm property (which is crucial to ensure numerical solvability) of the operator in (6.75) or (6.73).

6.4.2 Comments on the Fredholm property and unique solvability of the integral equations

The basic idea of reduction of the integral equations (6.73) to those of the Fredholm type is fairly general and is known as semi-inversion. Namely, a general linear equation (including an integral equation)

$$A\,x = f$$

is sought to be represented in the form

$$(A_1 + K)\,x = f,$$

where A_1 is an invertible operator with a bounded inverse, and K is compact. Then the left regularization

$$\left(I + A_1^{-1}K\right)\,x = A_1^{-1}f$$

gives a classical equation of the second kind with compact operator $A_1^{-1}K$ in an appropriate functional space. It is remarkable that in our problem the semi-inversion can be executed explicitly by use of Dixon's resolvent, analogously to the scalar problem [6.10]; see also [6.7] and the previous chapter.

The demonstration of this fact is traditional, by taking into account that the matrix $W(\nu)$ has the asymptotics ($\nu \to \mathrm{i}\infty$)

$$W(\nu) = N + \mathrm{O}\left(\nu^{-1}\right), \tag{6.79}$$

$$N = \frac{1}{2}\begin{bmatrix} \eta+\eta^{-1} & \eta-\eta^{-1} \\ \eta-\eta^{-1} & \eta+\eta^{-1} \end{bmatrix},$$

since $w(\nu) = 1 + \mathrm{O}(\nu^{-1})$. The constant matrix N in (6.79) has the eigenvalues $\lambda_1 = \eta^{-1}$ and $\lambda_2 = \eta$ with positive real parts, provided $\operatorname{Re}\eta > 0$. This ensures the existence of the splitting into a boundedly invertible and compact operators in (6.73), which is discussed in [6.7]. Additional technical details can also be found in [6.2].

The homogeneous integral equations have only a trivial solution in the corresponding functional space, which is shown in [6.7]. Then, in view of the Fredholm property, the equations are uniquely solvable.

6.5 Electromagnetic diffraction coefficients in M' and numerical results

The far-field diffracted spherical wave $\boldsymbol{E}^{\mathrm{d}}$ and $\boldsymbol{H}^{\mathrm{d}}$ can be also written in the form:

$$\boldsymbol{E}^{\mathrm{d}} = -2\pi\frac{\mathrm{e}^{ikr}}{kr}D\cdot\boldsymbol{E}^{0}\Big|_{r=0}\left[1+\mathrm{O}\left(\frac{1}{kr}\right)\right], \tag{6.80}$$

$$\boldsymbol{H}^{\mathrm{d}} = \boldsymbol{e}_r \times \boldsymbol{E}^{\mathrm{d}}, \tag{6.81}$$

with the matrix diffraction coefficient D defined as

$$D = \begin{bmatrix} D_{\theta\theta_0} & D_{\theta\varphi_0} \\ D_{\varphi\theta_0} & D_{\varphi\varphi_0} \end{bmatrix}, \tag{6.82}$$

$$\begin{aligned} D_{\theta\{\theta_0,\varphi_0\}} &= \partial_\theta\, f_u(\omega,\omega_0)|_{\beta=\pi/2,0} \\ &\quad +(\sin\theta)^{-1}\partial_\varphi\, f_v(\omega,\omega_0)|_{\beta=\pi/2,0}, \end{aligned} \tag{6.83}$$

$$\begin{aligned} D_{\varphi\{\theta_0,\varphi_0\}} &= (\sin\theta)^{-1}\partial_\varphi\, f_u(\omega,\omega_0)|_{\beta=\pi/2,0} \\ &\quad -\partial_\theta\, f_v(\omega,\omega_0)|_{\beta=\pi/2,0}, \end{aligned} \tag{6.84}$$

where $f_{u,v}(\omega,\omega_0)$ are defined in (6.58). The spectra $g_{u,v}$ of the scattered field are known at the same time as $\rho(\nu)$ because

$$R_u(\nu) = [\rho_+(\nu) + \rho_-(\nu)]/2, \tag{6.85}$$

$$R_v(\nu) = \mathrm{i}\,[\rho_+(\nu) - \rho_-(\nu)]/(2\eta). \tag{6.86}$$

Considering the domain covered by the oasis, the Fourier series for the diffraction coefficients will be defined in particular as $\theta < \theta_s(\varphi_0) = 2\theta_1 - \theta_0 - \pi$.

It is recalled that analogous expressions for the diffraction coefficients have been obtained for perfectly conducting cones [6.4, 6.16–6.27].

6.5.1 Numerical Solution

Since the diffraction coefficients depend on the spectra $g_{u,v}(\nu)$ along the imaginary axis, one has to solve the integral equations (6.73) and then to calculate the integrals (6.58).[12]

It proves beneficial to account for the asymptotic behavior of $\rho(\nu)$ in an explicit way. An appropriate Ansatz in the present work takes the following form:

$$\rho(\nu) = \mathrm{i}^n \exp[-\mathrm{Im}\,\nu\,(\theta_1 - \theta_0)]\,\widetilde{\rho}(\nu). \tag{6.87}$$

Recalling that both the spectra and their Fourier coefficients are even in ν allows one to concentrate on the upper half of the imaginary axis, which can be transformed into a finite domain via

$$\nu = \frac{\mathrm{i}p}{\theta_1 - \theta_0} \ln\frac{1+x}{1-x}, \quad 0 \le x < 1. \tag{6.88}$$

Here p is a positive large real number.

The integral equations for $\widetilde{\rho}(\nu)$ are solved by the aid of the quadrature method, that is, evaluating the integrals with the N-abscissa Gauss-Legendre scheme, enforcing the approximated equations precisely at these N abscissae, and resolving the resulting matrix equation (see for example [6.28]).

[12] Surely, our "numerical" experience from Chapter 5 proves to be quite helpful.

The evenness and the asymptotic behavior of the integrands allows for casting the integrals into the form

$$\frac{1}{\pi\,[\theta_s(\varphi_0)-\theta]}\int_0^{+\infty} G(\boldsymbol{\omega},\boldsymbol{\omega}_0,\nu)\,\mathrm{e}^{-T}\,\mathrm{d}T, \tag{6.89}$$

with $\nu = \mathrm{i}\tau$, $T = [\theta_s(\varphi_0)-\theta]\,\tau$ and the smallest singular angle $\theta_s(\varphi_0) = 2\theta_1 - \theta_0 - \boldsymbol{\pi}$. $G(\boldsymbol{\omega},\boldsymbol{\omega}_0,\nu)$ depends on, among other things, the spectra $g_{u,v}(\boldsymbol{\omega},\boldsymbol{\omega}_0,\nu)$ and grows at most algebraically as $\operatorname{Im}\nu \to +\infty$. The values of $g_{u,v}(\boldsymbol{\omega},\boldsymbol{\omega}_0,\nu)$ within the strip $|\operatorname{Re}\nu| < 1$ are interpolated with (6.73). The previous expression suggests the Gauss-Laguerre scheme for its numerical evaluation.

To study the convergence rate and for comparison purposes two special cases, owing to the available exact spectra, are of particular interest:

$$R_u(\nu) = -R_u^0(\nu),\quad R_v(\nu) = -\frac{w_{\mathrm{i}}(\nu)}{w(\nu)}R_v^0(\nu),\quad \eta = 0, \tag{6.90}$$

$$R_v(\nu) = -R_v^0(\nu),\quad R_u(\nu) = -\frac{w_{\mathrm{i}}(\nu)}{w(\nu)}R_u^0(\nu),\quad \eta \to \infty. \tag{6.91}$$

6.5.2 Numerical examples

Owing to the previously described solution procedure, the accuracy of the obtained results depend on the following parameters: the number of abscissae of the Gauss-Legendre scheme N; the number of abscissae of the Gauss-Laguerre scheme M; the parameter p; and the number of terms in the Fourier series $2N_m + 1$.

By virtue of the exact spectra for perfect electric ($\eta = 0$) and magnetic ($\eta \to \infty$) cones, and as a result of detailed internal convergence studies it has been found that the following parameter set:

$$N = 100,\ M = 20,\ N_m = 50,\ \text{and } p = 20 \tag{6.92}$$

suffices for engineering purposes, that is, the deviation of the computed results for the normalized radar cross section (RCS) from the reference values is well below 0.001. Hence, all the following results have been obtained with this set of parameters.

Let an electromagnetic plane wave impinge on the cone along the latter's axis with $\theta_0 = \varphi_0 = \beta = 0°$, and consider the radar cross section as a function of the co-latitude θ in the half-plane $\varphi = 0°$.

Under this circumstance, the radar cross section is related to the diffraction coefficient $D_{\varphi\varphi_0}$ according to

$$\sigma = \lim_{r\to\infty} 4\pi r^2\,|\boldsymbol{E}|^2 / \left|\boldsymbol{E}^0\right|^2 = 4\pi\lambda^2\left|D_{\varphi\varphi_0}\right|^2. \tag{6.93}$$

Now let the surface impedance η vary from $\eta = 0$ (perfect electric conductor PEC) through finite values to $\eta \to \infty$ (perfect magnetic conductor, or PMC) and display the normalized radar cross section σ/λ^2 in Fig. 6.2. For the PEC cone, the results of [6.16] and [6.18] are also included for comparison purposes.

Clearly, the radar cross section attains its lowest value at the backward direction, and this minimum depends on the surface impedance η. Starting from the PEC case with $\eta = 0$, this

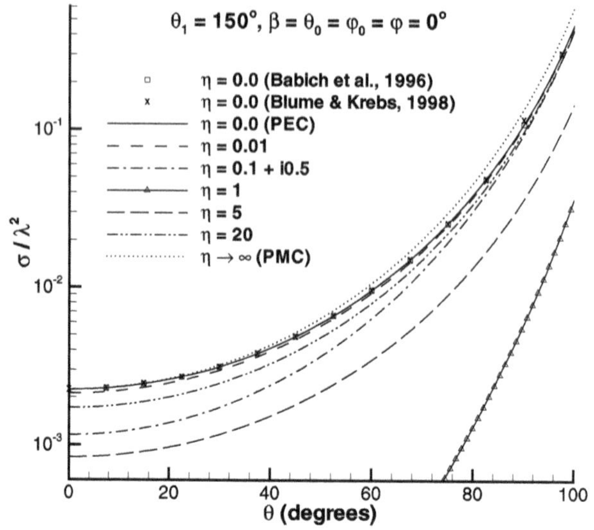

Figure 6.2 Normalized radar cross section of a circular impedance cone ($\theta_1 = 150°$) at axial incidence for different values of η.

minimum decreases with increasing modulus of η and disappears completely at $\eta = 1$, in agreement with Weston's theorem on the so-called zero backscattering for bodies of certain rotational symmetry [6.29]. Then this minimum grows with the modulus of η and attains its largest value at $\eta \to \infty$ (a PMC cone), which equals exactly, both theoretically and numerically, the backward normalized radar cross section for a PEC cone.

The radar cross section increases with the co-latitude θ, hinting at the physical character of the diffracted waves. The field at large distance changes its nature when approaching the region where nonspherical waves exist, and the "sources" of the diffracted waves can be considered as distributed on the shadow boundaries of the reflected waves which define singular directions. These "sources" are not located at the cone's tip even if (6.80) and (6.81) apply in the oasis region. In the words of Sommerfeld, it is a matter of optical delusion.

Next consider a nonaxial incidence with $\theta_0 = 40°$ (the cone with $\theta_1 = 160$ is not fully lit) and $\beta = 45°$ for a certain surface impedance and look at the radar cross sections σ_φ and σ_θ, which are now

$$\sigma_\varphi = 2\pi\lambda^2 \left| D_{\varphi\varphi_0} + D_{\varphi\theta_0} \right|^2, \tag{6.94}$$

$$\sigma_\theta = 2\pi\lambda^2 \left| D_{\theta\varphi_0} + D_{\theta\theta_0} \right|^2. \tag{6.95}$$

The normalized radar cross sections are depicted in Figs. 6.3 and 6.4, as a function of both the azimuth φ and co-latitude θ. They are conspicuously not symmetric with respect to $\varphi = 0°$. This phenomenon is due to the vectorial nature of the electromagnetic field, which is absent in acoustics, for example. Indeed, it can be demonstrated simply by Maxwell's equations, that, in the cases $\beta = 0°$ and $\beta = 90°$, the field can present some antisymmetry. This explains in particular why, by linear superposition, one obtains the asymmetry in modulus for the case $\beta = 45°$ shown in Figs. 6.3 and 6.4.

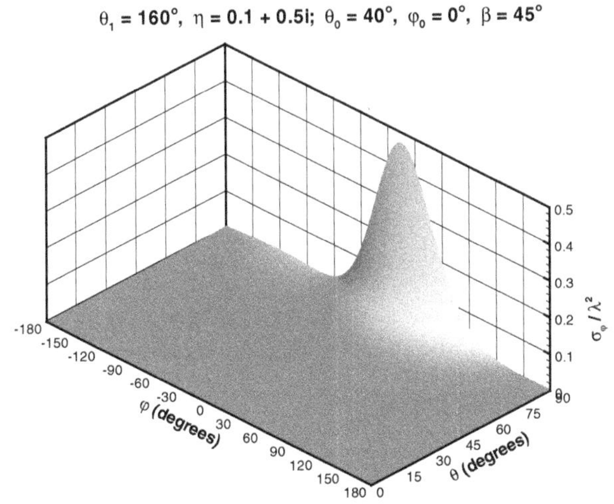

Figure 6.3 Normalized radar cross section of φ-polarization for a circular impedance cone ($\theta_1 = 160°$) at nonaxial incidence ($\theta_0 = 40°$).

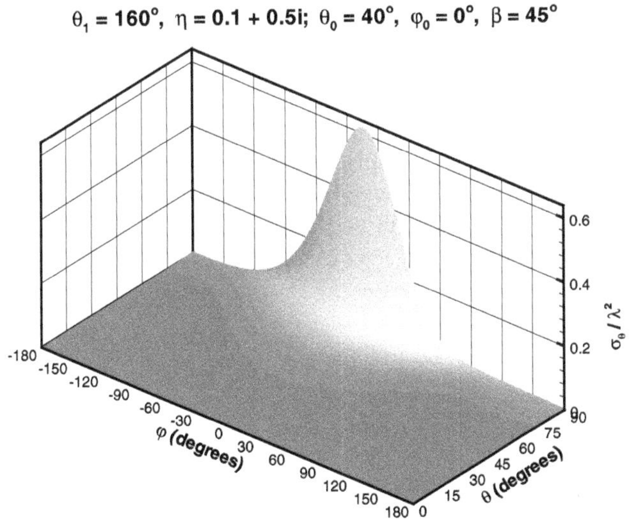

Figure 6.4 Normalized radar cross section of θ-polarization for a circular impedance cone ($\theta_1 = 160°$) at nonaxial incidence ($\theta_0 = 40°$).

6.6 Sommerfeld and Watson–Bessel (WB) integral representations

As mentioned earlier, the analytic continuation of the KL integral representation for the potentials for real k requires the use of the WB integral representation. On the other hand, this kind of representation will enable us to study the analytic properties of the Sommerfeld tranformant in the Sommerfeld integral representations for the potentials. The latter representations are most convenient for the evaluation of the far-field asymptotics. So the integral representations (6.24) can be transformed thereby to a new form that has certain advantages, in particular, for real wavenumbers $k > 0$. This form gives the desired "analytic continuation" of (6.24) valid within restrictions (6.25)–(6.26).

We exploit the functional relation between $\mathrm{K}_\nu(z)$ and the Bessel function J_ν (see, e.g., [6.13])

$$\sin(\nu\pi)\mathrm{K}_\nu(z) = (\pi/2)\left[\exp(\mathrm{i}\nu\pi/2)\,\mathrm{J}_{-\nu}(\mathrm{i}z) - \exp(-\mathrm{i}\nu\pi/2)\,\mathrm{J}_\nu(\mathrm{i}z)\right],$$

$-\pi < \arg z \leq \pi/2$, and the evenness of the spectral functions in ν, which transforms the KL integrals (6.24) into those of the so-called Watson–Bessel (WB) type integrals (see also Section 1.4.5):

$$\begin{aligned}(u, v) = -\frac{2}{k}\sqrt{\frac{2}{\pi}}\Bigg[&-\pi\int_{C_\phi}\frac{\nu\mathrm{e}^{-\mathrm{i}\pi\nu/2}}{\nu^2-1/4}\frac{\mathrm{J}_\nu(kr)}{\sqrt{-\mathrm{i}kr}}\,g_{u,v}(\omega,\omega_0,\nu)\,\mathrm{d}\nu \\ &+\mathrm{i}\pi\frac{\mathrm{K}_{1/2}(-\mathrm{i}kr)}{\sqrt{-\mathrm{i}kr}}\,g_{u,v}(\omega,\omega_0,1/2)\Bigg], \quad \phi\in(0,\pi/2]\,.\end{aligned} \tag{6.96}$$

The contour $C_\phi = (\infty\,\mathrm{e}^{-\mathrm{i}\phi}, \infty\,\mathrm{e}^{\mathrm{i}\phi})$ comprises the positive real axis and all the singularities of the spectral functions located in some strip along the positive real axis, and $C_{\pi/2} = (-\mathrm{i}\infty, \mathrm{i}\infty)$ is the traditional contour in the KL integrals (6.24).

It is worth noting that, provided the conditions (6.25)–(6.26) hold, the integration contour C_ϕ may coincide with $C_{\pi/2}$ and the representation (6.96) is then equivalent to the KL integrals. However, for $\phi \in [0, \pi/2)$ the WB integrals in (6.96) converge exponentially for fixed kr without the limitations (6.25)–(6.26), in particular, for real $k > 0$. This follows from the asymptotics of the Bessel functions $\mathrm{J}_\nu(z) \sim (kr/2)^\nu/\Gamma(\nu+1)$ ($\nu \to \infty$, kr is fixed) and from the estimate

$$\begin{aligned}\left|\frac{\nu\,\mathrm{e}^{-\mathrm{i}\pi\nu/2}}{\nu^2-1/4}g_{u,v}(\omega,\omega_0,\nu)\,\mathrm{J}_\nu(kr)\right| \leq C\exp\Big[&-|\nu|\log|\nu|\cos\phi \\ -|\nu|\big(\sin\phi(\arg k - \pi/2 - \phi) + |\sin\phi|\widehat{\theta}'(\omega,\omega_0) &- \cos\phi[1+\log(|k|r/2)]\big)\Big],\end{aligned}$$

on the contour C_ϕ as $|\nu| \to \infty$. This ensures that (6.96), always converging for $\phi > 0$, provides the analytic continuation of the KL integral (6.24).

6.6.1 Sommerfeld integral representations

It turns out that the Sommerfeld integral representations are well suited for the asymptotic evaluation of the potentials at infinity. Exploit the integral representation

$$\mathrm{K}_\nu(-\mathrm{i}kr) = \frac{1}{4\sin\pi\nu}\int_\gamma \mathrm{e}^{-\mathrm{i}kr\cos\alpha}\,\frac{\mathrm{e}^{\mathrm{i}\nu\alpha}-\mathrm{e}^{-\mathrm{i}\nu\alpha}}{2}\,\mathrm{d}\alpha, \tag{6.97}$$

where $\gamma = \gamma_+ \cup \gamma_-$ is the double-loop Sommerfeld contour (Fig. 6.5).

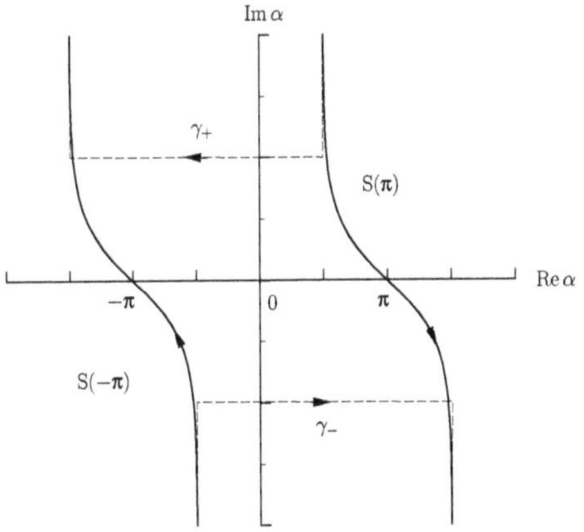

Figure 6.5 The Sommerfeld double-loop contour γ and the steepest-descent paths.

Substituting this representation into the KL integrals (6.24) and interchanging the order of integration (which can be justified), we arrive at the formulas

$$(u\ ,\ v) = \frac{1}{\sqrt{-\mathrm{i}kr}}\,\frac{1}{2\pi\mathrm{i}}\int\limits_{\gamma} \mathrm{e}^{-ikr\cos\alpha}\,\Phi_{u,v}(\alpha,\omega,\omega_0)\,\mathrm{d}\alpha, \tag{6.98}$$

where

$$\Phi_{u,v}(\alpha,\omega,\omega_0) = -\frac{\mathrm{i}\sqrt{2\pi}}{k}\left[\int\limits_{\mathrm{i}\mathbb{R}} \frac{\nu\, g_{u,v}(\omega,\omega_0,\nu)}{\nu^2-1/4}\,\frac{\mathrm{e}^{\mathrm{i}\nu\alpha}-\mathrm{e}^{-\mathrm{i}\nu\alpha}}{2}\,\mathrm{d}\nu \right.$$
$$\left. +\ \mathrm{i}\pi\, g_{u,v}(\omega,\omega_0,1/2)\,\frac{\mathrm{e}^{\mathrm{i}\alpha/2}-\mathrm{e}^{-\mathrm{i}\alpha/2}}{2}\right]. \tag{6.99}$$

It is obvious that, in view of the estimates (6.33), the Fourier-type integrals in (6.99) give the representations of the analytic functions that are regular (see (6.34)) in the strip:

$$\Pi_{\widehat{\theta}'} = \left\{\alpha : |\mathrm{Re}\,\alpha| < \widehat{\theta}'(\omega,\omega_0)\right\}. \tag{6.100}$$

Via integration by parts we also have

$$(u,v) = \frac{1}{(-\mathrm{i}kr)^{3/2}}\,\frac{1}{2\pi\mathrm{i}}\int\limits_{\gamma} \mathrm{e}^{-ikr\cos\alpha}\,\frac{\mathrm{d}}{\mathrm{d}\alpha}\left[\frac{\Phi_{u,v}(\alpha,\omega,\omega_0)}{\sin\alpha}\right]\mathrm{d}\alpha, \tag{6.101}$$

$$\widetilde{\Phi}_{u,v}(\alpha,\omega,\omega_0) := \frac{\mathrm{d}}{\mathrm{d}\alpha}\left[\frac{\Phi_{u,v}(\alpha,\omega,\omega_0)}{\sin\alpha}\right]$$
$$= -\frac{\mathrm{i}\sqrt{2\pi}}{k}\left[\int\limits_{\mathrm{i}\mathbb{R}} \frac{\nu\, g_{u,v}(\omega,\omega_0,\nu)}{\nu^2-1/4}\,\frac{\mathrm{d}}{\mathrm{d}\alpha}\left(\frac{\mathrm{e}^{\mathrm{i}\nu\alpha}-\mathrm{e}^{-\mathrm{i}\nu\alpha}}{2\sin\alpha}\right)\mathrm{d}\nu \right.$$
$$\left. +\,\mathrm{i}\pi\, g_{u,v}(\omega,\omega_0,1/2)\,\frac{\mathrm{d}}{\mathrm{d}\alpha}\,\frac{\mathrm{e}^{\mathrm{i}\alpha/2}-\mathrm{e}^{-\mathrm{i}\alpha/2}}{2\sin\alpha}\right].$$

The latter form is more convenient when studying surface waves. The Sommerfeld transformants $\widetilde{\Phi}_{u,v}$ vanish exponentially like $\exp[-(m+3/2)|\mathrm{Im}\,\alpha|]$, $(m > -3/2)$ as $|\mathrm{Im}\,\alpha| \to \infty$. When α is real and viewed as time, the transformants solve hyperbolic equations with appropriate boundary and initial conditions [6.4].

6.6.2 Regularity domains for the Sommerfeld transformants

We turn to discussion of the regularity of $\widetilde{\Phi}_{u,v}(\alpha, \omega, \omega_0)$, $\Phi_{u,v}(\alpha, \omega, \omega_0)$ and their analytic continuations to a larger domain.

Provided the spectral function $g_{u,v}$ admits the estimate (6.33) in M' and (6.34) in M'', from the properties of the Fourier integrals one has that $\widetilde{\Phi}_{u,v}(\alpha, \omega, \omega_0)$ and $\Phi_{u,v}(\alpha, \omega, \omega_0)$ are regular in $\Pi_{\widehat{\theta}'}$ as $\widehat{\theta}' \le \pi$ and in the strip $\Pi_{\pi+\epsilon}$, $(\epsilon > 0$ is small) when $\widehat{\theta}' > \pi$, that is, in the strip wider than 2π. Actually the width of the regularity strip is specified by location of the singularities of $\widetilde{\Phi}_{u,v}(\alpha, \omega, \omega_0)$ and $\Phi_{u,v}(\alpha, \omega, \omega_0)$ on the real axis. In particular, when the observation point is in the domain M' not illuminated by the rays reflected from the cone, there are no singularities in the strip $\Pi_{\pi+\epsilon}$.

Consider the domain $D_C = \{\alpha \in \mathbb{C} : \ \mathrm{Im}\,\alpha < -C\}$ for some positive constant C. Note that for this constant one can get the estimate from the following equation, $C \ge |\mathrm{Im}\,\zeta_{u,v}|$, where

$$\sin\zeta_u = \eta, \quad \sin\zeta_v = 1/\eta, \quad 0 < \mathrm{Re}\,\zeta_{u,v} \le \pi/2.$$

Let $D^{\star}_C$ be the domain symmetric with respect to the origin. We wish to prove the following.

Lemma 6.1. *Let $g_{u,v}(\nu, \omega, \omega_0)$ satisfy the following estimate on C_ϕ for any $\phi \in [0, \pi/2]$ as $|\nu| \to +\infty$*

$$|g_{u,v}(\nu, \omega, \omega_0)| \le \mathrm{const}\, \mathrm{e}^{C|\nu|}. \tag{6.102}$$

Then the function $\Phi_{u,v}(\alpha, \omega, \omega_0)$ $(\widetilde{\Phi}_{u,v}(\alpha, \omega, \omega_0))$ admits analytic continuation as a regular function to the domain $D = \Pi_{\widehat{\theta}'} \cup D_C \cup D^{\star}_C$. [13]

In other words, the singularities of the transformants are located in the domain $C \setminus D$. To prove this we turn to the representation (6.96) and exploit the Sommerfeld integral representation for the Bessel functions

$$\mathrm{J}_\nu(kr) = \frac{1}{2\pi} \int_{\gamma_-} \mathrm{e}^{-ikr\cos\alpha}\, \mathrm{e}^{i\nu\pi/2 - i\nu\alpha}\, d\alpha, \tag{6.103}$$

where γ_- is the lower loop of the contour γ (Fig. 6.5). Then interchange the order of integration in (6.96), which can be justified. This gives

$$\left(u^s, v^s\right)(r, \omega, \omega_0) = \frac{1}{\sqrt{-ikr}} \frac{1}{2i\pi} \int_{\gamma_-} \mathrm{e}^{-ikr\cos\alpha}\, \Psi_{u,v}(\alpha, \omega, \omega_0)\, d\alpha, \tag{6.104}$$

[13] The estimate (6.102) is indeed true, which follows from the analysis of the spectral functions.

where

$$\Psi_{u,v}(\alpha,\omega,\omega_0) = 2\frac{\mathrm{i}\sqrt{2\pi}}{k}\int_{C_\phi}\frac{\nu}{\nu^2-1/4}\, g_{u,v}(\nu,\omega,\omega_0)\,\mathrm{e}^{-\mathrm{i}\nu\alpha}\,\mathrm{d}\nu$$
$$+2\frac{\sqrt{2\pi}}{k}\,\pi\, g_{u,v}(\omega,\omega_0,1/2)\,\frac{\mathrm{e}^{\mathrm{i}\alpha/2}-\mathrm{e}^{-\mathrm{i}\alpha/2}}{2}. \qquad (6.105)$$

The contour C_ϕ, $\phi \in [0,\pi/2)$ comprises all the singularities of the meromorphic function $g_{u,v}(\nu,\omega,\omega_0)$ in the right complex half-plane, k can be taken real positive, $\omega \in M$. Recall that these singularities are in a strip parallel to the real axis.

There exists a constant C, such that $\Psi(\alpha,\omega,\omega_0)$ is regular in the domain D_C, and this holds for any C_ϕ, $\phi \in [0,\pi/2)$ in (6.105).

Let $\alpha \in \Pi_{\widehat{\theta'}} \cap D_C$. Transform the representation (6.99) to the form

$$\Phi_{u,v}(\alpha,\omega,\omega_0) = \frac{\mathrm{i}\sqrt{2\pi}}{k}\int_{C_\phi}\frac{\nu}{\nu^2-1/4}\, g_{u,v}(\nu,\omega,\omega_0)\,\mathrm{e}^{-\mathrm{i}\nu\alpha}\,\mathrm{d}\nu$$
$$+\frac{\sqrt{2\pi}}{k}\,\pi\, g_{u,v}(\omega,\omega_0,1/2)\,\frac{\mathrm{e}^{\mathrm{i}\alpha/2}-\mathrm{e}^{-\mathrm{i}\alpha/2}}{2}, \qquad (6.106)$$

where the contour can be deformed into $C_{\phi=0}$, taking into account the estimates for the integrands. Therefore, taking into account the symmetry of the contour $\gamma = \gamma_+ \cup \gamma_-$ in (6.98), (see also (6.104)), we have

$$2\Phi_{u,v}(\alpha,\omega,\omega_0) = \Psi_{u,v}(\alpha,\omega,\omega_0) = \frac{2\mathrm{i}\sqrt{2\pi}}{k}\left[\int_{C_0}\frac{\nu}{\nu^2-1/4}\, g_{u,v}(\nu,\omega,\omega_0)\mathrm{e}^{-\mathrm{i}\nu\alpha}\,\mathrm{d}\nu\right.$$
$$\left.+\,(-\mathrm{i})\,\pi\, g_{u,v}(\omega,\omega_0,1/2)\,\frac{\mathrm{e}^{\mathrm{i}\alpha/2}-\mathrm{e}^{-\mathrm{i}\alpha/2}}{2}\right]. \qquad (6.107)$$

The functions $\Phi_{u,v}(\alpha,\omega,\omega_0)$ admit continuation as a regular function into the domain $\Pi_{\widehat{\theta'}} \cup D_C$, and using the oddness of $\Phi_{u,v}(\alpha,\omega,\omega_0)$, also into $D = \Pi_{\widehat{\theta'}} \cup D_C \cup D_C^\star$.

Remark. In the domain $D = \Pi_{\widehat{\theta'}} \cup D_C^\star$ we have the representation

$$\Phi_{u,v}(\alpha,\omega,\omega_0) = \frac{1}{2}\Psi_{u,v}(\alpha,\omega,\omega_0) = -\frac{\mathrm{i}\sqrt{2\pi}}{k}\int_{C_\phi}\frac{\nu}{\nu^2-1/4}\, g_{u,v}(\nu,\omega,\omega_0)\,\mathrm{e}^{\mathrm{i}\nu\alpha}\,\mathrm{d}\nu$$
$$+\frac{\sqrt{2\pi}}{k}\,\pi\, g_{u,v}(\omega,\omega_0,1/2)\,\frac{\mathrm{e}^{\mathrm{i}\alpha/2}-\mathrm{e}^{-\mathrm{i}\alpha/2}}{2}. \qquad (6.108)$$

6.6.3 Diffraction coefficients and Sommerefeld transformants

The Sommerfeld integrals give the representations for the potentials also for positive real values of k, which is of main interest in electromagnetic scattering [6.2]. Provided the observation point is in the domain M' not illuminated by the reflected rays $\widehat{\theta}'(\omega,\omega_0) > \pi$, $k > 0$, the real singularities of the Sommerfeld transformants are located outside the segment $[-\pi,\pi]$ of the real axis. We can apply the saddle-point techniques to (6.98) and obtain for the spherical wave propagating from the vertex ($kr \to \infty$)

$$(u^{\mathrm{s}}, v^{\mathrm{s}})(kr, \omega, \omega_0) = D_{u,v}(\omega, \omega_0) \frac{\mathrm{e}^{\mathrm{i}kr}}{-\mathrm{i}kr}\left[1 + \mathrm{O}\left(\frac{1}{kr}\right)\right], \tag{6.109}$$

$$D_{u,v}(\omega, \omega_0) = -\sqrt{\frac{2}{\pi}}\Phi_{u,v}(+\pi, \omega, \omega_0), \quad \text{or}$$

$$D_{u,v}(\omega, \omega_0) = -\frac{2}{k}\left[\int_{\mathrm{i}\mathbb{R}} \frac{\nu \sin \pi\nu}{\nu^2 - 1/4}\, g_{u,v}(\omega, \omega_0, \nu)\, \mathrm{d}\nu + \mathrm{i}\pi\, g_{u,v}(\omega, \omega_0, 1/2)\right]. \tag{6.110}$$

The actual electromagnetic diffraction coefficients are then deduced from (6.109)–(6.110) in the same manner as in (6.57)–(6.59). Note that, contrary to (6.56), expressions (6.109) and (6.110) were deduced for positive real wavenumbers.

Remark. For a narrow cone the diffraction coefficients can be asymptotically evaluated, the explicit formulas in the leading approximation are given in [6.7]. They have the form

$$\mathcal{E}_\theta = \frac{\mathrm{i}|l|}{8\pi^2}\frac{\sin\beta}{\eta}\frac{\sin\theta\sin\theta_0}{(\cos\theta + \cos\theta_0)^2} + \mathrm{O}(|l|^2\log|l|),$$

$$\mathcal{E}_\varphi = \frac{-\mathrm{i}|l|}{8\pi^2}\,\eta\sin\beta\,\frac{\sin\theta\sin\theta_0}{(\cos\theta + \cos\theta_0)^2} + \mathrm{O}(|l|^2\log|l|),$$

where $|l| \ll 1$ is the length of the contour l.

The integral in (6.110) diverges for $\omega \in M''$, and we need to derive alternative expressions for the diffraction coefficients. One could use the analytic continuation of $\Phi_{u,v}(\alpha, \omega, \omega_0)$ with respect to α to derive an expression for the diffraction coefficients. It is convenient to denote the Sommerfeld transformant coresponding to the incident wave by $\Phi^0_{u,v}(\alpha, \omega, \omega_0)$. It can be computed

$$\Phi^0_{u,v}(\alpha, \omega, \omega_0) = -\frac{\mathrm{i}\sqrt{2\pi}}{k}\left[\int_{\mathrm{i}\mathbb{R}} \frac{\nu\, g^0_{u,v}(\omega, \omega_0, \nu)}{\nu^2 - 1/4}\frac{\mathrm{e}^{\mathrm{i}\nu\alpha} - \mathrm{e}^{-\mathrm{i}\nu\alpha}}{2}\,\mathrm{d}\nu \right.$$
$$\left. + \mathrm{i}\pi\, g_{u,v}(\omega, \omega_0, 1/2)\frac{\mathrm{e}^{\mathrm{i}\alpha/2} - \mathrm{e}^{-\mathrm{i}\alpha/2}}{2}\right]. \quad |\mathrm{Re}\,\alpha| < \widehat{\theta}(\omega, \omega_0),$$

where the integral could be computed explicitly; however, the corresponding expressions will not be exploited herein.

Outside the oasis as $\omega \in M''$ the expression in (6.110) must be replaced by

$$D_{u,v}(\omega, \omega_0) = -\sqrt{\frac{2}{\pi}}\lim_{\epsilon\to 0+}\frac{\Phi_{u,v}(\pi + \mathrm{i}\epsilon, \omega, \omega_0) + \Phi_{u,v}(\pi - \mathrm{i}\epsilon, \omega, \omega_0)}{2}. \tag{6.111}$$

A similar expression has been discussed for the scalar case (5.85). Its origin also becomes obvious from the analysis of the Sommerfeld transformants given below and is based on the

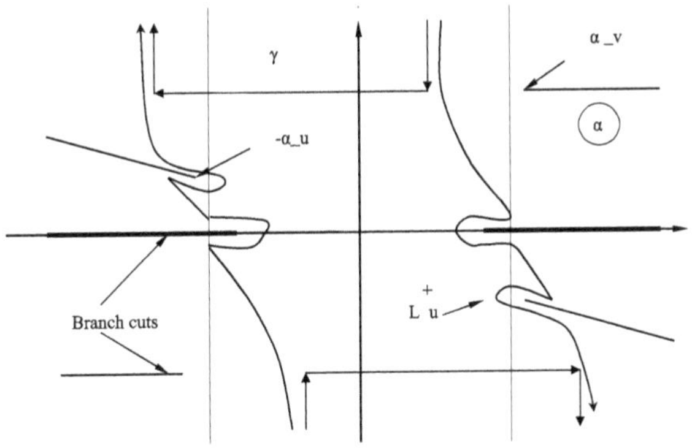

Figure 6.6 The Sommerfeld double-loop contour γ and its deformation into the steepest-descent paths, $\operatorname{Im}\eta < 0$.

saddle-point technique applied to the Sommerfeld integrals. In this case, the deformation of the integration contours are shown in Fig. 6.6. The saddle points are located on the real branch cuts, which leads to the expression (6.111).

We observe that the diffraction coefficients are connected with the value of the analytic function $\Phi_{u,v}(\alpha, \omega, \omega_0)$ at $\alpha = \pi$. In the case of $\omega \in M'$ (in the oasis) this function is regular in the vicinity of π. When the observation point belongs to the domain illuminated by the rays reflected from the cone, there are real branch points in the strip $\Pi_{\pi+\epsilon}$ and other singularities in $\Pi_{3\pi/2}$. While deforming the Sommerfeld contour into the steepest-descent paths $S(\pm\pi)$ (Fig. 6.6), these singularities can be captured, and therefore contribute to the asymptotics describing the reflected waves (the branch points at $\pm\widehat{\theta}'$) and the surface waves (the complex singularities at $\pm\alpha_{sw}$). However, if these singularities are not in the close vicinity of the saddle points $\pm\pi$, one can expect that the contributions from the saddle points to the high-frequency (far-field) asymptotics can be isolated, although the integrals representing the diffraction coefficients (6.110) should be appropriately modified.

The calculations for the expressions of the reflected waves are similar to those in the scalar acoustic case and are not presented herein. On the other hand, they could be found by means of the traditional ray method of the geometrical optics.

6.7 The diffraction coefficients outside the oasis as $\omega \in M''$

The analytic continuation of $\Phi_{u,v}(\alpha, \omega, \omega_0)$ given by (6.99) from the domain $|\operatorname{Re}\alpha| < \widehat{\theta}'(\omega, \omega_0)$ of definition onto a broader strip is performed using the formulas of the Fourier transform given in Section 1.4.1 as well as by use of the Fourier transform of the convolution. Remark that our calculations are close to the analog in Chapter 5.

Comparing with $\Phi_{u,v}(\omega, \omega_0)$, we observe that the auxiliary function

$$u_\nu^{\mathrm{r}}(\omega, \omega_0) = -\frac{\mathrm{P}_{\nu-1/2}\left(-\cos\widehat{\theta}'(\omega, \omega_0)\right)}{4\cos(\pi\nu)}, \tag{6.112}$$

up to a constant factor, has the same asymptotic behavior $C_{\pm}\nu^{-1/2}\mathrm{e}^{\pm\mathrm{i}\nu\widehat{\theta}'}$ as $\nu \to \pm\mathrm{i}\infty$. Assuming that $|\mathrm{Re}\,\alpha| < \widehat{\theta}'(\omega, \omega_0)$, one has

$$\begin{aligned}
\Phi_{u,v}(\alpha,\omega,\omega_0) &= -\frac{\mathrm{i}\sqrt{2\pi}}{k}\left[\int\limits_{\mathrm{i}\mathbb{R}} \frac{\nu\, g_{u,v}(\omega,\omega_0,\nu)}{\nu^2-1/4}\,\frac{\mathrm{e}^{\mathrm{i}\nu\alpha}-\mathrm{e}^{-\mathrm{i}\nu\alpha}}{2}\,\mathrm{d}\nu \right. \\
&\quad \left. +\mathrm{i}\pi\, g_{u,v}(\omega,\omega_0,1/2)\,\frac{\mathrm{e}^{\mathrm{i}\alpha/2}-\mathrm{e}^{-\mathrm{i}\alpha/2}}{2}\right] \\
&= \frac{\sqrt{2\pi}}{k}\frac{\mathrm{d}}{\mathrm{d}\alpha}\left(\frac{\mathrm{d}^2}{\mathrm{d}\alpha^2}+a^2\right)\frac{1}{2\pi}\int\limits_{\mathrm{i}\mathbb{R}} P(\alpha-\tau,\omega,\omega_0)\, r_{u,v}(\tau,\omega,\omega_0)\,\mathrm{d}\tau \\
&\quad + \frac{\sqrt{2\pi}}{k}\pi\, g_{u,v}(\omega,\omega_0,1/2)\,\frac{\mathrm{e}^{\mathrm{i}\alpha/2}-\mathrm{e}^{-\mathrm{i}\alpha/2}}{2},
\end{aligned} \tag{6.113}$$

where

$$\begin{aligned}
P(\alpha,\omega,\omega_0) &= \frac{1}{\mathrm{i}}\int\limits_{\mathrm{i}\mathbb{R}} u_\nu^{\mathrm{r}}(\omega,\omega_0)\,\mathrm{e}^{\mathrm{i}\nu\alpha}\,\mathrm{d}\nu, \\
r_{u,v}(\alpha,\omega,\omega_0) &= \frac{1}{\mathrm{i}}\int\limits_{\mathrm{i}\mathbb{R}} \frac{g_{u,v}(\omega,\omega_0,\nu)/u_\nu^{\mathrm{r}}(\omega,\omega_0)}{\left(\nu^2-a^2\right)\left(\nu^2-1/4\right)}\,\mathrm{e}^{\mathrm{i}\nu\alpha}\,\mathrm{d}\nu
\end{aligned} \tag{6.114}$$

and $a > 0$ is some constant. Note that the integrals in (6.114) converge absolutely and uniformly on $\alpha \in \mathrm{i}\mathbb{R}$.

Making use of the formula (7.216) from [6.13], one obtains

$$-\frac{1}{2}\int\limits_0^\infty \frac{\mathrm{P}_{\mathrm{i}\tau-1/2}\left(\cos\left(\pi-\widehat{\theta}'\right)\right)}{\cos(\mathrm{i}\pi\tau)}\cos(\mathrm{i}\alpha\tau)\,\mathrm{d}\tau = -\frac{1}{2\sqrt{2}}\left(\cos\alpha-\cos\widehat{\theta}'\right)^{-1/2}, \quad \mathrm{i}\alpha > 0. \tag{6.115}$$

Now we use argumentation of the analytic continuation, first, onto the strip $|\mathrm{Re}\,\alpha| < \widehat{\theta}'(\omega,\omega_0)$. To fix the regular branch of the right-hand side in (6.115) we conduct the branch cuts from $\pm\widehat{\theta}'$ to $\pm\infty$ correspondingly and assume that $\sqrt{\cos\alpha-\cos\widehat{\theta}'} > 0$ as $-\widehat{\theta}' < \alpha < \widehat{\theta}'$. The right-hand side of the formula (6.115) remains valid for $|\mathrm{Re}\,\alpha| < \pi+\delta$ for some small positive δ outside the branch cuts if $\widehat{\theta}' < \pi$; therefore, it also gives the analytic continuation of the left-hand side. The function

$$P(\alpha,\omega,\omega_0) = -\frac{1}{2\sqrt{2}}\left(\cos\alpha-\cos\widehat{\theta}'\right)^{-1/2}$$

is continuous on the sides of the branch cuts in the strip except the real points $\pm\widehat{\theta}'$ (Fig. 6.6).

Finally, introducing the notation

$$\begin{aligned}
\mathcal{P}(\alpha,\omega,\omega_0) &= -\frac{\mathrm{d}}{\mathrm{d}\alpha}\left(\frac{\mathrm{d}^2}{\mathrm{d}\alpha^2}+a^2\right)P(\alpha,\omega,\omega_0) \\
&= \left(\frac{\mathrm{d}^2}{\mathrm{d}\alpha^2}+a^2\right)\left[\frac{1}{4\sqrt{2}}\frac{\sin\alpha}{\left(\cos\alpha-\cos\widehat{\theta}'\right)^{3/2}}\right],
\end{aligned}$$

from (6.113) and (6.111) one has the formula for the scattering diagram

$$D_{u,v}(\omega,\omega_0) = -\frac{1}{2\pi k}\int_{\mathrm{i}\mathbb{R}} [\mathcal{P}(\pi+\mathrm{i}0-\tau,\omega,\omega_0)+\mathcal{P}(\pi-\mathrm{i}0-\tau,\omega,\omega_0)]\, r_{u,v}(\tau,\omega,\omega_0)\mathrm{d}\tau \tag{6.116}$$

$$+\mathrm{i}\pi\, g_{u,v}(\omega,\omega_0,1/2)\sqrt{2\pi}/k.$$

The electromagnetic diffraction coefficients are given by (6.59). So we can write

$$\begin{aligned} 2\pi\,\mathcal{E}_\theta &= -\frac{1}{\sin\theta}\frac{\partial D_v}{\partial\varphi} - \frac{\partial D_u}{\partial\theta}, \\ 2\pi\,\mathcal{E}_\varphi &= -\frac{1}{\sin\theta}\frac{\partial D_u}{\partial\varphi} + \frac{\partial D_v}{\partial\theta}. \end{aligned} \tag{6.117}$$

The numerical efficiency of these formulas is to be additionally studied and they could be modified appropriately.

It is worth commenting on the formula (6.113) giving analytic continuation of $\Phi_{u,v}(\alpha,\omega,\omega_0)$. In view of the assumption $\operatorname{Re}\zeta_{u,v} > 0$ (see Section 6.6.2 before Lemma 1) the right-hand side in (6.113) is regular in the strip $|\operatorname{Re}\alpha| < \pi+\delta$ with the above-mentioned branch cuts conducted from the real points $\pm\widehat{\theta}'(\omega,\omega_0)$, where $\delta > 0$ is such that $\operatorname{Re}\zeta_{u,v} \geq \delta$. The complex singularities at $\pm\alpha_{u,v}$, corresponding to the surface wave, may be located on the boundary of the strip $|\operatorname{Re}\alpha| < \pi+\delta$. In this case the formula (6.116) for the diffraction coefficient is valid in the whole domain M'', that is, as

$$\theta_1 \leq \theta < \theta_s(\varphi),$$

where $\theta = \theta_s(\varphi)$ is the equation of the line σ of singular directions on the unit sphere. However, provided $\operatorname{Re}\zeta_{u,v} = 0$, the complex singularities at $\pm\alpha_{u,v}$ are on the boundary of the strip $|\operatorname{Re}\alpha| < \pi$ as $\theta = \theta_1$ and the formula (6.116) for the diffraction coefficient is, generally speaking, valid as

$$\theta_1 - \delta < \theta < \theta_s(\varphi).$$

As $\theta_1 \leq \theta \leq \theta_1 - \delta$, applicability of (6.116) requires additional study. A procedure of the limiting absorption type, as $\operatorname{Re}\zeta_{u,v} \to 0$, may be efficient for such a domain.

We note that the Abel-Poisson summation exploited in [6.5] cannot be applied in the case of the impedance boundary conditions because of the more complex behavior of the spectral functions than those in the ideal case.

6.8 Problems for the Sommerfeld transformants and some complex singularities

6.8.1 Problems for the Sommerfeld transformants

Starting from the problems for the potentials u and v, we now formulate the equations and the boundary conditions in terms of $\Phi_{u,v}$, such that, provided one has solution of the latter, then, using the Sommerfeld integrals, the solution of the former can be obtained.

Lemma 6.2. *Let $\Phi_{u,v}(\alpha, \omega, \omega_0)$ be regular in the domain D from Lemma 6.1 (for any direction ω outside the cone, ω_0 is fixed) be smooth on ω and grow not faster than* $\exp(\mp \mathrm{i}a\alpha)$ *as* $\operatorname{Im}\alpha \to \pm\mathrm{i}\infty$, $a > 0$. *Besides, $\Phi_{u,v}(\alpha, \omega, \omega_0)$ satisfies the equation*

$$(\Delta_\omega - \partial_\alpha^2 - 1/4)\Phi_{u,v}(\alpha, \omega, \omega_0) = 0, \tag{6.118}$$

where Δ_ω is the Laplace–Beltrami operator on the unit sphere; then the integral (6.98) solves the Helmholtz equation.

Lemma 6.2 can be proved in a similar way as in Chapter 5. Now, under the conditions of Lemma 6.1, we turn to the boundary conditions for the Sommerefeld transformants.

Lemma 6.3. *Let $\Psi_{u,v}(\alpha, \omega, \omega_0) = \Phi_{u,v}(\alpha, \omega, \omega_0) + \Phi^0_{u,v}(\alpha, \omega, \omega_0)$ fulfill the boundary conditions*

$$\begin{aligned}
&\left(\frac{\partial \Psi_u}{\partial \theta} + \frac{\sin\alpha}{\sin\zeta_u}\frac{\partial \Psi_u}{\partial \alpha} + \frac{\sin\alpha}{4\sin\zeta_u}\widetilde{\Psi}_u\right)_{\theta=\theta_1} - \left(\frac{\sin\alpha}{2\sin\theta}\frac{\partial \widetilde{\Psi}_v}{\partial \varphi} + \frac{\cos\alpha}{\sin\theta}\frac{\partial \Psi_v}{\partial \varphi}\right)_{\theta=\theta_1} = 0,\\
&\left(\frac{\partial \Psi_v}{\partial \theta} + \frac{\sin\alpha}{\sin\zeta_v}\frac{\partial \Psi_v}{\partial \alpha} + \frac{\sin\alpha}{4\sin\zeta_v}\widetilde{\Psi}_v\right)_{\theta=\theta_1} + \left(\frac{\sin\alpha}{2\sin\theta}\frac{\partial \widetilde{\Psi}_u}{\partial \varphi} + \frac{\cos\alpha}{\sin\theta}\frac{\partial \Psi_u}{\partial \varphi}\right)_{\theta=\theta_1} = 0,\\
&\Psi_{u,v} = \partial_\alpha \widetilde{\Psi}_{u,v}.
\end{aligned} \tag{6.119}$$

Then the potentials u and v satisfy the boundary conditions (6.21) and (6.22).

To prove Lemma 6.3 is tedious; however, it is direct. We substitute the Sommerfeld integral representations for ru and rv (see (6.98) and (6.101)) into the boundary condition (6.21) and obtain

$$\begin{aligned}
0 = \frac{1}{2\pi\mathrm{i}}\int_\gamma \Bigg[&\frac{-\mathrm{i}k\mathrm{e}^{-\mathrm{i}kr\cos\alpha}}{\sqrt{-\mathrm{i}kr}}\left(\frac{\partial \Psi_u}{\partial \theta}\right)_{\theta=\theta_1} - \frac{\mathrm{i}k}{\eta}\frac{\mathrm{e}^{-\mathrm{i}kr\cos\alpha}}{\sqrt{-\mathrm{i}kr}}\frac{\partial}{\partial\alpha}\left(\frac{\Psi_u}{\sin\alpha}\right)_{\theta=\theta_1}\\
&+\frac{1}{\mathrm{i}k\eta}\frac{\partial^2}{\partial r^2}\left(\sqrt{-\mathrm{i}kr}\,\mathrm{e}^{-\mathrm{i}kr\cos\alpha}\Psi_u\right)_{\theta=\theta_1} + \frac{1}{\sin\theta}\frac{\partial}{\partial\varphi}\frac{1}{\mathrm{i}kr}\frac{\partial}{\partial r}\left(\sqrt{-\mathrm{i}kr}\,\mathrm{e}^{-\mathrm{i}kr\cos\alpha}\Psi_v\right)_{\theta=\theta_1}\Bigg]\mathrm{d}\alpha\\
= \frac{1}{2\pi\mathrm{i}}\int_\gamma \Bigg\{ &\frac{\mathrm{e}^{-\mathrm{i}kr\cos\alpha}}{\sqrt{-\mathrm{i}kr}}\left[(-\mathrm{i}k)\left(\frac{\partial \Psi_u}{\partial \theta}\right)_{\theta=\theta_1} - \frac{\mathrm{i}k}{\eta}\frac{\partial}{\partial\alpha}\left(\frac{\Psi_u}{\sin\alpha}\right)_{\theta=\theta_1}\right]\\
&+\frac{(-\mathrm{i}k)^2}{\mathrm{i}k\eta}\frac{\mathrm{e}^{-\mathrm{i}kr\cos\alpha}}{\sqrt{-\mathrm{i}kr}}(\cos\alpha\Psi_u)_{\theta=\theta_1} - \frac{(-\mathrm{i}k)^2}{4\mathrm{i}k\eta}\frac{\mathrm{e}^{-\mathrm{i}kr\cos\alpha}}{\left(\sqrt{-\mathrm{i}kr}\right)^3}\\
&+\frac{(-\mathrm{i}k)^2}{\mathrm{i}k\eta}(-\mathrm{i}kr)\frac{\mathrm{e}^{-\mathrm{i}kr\cos\alpha}}{\sqrt{-\mathrm{i}kr}}(\cos^2\alpha\,\Psi_u)_{\theta=\theta_1}\\
&+\frac{-\mathrm{i}k\,\mathrm{e}^{-\mathrm{i}kr\cos\alpha}}{2\mathrm{i}kr\sqrt{-\mathrm{i}kr}}\frac{1}{\sin\theta}\left(\frac{\partial \Psi_v}{\partial\varphi}\right)_{\theta=\theta_1} + \mathrm{i}k\frac{\mathrm{e}^{-\mathrm{i}kr\cos\alpha}}{\sqrt{-\mathrm{i}kr}}\frac{\cos\alpha}{\sin\theta}\left(\frac{\partial \Psi_v}{\partial\varphi}\right)_{\theta=\theta_1}\Bigg\}\mathrm{d}\alpha.
\end{aligned}$$

We then take into account that

$$(-ikr)\,e^{-ikr\cos\alpha} = -\frac{1}{\sin\alpha}\frac{\partial}{\partial\alpha}e^{-ikr\cos\alpha},$$

and $\Psi_{u,v} = \partial_\alpha \widetilde{\Psi}_{u,v}$ and integrate by parts, where it is necessary, and thus have

$$\frac{1}{2\pi i}\int_\gamma \left\{ \frac{e^{-ikr\cos\alpha}}{\sqrt{-ikr}} \left[-ik\left(\frac{\partial\Psi_u}{\partial\theta}\right)_{\theta=\theta_1} - \frac{ik}{\sin\zeta_u}\frac{\partial}{\partial\alpha}\left(\frac{\Psi_u}{\sin\alpha}\right)_{\theta=\theta_1} + \frac{ik\cos\alpha}{\sin\zeta_u}(\Psi_u)_{\theta=\theta_1}\right]\right.$$
$$-\frac{ik}{\sin\zeta_u}\left(\frac{\cos^2\alpha}{\sin\alpha}\right)\frac{\partial}{\partial\alpha}\left(\frac{e^{-ikr\cos\alpha}}{\sqrt{-ikr}}\right)(\Psi_u)_{\theta=\theta_1} + \frac{ik}{4\sin\zeta_u}\frac{1}{ikr}\frac{e^{-ikr\cos\alpha}}{\sqrt{-ikr}}\left(\frac{\partial\widetilde{\Psi}_u}{\partial\alpha}\right)_{\theta=\theta_1}$$
$$\left. + \frac{-ik}{2ikr}\frac{e^{-ikr\cos\alpha}}{\sqrt{-ikr}}\frac{1}{\sin\theta_1}\left[\frac{\partial}{\partial\varphi}\left(\frac{\partial\widetilde{\Psi}_v}{\partial\alpha}\right)\right]_{\theta=\theta_1} + ik\frac{e^{-ikr\cos\alpha}}{\sqrt{-ikr}}\frac{\cos\alpha}{\sin\theta_1}\left(\frac{\partial\Psi_v}{\partial\varphi}\right)_{\theta=\theta_1}\right\}d\alpha$$
$$= \frac{-ik}{2\pi i}\int_\gamma \left\{ \frac{e^{-ikr\cos\alpha}}{\sqrt{-ikr}} \left[\left(\frac{\partial\Psi_u}{\partial\theta}\right)_{\theta=\theta_1} + \frac{1}{\sin\zeta_u}\frac{\partial}{\partial\alpha}\left(\frac{\Psi_u}{\sin\alpha}\right)_{\theta=\theta_1} - \frac{\cos\alpha}{\sin\zeta_u}(\Psi_u)_{\theta=\theta_1}\right.\right.$$
$$-\frac{1}{\sin\zeta_u}\frac{\partial}{\partial\alpha}\left(\frac{\cos^2\alpha}{\sin\alpha}\Psi_u\right)_{\theta=\theta_1} + \left(\frac{\sin\alpha}{4\sin\zeta_u}\widetilde{\Psi}_u\right)_{\theta=\theta_1} - \left[\frac{\sin\alpha}{2\sin\theta}\frac{\partial\widetilde{\Psi}_v}{\partial\varphi}\right.$$
$$\left.\left.\left. + \frac{\cos\alpha}{\sin\theta}\frac{\partial\Psi_v}{\partial\varphi}\right]_{\theta=\theta_1}\right\} d\alpha = 0,$$

which completes the proof after elementary reductions and cancellations. The second boundary condition is treated in the same manner. Recall that $\Psi_{u,v}$ are assumed to be analytically continued as it is implied in the proofs of the lemmas.

6.8.2 Local behavior of the Sommerfeld transformants near complex singularities

As previously mentioned, the real singularities of the Sommerfeld transformants give rise to the reflected wave in the far-field asymptotics. In this case, the observation point is located in the nonoasis zone and the Sommerfeld integrals are evaluated by use of the saddle-point technique. In the process of deformation of the integration contour into the steepest-descent paths not only real but also complex singularities of the transformants are captured. The complex singularities are responsible for the surface waves. We turn to study of the complex singularities.

First, taking into account the rotational symmetry of the scatterer, we expand the transformants into the Fourier series with respect to φ

$$\Psi_{u,v}(\alpha,\omega,\omega_0) = \sum_{n=-\infty}^{\infty} i^n\, e^{-in\varphi}\Psi_{u,v}(\alpha,\theta,n) \tag{6.120}$$
$$= \sum_{n=-\infty}^{\infty} i^n\, e^{-in\varphi}[\Phi_{u,v}(\alpha,\theta,n) + \Phi^0_{u,v}(\alpha,\theta,n)].$$

The experience gained in the scalar problem shows that the complex singularities responsible for the surface waves are located at the points[14]

$$\alpha_{u,v}(\theta) = \pi + \zeta_{u,v} + \theta_1 - \theta.$$

We shall study contribution of $\alpha_u(\theta)$ provided $\mathrm{Im}\,\zeta_u < 0$. In this case, $\mathrm{Im}\,\zeta_v > 0$, and, in the leading approximation, only the singularities at $\pm\alpha_u(\theta)$ of $\Phi_{u,v}(\alpha, \omega, \omega_0)$ may contribute and give rise to the expression of the surface wave in the framework of the saddle-point technique applied to the Sommerfeld integrals.

We look for the local expansion for the Fourier coefficients in the form[15]

$$\Phi_u(\alpha, \theta, n) = \{A_{0u}(\theta, n) + A_{1u}(\theta, n)\,[\alpha - \alpha_u(\theta)] + \cdots\}\,[\alpha - \alpha_u(\theta)]^{\lambda_u} + \cdots, \qquad (6.121)$$

and

$$\Phi_v(\alpha, \theta, n) = \{A_{0v}(\theta, n) + A_{1v}\theta, n)\,[\alpha - \alpha_u(\theta)] + \cdots\}\,[\alpha - \alpha_u(\theta)]^{\lambda_u+1} + \cdots. \qquad (6.122)$$

The dots denote regular terms in the series near the nonregular point α_u. Remark that $\Phi_v(\alpha, \theta, n)$ has the lower order of singularity at α_u. Actually, the expansion (6.122) will not be exploited for the construction of the leading terms of the surface waves; however, we take into account the order of the singularity only. We substitute the Fourier series (6.120) into the equation (6.118) then exploit the expansion (6.121) and equate the terms of the same order of singularity at $\alpha_u(\theta)$. Thus we obtain in the leading approximations

$$2\lambda_u \frac{\mathrm{d}A_{0u}(\theta, n)}{\mathrm{d}\theta} + \lambda_u \cot\theta\, A_{0u}(\theta, n) = 0,$$

where λ_u is still undetermined. The solution of this equation is

$$A_{0u}(\theta, n) = C_{0u}(n)\sqrt{\frac{\sin\theta_1}{\sin\theta}}, \qquad (6.123)$$

whereas the constant $C_{0u}(n)$ cannot be determined from the local considerations of the present section and is discussed next. The exponent λ_u is determined from the boundary conditions (6.119) written for the Fourier coefficients $\Phi^s_{u,v}(\alpha, \theta, n)$. Taking into account only leading nontrivial singular terms (of $O([\alpha - \alpha_u]^{\lambda_u})$) in the first boundary condition, we find

$$\left.\frac{\mathrm{d}A_{0u}(\theta, n)}{\mathrm{d}\theta}\right|_{\theta=\theta_1} - \lambda_u \cot\zeta_u\, A_{0u}(\theta_1, n) = 0.$$

Since it is assumed that $C_{0u}(n) \neq 0$, we obtain

$$\lambda_u = -(1/2)\cot\theta_1 \tan\zeta_u. \qquad (6.124)$$

Remark that λ_u is purely imaginary and the singularity of $\Phi_{u,v}(\alpha, \theta, n)$ is of the logarithmic branch-point type. To fix its regular branch we conduct the branch cuts from the points $\alpha_{u,v}(\theta)$ as shown in Fig. 6.6. The branch is defined in the following section.

---

[14] Recall that the Sommerfeld transformants are odd with respect to α so that the singularities at $-\alpha_{u,v}(\theta)$ are analogously treated.

[15] We can consider the transformant Φ_u instead of Ψ_u because $\Phi^0_{u,v}(\alpha, \theta, n)$ are regular at this point.

We emphasize that the singularities of $\Phi_{u,v}$ at the points $\pm\alpha_v$ are not captured when deforming the Sommerfeld contour as $\operatorname{Im}\zeta_v > 0$.

6.9 Asymptotics of the Sommerfeld integrals and the electromagnetic surface waves

In the process of deformation of the Sommerfeld contour γ into the steepest-descent paths several branch cuts can be captured as shown on Fig. 6.6. The asymptotic contribution of the integrals over the sides of the real branch cuts give rise to the GO reflected electromagnetic wave, which is derived similarly to the scalar acoustic case and is omitted herein. We are interested in the contribution of complex singularities captured. They are responsible for the surface waves. We evaluate asymptotically the integral for the potentials over the contour l_u^+ comprising the branch cut from the point $\alpha_u(\theta)$[16]

$$u^{\text{sw}}(kr,\theta,\varphi) = \frac{1}{\pi \mathrm{i}} \int\limits_{l_u^+} \frac{\mathrm{e}^{-ikr\cos\alpha}}{\sqrt{-ikr}} \Phi_u(\alpha,\omega,\omega_0)\,\mathrm{d}\alpha. \tag{6.125}$$

Remark that contribution of l_u^- is also taken into account in (6.125) because $\Phi_u(\alpha,\omega,\omega_0)$ is odd with respect to α. In our conditions ($kr \to \infty$) it suffices to restrict ourselves to the local considerations, taking into account the expansion

$$\cos\alpha = \cos\alpha_u(\theta) - \sin\alpha_u(\theta)[\alpha - \alpha_u(\theta)] + \mathrm{O}([\alpha - \alpha_u(\theta)]^2), \quad \alpha - \alpha_u(\theta) \sim \mathrm{O}(1/kr)$$

near the singular point and the local behavior of Φ_u in the integrand, thus having

$$u^{\text{sw}}(kr,\theta) = \frac{C_{0u}(\varphi)}{\pi\mathrm{i}} \sqrt{\frac{\sin\theta_1}{\sin\theta}} \frac{\mathrm{e}^{-ikr\cos\alpha_u(\theta)}}{\sqrt{-ikr}}$$
$$\times \int\limits_{l_u^+} \mathrm{e}^{ikr\sin\alpha_\zeta(\theta)[\alpha-\alpha_u(\theta)]}[\alpha - \alpha_u(\theta)]^{-1+\mu}\,\mathrm{d}\alpha \left[1 + \mathrm{O}\left(\frac{1}{kr}\right)\right]. \tag{6.126}$$

with

$$C_{0u}(\varphi) = \sum_{n=-\infty}^{\infty} \mathrm{i}^n \mathrm{e}^{-in\varphi} C_{0u}(n)$$

and $\lambda_u = \mu - 1$.

Now it is reasonable to introduce a new variable of integration

$$\tau = -ikr\sin\alpha_u(\theta)[\alpha - \alpha_\zeta(\theta)]$$

and choose the branch of $[\alpha - \alpha_u(\theta)]^\mu$ or equivalently of τ^μ as follows

$$a_0 < \arg(\alpha - \alpha_u(\theta)) < 2\pi + a_0, \quad a_0 := -\arg(-ikr\sin\alpha_u(\theta)) = \arg(\alpha - \alpha_u(\theta))|_{l_u^{+,\text{up}}},$$

[16]The contour l_u^+ is located in a small vicinity of the branch point, goes along the upper side of the branch cut, enveloping the point $\alpha_u(\theta)$, and then passes along the lower side of the cut (Fig. 6.6).

where $l_u^{+,\mathrm{up}}$ is the upper side of the branch cut from $\alpha_u(\theta)$ to infinity,

$$\tau^{\mu} = \exp(\mu \log|\tau| + \mathrm{i}\,\mu \arg\tau), \quad \arg\tau = a_0 + \arg[\alpha - \alpha_u(\theta)]\,.$$

We evaluate asymptotically the integral in (6.126)

$$\int_{l_\zeta^+} \mathrm{e}^{\mathrm{i}kr\sin\alpha_u(\theta)[\alpha-\alpha_u(\theta)]}[\alpha - \alpha_u(\theta)]^{-1+\mu}\,\mathrm{d}\alpha$$

$$= [-\mathrm{i}kr\sin\alpha_u(\theta)]^{-\mu}\left(\mathrm{e}^{2\pi\mathrm{i}\mu} - 1\right)\int_0^{Ckr} \mathrm{e}^{-\tau}\tau^{-1+\mu}\,\mathrm{d}\tau$$

$$= [-\mathrm{i}kr\sin\alpha_u(\theta)]^{-\mu}\left(\mathrm{e}^{2\pi\mathrm{i}\mu} - 1\right)\Gamma(\mu)\,[1 + \mathrm{o}(1)]\,,$$

where $\Gamma(z)$ is the Euler gamma-function. Collecting these formulas, we obtain from (6.126)

$$u^{\mathrm{sw}}(kr,\theta,\varphi) = \frac{C_{0,u}(\varphi)}{\pi\mathrm{i}}\,\frac{\left(\mathrm{e}^{2\pi\mathrm{i}\mu} - 1\right)\Gamma(\mu)}{[-\sin(\theta_1 - \theta + \zeta_u)]^{\mu}}$$
$$\times\sqrt{\frac{\sin\theta_1}{\sin\theta}}\,\frac{\mathrm{e}^{\mathrm{i}kr\cos[\theta_1-\theta+\zeta_u]}}{(-\mathrm{i}kr)^{3/2+\lambda_u}}\left[1 + \mathrm{O}\left(\frac{1}{kr}\right)\right], \tag{6.127}$$

where $\lambda_u = -(1/2)\cot\theta_1\tan\zeta_u$, ζ_u is purely imaginary, $\mu = 1 + \lambda_u$ and

$$v^{\mathrm{sw}}(kr,\theta,\varphi) = \mathrm{O}\left(\frac{\mathrm{e}^{\mathrm{i}kr\cos[\theta_1-\theta+\zeta]}}{(-\mathrm{i}kr)^{5/2+\lambda_u}}\right).$$

We notice that under the sufficient conditions

$$\theta - \theta_1 - \mathrm{Re}\,\zeta_u - \mathrm{gd}\,(\mathrm{Im}\,\zeta_u) > 0, \quad \mathrm{Im}\,\zeta_u < 0\,, \quad \mathrm{Re}\,\zeta_u = 0,$$

where $\mathrm{gd}(x)$ stands for the Gudermann function, the corresponding singularities are indeed captured; hence the surface waves are excited.

Now we substitute the asymptotic expression (6.127) into the formulas (6.54); then, after some reduction, in the leading approximation we obtain

$$E_r^{\mathrm{sw}}(kr,\theta,\varphi) = \mathrm{i}kT_u(\varphi,\theta)\sin^2(\theta_1 - \theta + \zeta_u)$$
$$\times\frac{\exp[\mathrm{i}kr\cos(\theta_1 - \theta + \zeta_u)]}{(-\mathrm{i}kr)^{1/2+\lambda_u}}\left[1 + \mathrm{O}\left(\frac{1}{kr}\right)\right],$$
$$E_\theta^{\mathrm{sw}}(kr,\theta,\varphi) = -\mathrm{i}kT_u(\varphi,\theta)\sin(\theta_1 - \theta + \zeta_u)$$
$$\times\cos(\theta_1 - \theta + \zeta_u)\,\frac{\exp[\mathrm{i}kr\cos(\theta_1 - \theta + \zeta_u)]}{(-\mathrm{i}kr)^{1/2+\lambda_u}}\left[1 + \mathrm{O}\left(\frac{1}{kr}\right)\right], \tag{6.128}$$
$$H_\varphi^{\mathrm{sw}}(kr,\theta,\varphi) = -\mathrm{i}kT_u(\varphi,\theta)\sin(\theta_1 - \theta + \zeta_u)$$
$$\times\frac{\exp[\mathrm{i}kr\cos(\theta_1 - \theta + \zeta_u)]}{(-\mathrm{i}kr)^{1/2+\lambda_u}}\left[1 + O\left(\frac{1}{kr}\right)\right],$$

with

$$T_u(\varphi, \theta) = \frac{C_{0u}(\varphi)}{\pi \mathrm{i}} \frac{\left(\mathrm{e}^{2\pi \mathrm{i}\mu} - 1\right) \Gamma(\mu)}{[-\sin(\theta_1 - \theta + \zeta)]^{\mu}} \sqrt{\frac{\sin\theta_1}{\sin\theta}}\,.$$

It is natural to call the wave (6.128) the electrical-type surface wave, which is excited provided $\operatorname{Im}\zeta_u < 0,\ \operatorname{Im}\zeta_v > 0$. Contrary to this case, provided $\operatorname{Im}\zeta_v < 0,\ \operatorname{Im}\zeta_u > 0$, the magnetic-type surface wave propagates along the impedance cone.

From (6.128) we arrive at some simple physical conclusions. The electromagnetic surface wave has not only transversal but also longitudinal components. It propagates with the velocity $c/\cosh(|\zeta_u|)$, that is, slower than the spherical wave having the velocity c. On the conical surface its amplitude decreases as $O(1/(kr)^{1/2+\lambda_u})$. The type of the surface wave (electrical or magnetic) depends on the sign of the imaginary part of the impedance η. The presence of the factor $(kr)^{-\lambda_u}$ written in the form $\exp[-\lambda_u \log(kr)]$ enables us to interpret it as being responsible for the geometrical phase $\Phi_g(kr) = \mathrm{i}\lambda_u \log(kr)$. Note that, in our case, the geometrical phase is specified by the mean curvature of the circular cone, which also follows from [6.30] and [6.31].

6.9.1 Derivation of the functionals $C_{0u}(n)$

The final step in construction of the expressions of the electromagnetic surface waves can be done by giving a procedure of derivation of the constants $C_{0u}(n)$ in the excitation coefficients of the surface waves. First, we introduce $\rho_{u,v}(\alpha, n)$ as follows:

$$\rho_{u,v}(\alpha, n) = \frac{1}{\mathrm{i}} \int_{\mathrm{i}\mathbb{R}} R_{u,v}(\nu, n)\, \mathrm{e}^{\mathrm{i}\nu\alpha} \mathrm{d}\nu, \qquad R_{u,v}(\nu, n) = \frac{1}{2\pi \mathrm{i}} \int_{\mathrm{i}\mathbb{R}} \rho_{u,v}(\tau, n)\, \mathrm{e}^{-\mathrm{i}\nu\tau} \mathrm{d}\tau,$$

where $R_{u,v}(\nu, n)$ are solutions of the problem for the functional difference equations (6.62), and α belongs to a strip including the imaginary axis. We take into account an obvious correlation

$$\rho_{u,v}(\alpha, n) = -\left(\frac{\mathrm{d}^2}{\mathrm{d}\alpha^2} + \frac{1}{4}\right) \frac{1}{\mathrm{i}} \int_{\mathrm{i}\mathbb{R}} \frac{R_{u,v}(\nu, n)}{\nu^2 - 1/4} \mathrm{e}^{\mathrm{i}\nu\alpha} \mathrm{d}\nu.$$

Using the previous formula and the first integral for $\Phi_{u,v}$ in (6.113), we can write

$$\rho_{u,v}(\alpha, n) = -\frac{ik}{\sqrt{2\pi}} \left(\frac{\mathrm{d}^2}{\mathrm{d}\alpha^2} + \frac{1}{4}\right) \tilde{\Phi}_{u,v}(\alpha, \theta_1, n) + \cdots,$$

where the dots denote some regular terms. We consider the formula (see also (6.121))

$$\Phi_u(\alpha, \theta_1, n) = C_{0u}(\theta_1, n)[\alpha - \alpha_u(\theta_1)]^{\lambda_u} + \cdots,$$

thus have

$$\tilde{\Phi}_u(\alpha, \theta_1, n) = \frac{C_{0u}(n)}{\lambda_u + 1}[\alpha - \alpha_u(\theta_1)]^{\lambda_u + 1} + \cdots.$$

We arrive at the expansion

$$\rho_{u,v}(\alpha, n) = -\frac{ik}{\sqrt{2\pi}} C_{0u}(n) \lambda_u [\alpha - \alpha_u(\theta_1)]^{\lambda_u - 1} + \cdots. \tag{6.129}$$

Our task is to get analytic continuation of $\rho_u(\alpha, n)$ into a vicinity of the point $\alpha_u(\theta_1) = \pi+\zeta_u$. The idea of such continuations is due to Bernard [6.10]. Remark that there are some other ways of the desired analytical continuation.

Recall that $R_{u,v}$ is the Fourier transform of $\rho_{u,v}$. We apply the Fourier transform to the functional equation for R_u (6.62) and write it in the form ($\Lambda = n/(2\sin\theta_1)$)

$$(\sin\alpha + \sin\zeta_u)\rho_u(\alpha) + \sin\zeta_u\, \Lambda\, \rho_v^*(\alpha)$$
$$= \sin\zeta_u\, V_u(\alpha) + \sin\zeta_u\, V_1^u(\alpha) + \sin\zeta_u\, V_2^u(\alpha) + \sin\zeta_u\, V_4^v(\alpha) - T_u^{\mathrm{i}}(\alpha), \tag{6.130}$$

where

$$V_{u,v}(\alpha) = \frac{1}{\mathrm{i}} \int_{\mathrm{i}\mathbb{R}} [w(\nu, n) + \mathrm{i}\tan(b\nu)]\; R_{u,v}(\nu)\, \mathrm{e}^{\mathrm{i}\nu\alpha}\, d\nu\,, \quad b > 2\pi,$$

$$V_1^{u,v}(\alpha) = \frac{1}{\mathrm{i}} \int_{\mathrm{i}\mathbb{R}} \left[\mathrm{sgn}(\mathrm{i}\nu) - \mathrm{i}\tan(b\nu)\right]\; R_{u,v}(\nu)\, \mathrm{e}^{\mathrm{i}\nu\alpha}\, d\nu,$$

$$V_2^{u,v}(\alpha) = \frac{1}{\mathrm{i}} \int_{\mathrm{i}\mathbb{R}} \left[1 - \mathrm{sgn}(\mathrm{i}\nu)\right]\; R_{u,v}(\nu)\, \mathrm{e}^{\mathrm{i}\nu\alpha}\, d\nu,$$

$$\rho_v^*(\alpha) = \frac{1}{\mathrm{i}} \int_{\mathrm{i}\mathbb{R}} \left[\frac{\mathrm{e}^{-\mathrm{i}\alpha/2}}{t - 1/2} + \frac{\mathrm{e}^{\mathrm{i}\alpha/2}}{t + 1/2}\right] R_v(t)\, \mathrm{e}^{\mathrm{i}t\alpha} dt,$$

$$V_4^u(\alpha) = -2\pi\, R_u(1/2)\Lambda(\mathrm{e}^{\mathrm{i}\alpha/2} - \mathrm{e}^{-\mathrm{i}\alpha/2}), \quad V_4^v(\alpha) = 2\pi\, R_v(1/2)\Lambda(\mathrm{e}^{\mathrm{i}\alpha/2} - \mathrm{e}^{-\mathrm{i}\alpha/2}),$$

$T_{u,v}^{\mathrm{i}}(\alpha)$ is the Fourier transform of $G_{u,v}^0(\nu)$.[17]

It is important to notice that the right-hand side in (6.130) represents an analytic continuation of the left-hand side on the strip $|\mathrm{Re}\,\alpha| < \pi+\delta$ for some positive δ, because the integrals V_u and V_j^u, $j = 1, 2$ converge there, and, as a result, to the vicinity of the singularity $\alpha_u(\theta_1) = \pi + \zeta_u$. It is not difficult to demonstrate that $\rho_v^*(\alpha) = \mathrm{O}([\alpha - \alpha_u(\theta_1)]^{\lambda_u+1})$ near the singular point $\alpha_u(\theta_1)$. Then we take into account (6.130) and the expansion (6.129) and thus obtain

$$C_{0u}(n) = \lim_{\alpha\to\alpha_u(\theta_1)} \left\{ \frac{\sqrt{2\pi}}{ik} \frac{\tan\zeta_u}{\lambda_u\, [\alpha - \alpha_u(\theta_1)]^{\lambda_u}} \right.$$
$$\left. \times \left[V_u(\alpha) + V_1^u(\alpha) + V_2^u(\alpha) + V_4^v(\alpha) - T_u^{\mathrm{i}}(\alpha)/\sin\zeta_u \right] \right\}. \tag{6.131}$$

It is worth noting that the constant $C_{0u}(n)$ is actually a functional of $R_{u,v}(\nu, n)$, which are solutions of the system (6.62). The numerical calculation of C_{0u} may not be a simple task, which follows from the experience gained in the works [6.32] and [6.33].

---

[17] Analytic continuation of $T_{u,v}^{\mathrm{i}}(\alpha)$ in a vicinity of α_u is implied.

6.9.2 Some comments on the asymptotics uniform with respect to the direction of observation

As we have already mentioned, the nonuniform asymptotics for the spherical scattered and reflected waves ($kr \gg 1$) fail to be valid in the neighborhood of the singular directions $\widehat{\theta}'(\omega, \omega_0) \approx \pi$. The uniform asymptotics are described by means of the parabolic cylinder functions and have a form which is very similar to that exploited in the case of diffraction of an acoustic wave in the previous chapter, see also [6.34]. Thus we can write (for the scattered field except the surface wave)

$$
\begin{aligned}
\boldsymbol{E}(\boldsymbol{x}) &= (kr)^{-1/4} \mathrm{e}^{ikl(\boldsymbol{x})} \\
&\times \left[\boldsymbol{A}\, \mathrm{D}_{-3/2}(k^{1/2} m(\boldsymbol{x}) \mathrm{e}^{-\mathrm{i}\pi/4}) + \mathrm{e}^{\mathrm{i}\pi/4} \boldsymbol{B}\, \mathrm{D}'_{-3/2}(k^{1/2} m(\boldsymbol{x}) \mathrm{e}^{-\mathrm{i}\pi/4})\right] \\
&+ (kr)^{-1} \mathrm{e}^{ikr} \boldsymbol{F} \\
\boldsymbol{H}(\boldsymbol{x}) &= (kr)^{-1/4} \mathrm{e}^{ikl(\boldsymbol{x})} \\
&\times \left[\boldsymbol{C}\, \mathrm{D}_{-3/2}(k^{1/2} m(\boldsymbol{x}) \mathrm{e}^{-\mathrm{i}\pi/4}) + \mathrm{e}^{\mathrm{i}\pi/4} \boldsymbol{D}\, \mathrm{D}'_{-3/2}(k^{1/2} m(\boldsymbol{x}) \mathrm{e}^{-\mathrm{i}\pi/4})\right] \\
&+ (kr)^{-1} \mathrm{e}^{ikr} \boldsymbol{G},
\end{aligned}
\tag{6.132}
$$

where the coefficients $\boldsymbol{A}$, $\boldsymbol{B}$, $\boldsymbol{C}$, $\boldsymbol{D}$, $\boldsymbol{F}$, and $\boldsymbol{G}$ are the asymptotic series

$$\boldsymbol{A} \sim \boldsymbol{A}_0 + (kr)^{-1} \boldsymbol{A}_1 + \cdots ,$$

$$\boldsymbol{B} \sim \boldsymbol{B}_0 + (kr)^{-1} \boldsymbol{B}_1 + \cdots ,$$

$$\boldsymbol{C} \sim \boldsymbol{C}_0 + (kr)^{-1} \boldsymbol{C}_1 + \cdots ,$$

$$\boldsymbol{D} \sim \boldsymbol{D}_0 + (kr)^{-1} \boldsymbol{D}_1 + \cdots ,$$

$$\boldsymbol{F} \sim \boldsymbol{F}_0 + (kr)^{-1} \boldsymbol{F}_1 + \cdots ,$$

$$\boldsymbol{G} \sim \boldsymbol{G}_0 + (kr)^{-1} \boldsymbol{G}_1 + \cdots$$

and the functions l and m are defined by

$$l(\boldsymbol{x}) = r \sin^2[\widehat{\theta}'(\omega, \omega_0)/2], \quad m(\boldsymbol{x}) = -2\sqrt{r} \cos[\widehat{\theta}'(\omega, \omega_0)/2].$$

The coefficients in the asymptotic expansion can be found by substitution into Maxwell's equations and by matching (6.132) with the local asymptotics similarly to the acoustic diffraction by an impedance cone.

CHAPTER 7

Epilogue

In this monograph we tried to give a systematic and detailed exposition of different analytical techniques employed in the study of diffraction phenomena for a limited number of the so-called canonical problems, namely, the wave scattering by wedges or cones with impedance boundary conditions.

The key steps consist of representing the unknown solutions in terms of suitable integral transforms (the Sommerfeld–Malyuzhinets transform for wedge problems and the Kontorovich–Lebedev transform for cone problems), deriving difference equations for the spectral functions from the respective boundary conditions, simplifying and then converting the resultant higher-order functional difference equation to a Fredholm integral equation of the second kind, solving numerically the integral equation, and lastly deducing the scattering diagram (diffraction coefficient) by evaluating asymptotically the respective integral representations. It is precisely the essential analytical efforts described in the foregoing chapters that have enabled an accurate and fast solution of these boundary value problems.

A few subproblems should be dealt with in the near future: to study all surface waves excited by diffraction of a dipole wave field by an impedance wedge (Chapter 3) and to verify the efficiency of the formulas for at first the reflected, tip-diffracted, and surface waves outside the oasis, and then the uniform formulas for the far-field wave fields in the case of diffraction of a plane acoustic or electromagnetic wave by an impedance cone (Chapters 5 and 6).

Although we did not intend on giving an exhaustive study of the problems at hand, we expect that the presented results may be efficiently applied in the research practice and could be further developed to be exploited also in other related problems, for example, for studying diffraction by penetrable wedges or cones (see, e.g., [6.35]).

References

Introduction

0.1 J. B. Keller, "Geometrical theory of diffraction," *J. Opt. Soc. Am.*, vol. 52, no. 2, pp. 116–130, Feb. 1962.

0.2 V. A. Fock, "New methods in diffraction theory," *Philosph. Mag.*, vol. 39 (ser. 7), pp. 149–155, 1948.

0.3 V. A. Fock, *Electromagnetic Diffraction and Propagation Problems* (Int. Series of Monographs on Electromagnetic Waves 1). Oxford, UK: Pergamon Press, 1965.

0.4 V. M. Babich and V. S. Buldyrev, *Asymptotic Methods in Short-Wavelength Diffraction Theory* (Alpha Science Series on Wave Phenomena). Oxford, UK: Alpha Science, 2009.

0.5 G. L. James, *Geometrical Theory of Diffraction for Electromagnetic Waves* (IEE Electromagnetic Wave Series 1). 3rd ed. London, UK: Peregrinus, 1986.

0.6 P. H. Pathak, "Techniques for high-frequency problems," in *Antenna Handbook: Theory, Applications, and Design*, Y. T. Lo and S. W. Lee, Eds. New York, NY: Van Nostrand Reinhold, 1988.

0.7 V. A. Borovikov and B. Ye. Kinber, *Geometrical Theory of Diffraction* (IEE Electromagnetic Wave Series 37). London, UK: Institution of Elect. Engineers, 1994.

0.8 Y. A. Kravtsov and N. Y. Zhu, *Theory of Diffraction. Heuristic Approaches* (Alpha Science Series on Wave Phenomena). Oxford, UK: Alpha Science, 2010.

0.9 P. Ya. Ufimtsev, *Method of Fringe Waves in the Physical Theory of Diffraction*. Moscow, USSR: Sovetskoe Radio, 1962.

0.10 P. Ya. Ufimtsev, *Theory of Edge Diffraction in Electromagnetics: Origination and Validation of the Physical Theory of Diffraction*. Raleigh, NC: SciTech, 2009.

0.11 P. Ya. Ufimtsev, *Fundamentals of the Physical Theory of Diffraction*. Hoboken, NJ: Wiley, 2007.

0.12 T. N. Galishnikova and A.S. Il'inskiy, *Numerical Methods in Diffraction Problems*. Moscow, USSR: Moscow Univ. Press, 1987.

0.13 M. A. Leontovich, "Approximate boundary conditions for electromagnetic waves on surfaces of good conducting bodies," in *Investigation on Propagation of Radiowaves. Part II*, B. A. Vvedenskiĭ, Ed. Moscow, USSR: Publishing House of the USSR Academy of Sci., 1948, pp. 5–12.

0.14 G. Kirchhoff, "Zur Theorie der Lichtstrahlen," *Sitzgsber. Berl. Akad. Wiss.*, vol. 30, pp. 641–669, 1882; *Ann. Phys.*, vol. 18, no. 4, pp. 663–695, 1883.

0.15 G. Kirchhoff, *Vorlesungen über Mathematische Optik*. Leipzig, Germany: Teubner, 1891.

0.16 H. Poincaré, "Sur la polarisation par diffraction," *Acta Mathematica*, vol. 16, no. 1, pp. 297–338, 1892.

0.17 A. Sommerfeld, "Mathematische Theorie der Diffraction," *Math. Ann.*, vol. 47, pp. 317–374, 1896 (Transl.: *Mathematical Theory of Diffraction* (Progress in Mathematical Physics 35). Boston, MA: Birkhäuser, 2004).

0.18 H. M. Macdonald, *Electric Waves*. Cambridge, UK: Cambridge Univ. Press, 1902.

0.19 G. D. Maliuzhinets [Malyuzhinets], "Inversion formula for the Sommerfeld integral," *Soviet Phys.: Doklady*, vol. 3, no. 1, pp. 52–56, 1958.

0.20 G. D. Maliuzhinets [Malyuzhinets], "Excitation, reflection and emission of surface waves from a wedge with given face impedances," *Soviet Phys.: Doklady*, vol. 3, no. 4, pp. 752–755, 1958.

0.21 A. Sommerfeld, "Theorie der Beugung," in *Die Differential- und Integralgleichungen der Mechanik und Physik: II. Physikalischer Teil*, Ph. Frank and R. v. Mises, Eds. New York, NY: Dover, 1961, ch. 20.

0.22 A. Sommerfeld, *Vorlesungen über Theoretische Physik. Band IV: Optik*, 3rd ed. Frankfurt am Main, Germany: Harri Deutsch, 1989.

0.23 G. D. Malyuzhinets, *Sommerfeld Integrals and Their Applications*. Leningrad, USSR: Rumb, 1981.

0.24 T. B. A. Senior, "Diffraction by an imperfectly conducting wedge," *Commun. Pure Appl. Math.*, vol. 12, no. 2, pp. 337–372, 1959.

0.25 W. E. Williams, "Diffraction of an E-polarized plane wave by an imperfectly conducting wedge," *Proc. Roy. Soc. Lond. A*, vol. 252, no. 1270, pp. 376–393, 1959.

0.26 E. W. Barnes, "The genesis of the double gamma function," *Proc. Lond. Math. Soc.*, vol. 31, no. 1, pp. 358–381, 1899.

0.27 A. Rubinowicz, *Die Beugungswelle in der Kirchhoffschen Theorie der Beugung*. 2nd ed. Berlin, Germany: Springer, 1966.

0.28 J. J. Bowman, T. B. A. Senior and P. L. E. Uslenghi, Eds., *Electromagnetic and Acoustic Scattering by Simple Shapes*, rev. printing. New Yok, NY: Hemisphere, 1987.

0.29 B. V. Budaev, *Diffraction by Wedges* (Pitman Research Notes in Math. 322). Essex, UK: Longman Scientific & Tech., 1995.

0.30 T. B. A. Senior and J. L. Volakis, *Approximate Boundary Conditions in Electromagnetics* (IEE Electromagnetic Wave Series 41). London, UK: Institution of Elect. Engineers, 1995.

0.31 V. S. Buldyrev and M. A. Lyalinov, *Mathematical Methods in Modern Electromagnetic Diffraction Theory* (Int. Series of Monographs on Advanced Electromagnetics 1). Tokyo, Japan: Science House, 2001.

0.32 V. M. Babich, M. A. Lyalinov and V. E. Grikurov, *Diffraction Theory: the Sommerfeld-Malyuzhinets Technique* (Alpha Science Series on Wave Phenomena). Oxford, UK: Alpha Science, 2008.

0.33 C. DeWitt-Morette, S. G. Low, L. S. Schulman and A. Y. Shiekh, "Wedges I," *Found. Phys.*, vol. 16, no. 4, pp. 311–349, 1986.

0.34 A. V. Osipov and A. N. Norris, "The Malyuzhinets theory for scattering from wedge boundaries: a review," *Wave Motion*, vol. 29, no. 4, pp. 313–340, May 1999.

0.35 A. A. Tuzhilin, "The theory of Malyuzhinets' inhomogeneous functional difference equations," (in Russian), *Different. Uravn.*, vol. 9, no. 11, pp. 2058–2064, Nov. 1973.

0.36 M. S. Bobrovnikov and V. V. Fisanov, *Diffraction of Waves in Angular Regions*. Tomsk, USSR: Tomsk Univ. Press, 1988.

0.37 A. D. Avdeev, "On a special function that enters in the problem of diffraction by a wedge in an anisotropic plasma," *J. Commun. Technol. Electron.*, vol. 39, no. 10, pp. 70–78, 1994.

0.38 O. M. Bucci and G. Franceschetti, "Electromagnetic scattering by a half-plane with two face impedances," *Radio Sci.*, vol. 11, no. 1, pp. 49–59, 1976.

0.39 V. G. Vaccaro, "The generalized reflection method in electromagnetism," *Archiv für Elektronik und Übertragungstechnik*, vol. 34, no. 12, pp. 493–500, 1980.

0.40 V. G. Vaccaro, "Electromagnetic diffraction from a right-angled wedge with soft conditions on one face," *Optica Acta*, vol. 28, no. 3, pp. 293–311, 1981.

0.41 N. E. Nörlund, *Vorlesungen über Differenzenrechnung* (Die Grundlehren der mathematischen Wissenschaften). Berlin, Germany: Springer, 1924.

0.42 K. Lipszyc, "On the application of the Sommerfeld-Maluzhinetz transformation to some one-dimensional three-particle problems," *J. Math. Phys.*, vol. 21, no. 5, pp. 1092–1102, 1980.

0.43 L. A. Dmitrieva, Y. A. Kuperin and G. E. Rudin, "Extended class of Dubrovin's equations related to the one-dimensional quantum three-body problem," *Comput. and Math. with Applicat.*, vol. 34, no. 5/6, pp. 571–585, 1997.

0.44 R. Jost, "Lineare Differenzengleichungen mit periodischen Koeffizienten," *Commentari Mathematici Helvetici*, vol. 28, pp. 173–185, 1954.

0.45 R. Jost, "Mathematical analysis of a simple model for the stripping reaction," *Z. angew. Math. Phys.*, vol. 6, pp. 316–326, 1955.

0.46 S. Albeverio, "Analytische Lösung eines idealisierten Stripping- oder Beugungsproblems," *Helvetica Physica Acta*, vol. 40, pp. 135–184, 1967.

0.47 M. Gaudin and B. Derrida, "Solution exacte d'un probléme modéle a trois corps. État lié," *Journal de Physique*, vol. 36, no. 12, pp. 1183–1197, 1975.

0.48 S. Albeverio and P. Kurasov, *Singular Perturbations of Differential Operators. Solvable Schrödinger Type Operators* (London Math. Soc. Lecture Note Series 271). Cambridge, UK: Cambridge Univ. Press, 2000.

0.49 V. Buslaev and A. Fedotov, "On the difference equations with periodic coefficients," *Adv. Theor. Math. Phys.*, vol. 5, no. 6, pp. 1105–1168, 2001.

0.50 I. V. Komarov, "Various approaches to spectral problems for integrable systems in the QISM," *Int. J. Mod. Phys. A*, vol. 40, pp. 79–87, 1997.

0.51 C. Demetrescu, C. C. Constantinou, M. J. Mehler and B. V. Budaev, "Diffraction by a resistive sheet attached to a two-sided impedance plane," *Electromagnetics*, vol. 18, no. 3, pp. 315–332, 1998.

0.52 C. Demetrescu, C. C. Constantinou and M. J. Mehler, "Diffraction by a right-angled resistive wedge," *Radio Sci.*, vol. 33, no. 1, pp. 39–53, 1998.

0.53 M. A. Lyalinov and N. Y. Zhu, "Diffraction of a skew incident plane electromagnetic wave by an impedance wedge," *Wave Motion*, vol. 44, no. 1, pp. 21–43, Nov. 2006.

0.54 M. A. Lyalinov and N. Y. Zhu, "Diffraction of a skew incident plane electromagnetic wave by a wedge with axially anisotropic impedance faces," *Radio Sci.*, vol. 42, no. 6, RS6S03, 2007.

0.55 T. B. A. Senior and S. R. Legault, "Second-order difference equations in diffraction theory," *Radio Sci.*, vol. 35, no. 3, pp. 683–690, 2000.

0.56 T. B. A. Senior, S. R. Legault and J. L. Volakis, "A novel technique for the solution of second-order difference equations," *IEEE Trans. Antennas Propag.*, vol. 49, no. 12, pp. 1612–1617, 2001.

0.57 S. R. Legault and T. B. A. Senior, "Solution of a second order difference equation using the bilinear relations of Riemann," *J. Math. Phys.*, vol. 43, no. 3, pp. 1598–1621, 2002.

0.58 Y. A. Antipov and V. V. Silvestrov, "Vector functional-difference equation in electromagnetic scattering," *IMA J. Appl. Math.*, vol. 69, no. 1, pp. 27–69, 2004.

0.59 Y. A. Antipov and V. V. Silvestrov, "Second-order functional-difference equations. I: Method of the Riemann-Hilbert problem on Riemann surfaces," *Quart. J. Mech. Appl. Math.*, vol. 57, no. 2, pp. 245–265, 2004.

0.60 Y. A. Antipov and V. V. Silvestrov, "Second-order functional-difference equations. II: Scattering from a right-angled conductive wedge for E-polarization," *Quart. J. Mech. Appl. Math.*, vol. 57, no. 2, pp. 267–313, 2004.

0.61 H. S. Carslaw, "The scattering of sound waves by a cone," *Math. Ann.*, vol. 75, pp. 133–147, 1914. (correction on p. 592)

0.62 L. B. Felsen, "Electromagnetic properties of wedge and cone surfaces with a linearly varying surface impedance," *IRE Trans. Antennas Propagat.*, vol. 7, no. 5, pp. 231–243, 1959.

0.63 L. B. Felsen, "Plane wave scattering by small-angle cones," *IRE Trans. Antennas Propag.*, vol. 5, no. 1, pp. 121–129, 1957.

0.64 V. A. Borovikov, *Diffraction by Polygons and Polyhedrons*. Moscow, USSR: Nauka, 1966.

0.65 D. S. Jones, *The Theory of Electromagnetism*. Oxford, UK: Pergamon, 1964.

0.66 D. S. Jones, "Scattering by a cone," *Quart. J. Mech. Appl. Math.*, vol. 50, no. 4, pp. 499–523, Nov. 1997.

0.67 S. Blume and V. Krebs, "Numerical evaluation of dyadic diffraction coefficients and bistatic radar cross sections for a perfectly conducting semi-infinite elliptic cone," *IEEE Trans. Antennas Propag.*, vol. 46, no. 3, pp. 414–424, 1998.

0.68 L. Klinkenbusch, "Electromagnetic scattering by semi-infinite circular and elliptic cones," *Radio Sci.*, vol. 42, no. 6, RS6S10, 2007.

0.69 S. Blume and U. Uschkerat, "The radar cross section of the semi-infinite elliptic cone: numerical evaluation," *Wave Motion*, vol. 22, no. 3, pp. 311–326, Nov. 1995.

0.70 J. Cheeger and M. E. Taylor, "On the diffraction of waves by conical singularities. I," *Commun. Pure Appl. Math.*, vol. 35, no. 3, pp. 275–331, 1982.

0.71 J. Cheeger and M. E. Taylor, "On the diffraction of waves by conical singularities. II," *Common. Pure Appl. Math.*, vol. 35, no. 4, pp. 487–529, 1982.

0.72 V. P. Smyshlyaev, "Diffraction by conical surfaces at high frequencies," *Wave Motion*, vol. 12, no. 4, pp. 329–339, 1990.

0.73 V. P. Smyshlyaev, "The high frequency diffraction of electromagnetic waves by cones of arbitrary cross section," *SIAM J. Appl. Math.*, vol. 53, no. 3, pp. 670–688, 1993.

0.74 V. M. Babich, D. B. Dement'ev, B. A. Samokish and V. P. Smyshlyaev, "On evaluation of the diffraction coefficients for arbitrary 'non-singular' directions of a smooth convex cone," *SIAM J. Appl. Math.*, vol. 60, no. 2, pp. 536–573, 2000.

0.75 V. M. Babich, V. P. Smyshlyaev, D. B. Dement'ev and B. A. Samokish, "Numerical calculation of the diffraction coefficients for an arbitrary shaped perfectly conducting cone," *IEEE Trans. Antennas Propag.*, vol. 44, no. 5, pp. 740–747, 1996.

0.76 B. D. Bonner, I. G. Graham and V. P. Smyshlyaev, "The computation of the conical diffraction coefficients in high-frequency acoustic wave scattering," *SIAM J. Numer. Anal.*, vol. 43, no. 3, pp. 1202–1230, 2005.

0.77 J.-M. L. Bernard, "Méthode analytique et transformées fonctionnelles pour la diffraction d'ondes par une singularité conique: équation intégrale de noyau non oscillant pour le cas d'impédance constante," CEA, Saclay, France, Tech. Rep. CEA-R-5764, Sept. 1997. (erratum in J. Phys. A, vol. 32, p. L45)

0.78 J.-M. L. Bernard and M. A. Lyalinov, "Diffraction of acoustic waves by an impedance cone of an arbitrary cross-section," *Wave Motion*, vol. 33, no. 2, pp. 155–181, 2001. (erratum: p. 177 replace $O(1/\cos(\pi(\nu - b)))$ by $O(\nu^d \sin(\pi\nu)/\cos(\pi(\nu - b)))$)

0.79 J.-M. L. Bernard and M. A. Lyalinov, "Electromagnetic scattering by a smooth convex impedance cone," *IMA J. Appl. Math.*, vol. 69, no. 3, pp. 285–333, Jun. 2004. (multiply $\sin(\zeta)$ by $n/|n|$ in (D.20) of appendix D)

0.80 J.-M. L. Bernard, M. A. Lyalinov and N. Y. Zhu, "Analytical-numerical calculation of diffraction coefficients for a circular impedance cone," *IEEE Trans. Antennas Propag.*, vol. 56, no. 6, pp. 1616–1622, 2008.

0.81 M. A. Lyalinov, N. Y. Zhu and V. P. Smyshlyaev, "Scattering of a plane electromagnetic wave by a hollow circular cone with thin semi-transparent walls," *IMA J. Appl. Math.*, vol. 75, no. 5, pp. 676–719, Oct. 2010.

0.82 J.-M. L. Bernard, M. A. Lyalinov and N. Y. Zhu, "Chapter 5. Diffraction of acoustic and electromagnetic waves by impedance cones," in *Electromagnetics and Network Theory and Their Microwave Technology Applications*, S. Lindenmeier and R. Weigel, Eds. Berlin, Germany: Springer, 2011, pp. 65–73.

Chapter 1

1.1 A. D. Pierce, *Acoustics*. New York, NY: Acoust. Soc. Am., 1989.

1.2 A. D. Pierce, "Basic linear acoustics," in *Springer Handbook of Acoustics*, Part A, T. D. Rossing Ed. Berlin, Germany: Springer, 2007, ch. 3, pp. 25–111.

1.3 D. S. Jones, *The Theory of Electromagnetism*. Oxford, UK: Pergamon, 1964.

1.4 J. D. Jackson, *Classical Electrodynamics*, 3rd ed. New York, NY: Wiley, 1999.

1.5 S. A. Nazarov and B. A. Plamenevskii, *Elliptic Problems in Domains with Piecewise Smooth Boundaries*. Berlin, Germany: de Gruyter, 1994.

1.6 J. Meixner,"The behavior of electromagnetic fields at edges," *IEEE Antennas Propag.*, vol. 20, no. 4, pp. 442–446, 1972.

1.7 V. M. Babich, M. A. Lyalinov and V. E. Grikurov, *Diffraction Theory: the Sommerfeld-Malyuzhinets Technique* (Alpha Science Series on Wave Phenomena). Oxford, UK: Alpha Science, 2008.

1.8 R. Courant and D. Hilbert, *Methods of Mathematical Physics. Volume 2: Partial Differential Equations*. New York, NY: Wiley, 1962.

1.9 V. V. Kamotski and G. Lebeau, "Diffraction by an elastic wedge with stress-free boundary: existence and uniquiness," *Proc. R. Soc. Lond. A*, vol. 462, no. 2065, pp. 289–317, 2006.

1.10 M. A. Lyalinov, "Scattering of an acoustic axially symmetric surface wave propagating to the vertex of a right-circular impedance cone," *Wave Motion*, vol. 47, no. 4, pp. 241–252, 2010.

1.11 E. M. Stein and G. Weiss, *Introduction to Fourier Analysis on Euclidean Spaces* (Princeton Mathematical Series 32). Princeton, NJ: Princeton Univ. Press, 1990.

1.12 A. Sommerfeld, "Mathematische Theorie der Diffraction," *Math. Ann.*, vol. 47, pp. 317–374, 1896 (Transl.: *Mathematical Theory of Diffraction* (Progress in Mathematical Physics 35). Boston, MA: Birkhäuser, 2004).

1.13 G. D. Maliuzhinets [Malyuzhinets], "Excitation, reflection and emission of surface waves from a wedge with given face impedances," *Soviet Phys.: Doklady*, vol. 3, no. 4, pp. 752–755, 1958.

1.14 G. D. Maliuzhinets [Malyuzhinets], "Inversion formula for the Sommerfeld integral," *Soviet Phys.: Doklady*, vol. 3, no. 1, pp. 52–56, 1958.

1.15 A. N. Norris and A. V. Osipov, "Far-field analysis of the Malyuzhinets solution for plane and surface waves diffraction by an impedance wedge," *Wave Motion*, vol. 30, no. 1, pp. 69–89, Jul. 1999.

1.16 A. V. Osipov and A. N. Norris, "The Malyuzhinets theory for scattering from wedge boundaries: a review," *Wave Motion*, vol. 29, no. 4, pp. 313–340, May 1999.

1.17 B. V. Budaev, *Diffraction by Wedges* (Pitman Research Notes in Math. 322). Essex, UK: Longman Scientific & Tech., 1995.

1.18 V. S. Buldyrev and M. A. Lyalinov, *Mathematical Methods in Modern Electromagnetic Diffraction Theory* (Int. Series of Monographs on Advanced Electromagnetics 1). Tokyo, Japan: Science House, 2001.

1.19 M. J. Kontorowich [Kontorovich] and N. N. Lebedev, "On a method of solution of some problems of the diffraction theory," *J. Phys. (Moscow)*, vol. 1, no. 3, pp. 229–241, 1939.

1.20 G. D. Maliuzhinets [Malyuzhinets], "Relation between the inversion formulas for the Sommerfeld integral and the formulas of Kontorovich-Lebedev," *Soviet Phys.: Doklady*, vo. 3, pp. 266–268, 1958.

1.21 J.-M. L. Bernard and M. A. Lyalinov, "Diffraction of acoustic waves by an impedance cone of an arbitrary cross-section," *Wave Motion*, vol. 33, no. 2, pp. 155–181, 2001. (erratum: p. 177 replace $O(1/\cos(\pi(\nu - b)))$ by $O(\nu^d \sin(\pi\nu)/\cos(\pi(\nu - b)))$)

1.22 J.-M. L. Bernard and M. A. Lyalinov, "Electromagnetic scattering by a smooth convex impedance cone," *IMA J. Appl. Math.*, vol. 69, no. 3, pp. 285-333, Jun. 2004. (multiply $\sin(\zeta)$ by $n/|n|$ in (D.20) of appendix D)

1.23 J.-M. L. Bernard, "Méthode analytique et transformées fonctionnelles pour la diffraction d'ondes par une singularité conique: équation intégrale de noyau non oscillant pour le cas d'impédance constante," CEA, Saclay, France, Tech. Rep. CEA-R-5764, Sept. 1997. (erratum in J. Phys. A, vol. 32, p. L45)

1.24 I. S. Gradshteyn and I. M. Ryzhik, *Table of Integrals, Series and Products*, 4th ed. New York, NY: Academic Press, 1980.

1.25 M. Cessenat, *Mathematical Methods in Electromagnetism: Linear Theory and Applications* (Series on Advances in Mathematics for Applied Sciences 41). Singapore: World Scientific, 1996.

1.26 D. S. Jones, "The Kontorovich-Lebedev transform," *IMA J. Appl. Math.*, vol. 26, no. 2, pp. 133–141, 1980.

1.27 V. M. Babich, D. B. Dement'ev , B. A. Samokish and V. P. Smyshlyaev, "On evaluation of the diffraction coefficients for arbitrary 'non-singular' directions of a smooth convex cone," *SIAM J. Appl. Math.*, vol. 60, no. 2, pp. 536–573, 2000.

1.28 M. A. Lyalinov, N. Y. Zhu and V. P. Smyshlyaev, "Scattering of a plane electromagnetic wave by a hollow circular cone with thin semi-transparent walls," *IMA J. Appl. Math.*, vol. 75, no. 5, pp. 676–719, Oct. 2010.

1.29 G. D. Maliuzhinets [Malyuzhinets], "The radiation of sound by the vibrating boundaries of an arbitrary wedge. Part I," *Soviet Phys.: Acoust.*, vol. 1, pp. 152–174, 1955.

1.30 G. D. Maliuzhinets [Malyuzhinets], "Radiation of sound from the vibrating faces of an arbitrary wedge. Part II," *Soviet Phys.: Acoust.*, vol. 1, pp. 240–248, 1955

1.31 G. D. Malyughinetz [Malyuzhinets], "Das Sommerfeldsche Integral und die Lösung von Beugungsaufgaben in Winkelgebieten," *Ann. Phys.*, vol. 6 (ser. 7), pp. 107–112, 1960.

1.32 T. B. A. Senior and J. L. Volakis, *Approximate Boundary Conditions in Electromagnetics* (IEE Electromagnetic Wave Series 41). London, UK: Institution of Elect. Engineers, 1995.

1.33 R. G. Kouyoumjian and P. H. Pathak, "A uniform geometrical theory of diffraction for an edge in a perfectly conducting surface," *Proc. IEEE*, vol. 62, no. 11, pp. 1448–1461, Nov. 1974.

1.34 A. A. Tuzhilin, "The theory of Malyuzhinets' inhomogeneous functional difference equations," (in Russian), *Different. Uravn.*, vol. 9, no. 11, pp. 2058-2064, Nov. 1973.

Chapter 2

2.1 A. Sommerfeld, "Mathematische Theorie der Diffraction," *Math. Ann.*, vol. 47, pp. 317–374, 1896 (Transl.: *Mathematical Theory of Diffraction* (Progress in Mathematical Physics 35). Boston, MA: Birkhäuser, 2004).

2.2 G. D. Malyuzhinets, *Sommerfeld Integrals and Their Applications*. Leningrad, USSR: Rumb, 1981.

2.3 M. S. Bobrovnikov and V. V. Fisanov, *Diffraction of Waves in Angular Regions*. Tomsk, USSR: Tomsk Univ. Press, 1988.

2.4 B. V. Budaev, *Diffraction by Wedges* (Pitman Research Notes in Math. 322). Essex, UK: Longman Scientific & Tech., 1995.

2.5 T. B. A. Senior and J. L. Volakis, *Approximate Boundary Conditions in Electromagnetics* (IEE Electromagnetic Wave Series 41). London, UK: Institution of Elect. Engineers, 1995.

2.6 V. S. Buldyrev and M. A. Lyalinov, *Mathematical Methods in Modern Electromagnetic Diffraction Theory* (Int. Series of Monographs on Advanced Electromagnetics 1). Tokyo, Japan: Science House, 2001.

2.7 V. M. Babich, M. A. Lyalinov and V. E. Grikurov, *Diffraction Theory: the Sommerfeld-Malyuzhinets Technique* (Alpha Science Series on Wave Phenomena). Oxford, UK: Alpha Science, 2008.

2.8 A. D. Avdeev, "On a special function that enters in the problem of diffraction by a wedge in an anisotropic plasma," *J. Commun. Technol. Electron.*, vol. 39, no. 10, pp. 70–78, 1994.

2.9 A. A. Tuzhilin, "The theory of Malyuzhinets' inhomogeneous functional difference equations," (in Russian), *Different. Uravn.*, vol. 9, no. 11, pp. 2058–2064, Nov. 1973.

2.10 M. A. Lyalinov and N. Y. Zhu, "A solution procedure for second-order difference equations and its application to electromagnetic-wave diffraction in a wedge-shaped region," *Proc. R. Soc. Lond. A*, vol. 459, no. 2040, pp. 3159–3180, Dec. 2003.

2.11 N. Y. Zhu and M. A. Lyalinov, "Diffraction of a normally incident plane wave by an impedance wedge with its exterior bisected by a semi-infinite impedance sheet," *IEEE Trans. Antennas Propag.*, vol. 52, no. 10, pp. 2753–2758, 2004.

2.12 N. Y. Zhu and M. A. Lyalinov, "Diffraction by a wedge or by a cone with impedance-type boundary conditions and second-order functional difference equations," *Progress in Electromagnetcs Research B*, vol. 6, pp. 239–256, 2008.

2.13 M. A. Lyalinov and N. Y. Zhu, "Diffraction of a skew incident plane electromagnetic wave by an impedance wedge," *Wave Motion*, vol. 44, no. 1, pp. 21–43, Nov. 2006.

2.14 B. Budaev and D. B. Bogy, "Diffraction of a plane skew electromagnetic wave by a wedge with general anisotropic impedance boundary conditions," *IEEE Trans. Antennas Propag.*, vol. 54, no. 5, pp. 1559–1567, 2006.

2.15 V. G. Daniele and G. Lombardi, "Wiener-Hopf solution for impedance wedges at skew incidence," *IEEE Trans. Antennas Propag.*, vol. 54, no. 9, pp. 2472–2485, 2006.

2.16 A. V. Osipov and T. B. A. Senior, "Electromagnetic diffraction by arbitrary-angle impedance wedges," *Proc. R. Soc. Lond. A*, vol. 464, no. 2089, pp. 177–195, 2008.

2.17 M. A. Lyalinov and N. Y. Zhu, "Diffraction of a skew incident plane wave by an anisotropic impedance wedge — a class of exactly solvable cases," *Wave Motion*, vol. 30, no. 3, pp. 275–288, Oct. 1999.

2.18 M. A. Lyalinov and N. Y. Zhu, "Exact solution to diffraction problems by wedges with a class of anisotropic impedance faces: oblique incidence of a plane electromagnetic wave," *IEEE Trans. Antennas Propag.*, vol. 51, no. 6, pp. 1216–1220, 2003.

2.19 Y. A. Antipov and A. A. Silvestrov, "Electromagnetic scattering from an anisotropic impedance half-plane at oblique incidence: the exact solution," *Quart. J. Mech. Appl. Math.*, vol. 59, no. 2, pp. 211–251, 2006.

2.20 M. A. Lyalinov and N. Y. Zhu, "Diffraction of a skew incident plane electromagnetic wave by a wedge with axially anisotropic impedance faces," *Radio Sci.*, vol. 42, no. 6, RS6S03, 2007.

2.21 R. G. Kouyoumjian and P. H. Pathak, "A uniform geometrical theory of diffraction for an edge in a perfectly conducting surface," *Proc. IEEE*, vol. 62, no. 11, pp. 1448–1461, Nov. 1974.

2.22 K. E. Atkinson, *The Numerical Solution of Integral Equations of the Second Kind* (Cambridge Monographs on Appl. and Computational Math.). Cambridge, UK: Cambridge Univ. Press, 1997.

2.23 A. D. Polyanin and A. V. Manzhirov, *Handbook of Integral Equations*, 2nd ed. London, UK: Chapman & Hall, 2008.

2.24 E. J. Nyström, "Über die praktische Auflösung von Integralgleichungen mit Anwendungen auf Randwertaufgaben," *Acta Mathematica*, vol. 54, no. 1, pp. 185–204, 1930.

2.25 J.-L. Hu, S.-M. Lin and W.-B. Wang, "Calculation of Maliuzhinets function in complex region," *IEEE Trans. Antennas Propag.*, vol. 44, no. 8, pp. 1195–1196, Aug. 1996.

2.26 A. V. Kostyuk, "Calculation of the Malyuzhinets function in the problem of diffraction of electromagnetic waves by a wedge with impedance faces," *J. Commun. Technol. Electron.*, vol. 42, no. 10, pp. 1161–1168, Oct. 1997.

2.27 A. Osipov and V. Stein, "The theory and numerical computation of Maliuzhinets' special function," DLR Inst. Radio Frequency Technol., Oberpfaffenhofen, Germany, Tech. Rep. DLR-IB 551–5/1999, Aug. 23, 1999.

2.28 S. Zhang and J. Jin, *Computation of Special Functions*. New York, NY: Wiley, 1996.

2.29 A. V. Osipov, "Calculation of the Malyuzhinets function in a complex region," *Soviet Phys.: Acoust.*, vol. 35, no. 1, pp. 63–66, 1990.

Chapter 3

3.1 T. B. A. Senior, "The diffraction of a dipole field by a perfectly conducting half-plane," *Quart. J. Mech. Appl. Math.*, vol. 6, no. 1, pp. 101–114, 1953.

3.2 I. S. Gradshteyn and I. M. Ryzhik, *Table of Integrals, Series and Products*, 4th ed. New York, NY: Academic Press, 1980.

3.3 A. Sommerfeld, "Mathematische Theorie der Diffraction," *Math. Ann.*, vol. 47, pp. 317–374, 1896 (Transl.: *Mathematical Theory of Diffraction* (Progress in Mathematical Physics 35). Boston, MA: Birkhäuser, 2004).

3.4 H. Weyl, "Ausbreitung elektromagnetischer Wellen über einem ebenen Leiter," *Ann. Phys.*, vol. 62, no. 21, pp. 481–500, 1919.

3.5 R. Teisseyre, "New method of solving the diffraction problem for a dipole field," *Bulletin de l'Académie Polonaise des Sciences Cl. III*, vol. 4, no. 7, pp. 433–438, 1956.

3.6 G. D. Malyuzhinets and A. A. Tuzhilin, "The electromagnetic field excited by an electric dipole in a wedge-shaped region," *Soviet Phys.: Doklady*, vol. 7, no. 10, pp. 879–882, Apr. 1963.

3.7 J. J. Bowman, T. B. A. Senior and P. L. E. Uslenghi, Eds., *Electromagnetic and Acoustic Scattering by Simple Shapes*, rev. printing. New Yok, NY: Hemisphere, 1987.

3.8 M. A. Lyalinov and N. Y. Zhu, "Diffraction of a skew incident plane electromagnetic wave by an impedance wedge," *Wave Motion*, vol. 44, no. 1, pp. 21–43, Nov. 2006.

3.9 M. A. Lyalinov and N. Y. Zhu, "Diffraction of a skew incident plane electromagnetic wave by a wedge with axially anisotropic impedance faces," *Radio Sci.*, vol. 42, no. 6, RS6S03, 2007.

3.10 M. V. Fedoryuk, *Asymptotics, Series and Integrals*. Moscow, USSR: Nauka, 1987.

3.11 D. S. Ahluwalia, R. M. Lewis and J. Boersma, "Uniform asymptotic theory of diffraction by a plane screen," *SIAM J. Appl. Math.*, vol. 16, no. 4, pp. 783–807, Jul. 1968.

3.12 R. M. Lewis and J. Boersma, "Uniform asymptotic theory of edge diffraction," *J. Math. Phys.*, vol. 10, no. 12, pp. 2291–2305, Dec. 1969.

3.13 S. W. Lee and G. A. Deschamps, "A uniform asymptotic theory of electromagnetic diffraction by a curved wedge," *IEEE Trans. Antennas Propag.*, vol. 24, no. 1, pp. 25–34, Jan. 1976.

3.14 V. M. Babich, "On PC-Ansätze," *J. Math. Sci.*, vol. 132, no. 11, pp. 2–10, 2006.

3.15 M. V. Fedoryuk, "II. Asymptotic Methods in Analysis," in *Encyclopaedia of Mathematical Sciences, Analysis I, vol 13*. R.V. Gamkrelidze, Ed. Berlin, Germany: Springer, 1989, pp. 83–191.

3.16 T. B. A. Senior and J. L. Volakis, *Approximate Boundary Conditions in Electromagnetics* (IEE Electromagnetic Wave Series 41). London, UK: Institution of Elect. Engineers, 1995.

3.17 V. A. Borovikov and B. Ye. Kinber, *Geometrical Theory of Diffraction* (IEE Electromagnetic Wave Series 37). London, UK: Institution of Elect. Engineers, 1994.

3.18 G. D. Malyuzhinets, "Development in our concepts of diffraction phenomena (on the 130th aniversary of the death of Thomas Young)," *Soviet Phys. Uspekhi*, vol. 69 (ser. 2), no. 5, pp. 749–758, Sept.-Oct. 1959.

3.19 A. Erdélyi, *Asymptotic Expansion*. New York, NY: Dover, 1956.

3.20 E. T. Copson, *Asymptotic Expansions* (Cambridge Tracts in Mathematics 55). Cambridge, UK: Cambridge Univ. Press, 1965.

3.21 D. B. Paris, *Hadamard Expansions and Hyperasymptotic Evaluation. An Extension of the Method of Steepest Descents* (Encyclopedia of Mathematics and Its Applications 141). Cambridge, UK: Cambridge Univ. Press, 2011.

3.22 F. Ursell, "Integrals with a large parameter: a double complex integral with four nearly coincident saddle-points," *Math. Proc. Camb. Phil. Soc.*, vol. 87, no. 2, pp. 249–273, 1980.

3.23 L. C. Hsu, *Asymptotic Integration and Integral Approximation*. Beijing, China: Science Press, 1958.

3.24 N. Bleistein and R. A. Handelsman, *Asymptotic Expansions of Integrals*. New York, NY: Dover, 1986.

3.25 R. Wong, *Asymptotic Approximations of Integrals* (SIAM's Classics in Applied Mathematics 34). Philadelphia: SIAM, 2001.

3.26 M. V. Fedoryuk, *The Saddle-Point Method*. Moscow, USSR: Nauka, 1977.

3.27 C. J. Howls, "Hyperasymptotics for multidimensional integrals, exact remainder terms and the global connection problem," *Proc. Roy. Soc. Lond. A*, vol. 453, no. 1966, pp. 2271–2294, 1997.

3.28 E. Delabaere and C. J. Howls, "Global asymptotics for multiple integrals with boundaries," *Duke Math. J.*, vol. 112, no. 2, pp. 199–266, 2002.

Chapter 4

4.1 V. E. Grikurov and M. A. Lyalinov, "Diffraction of the surface H-polarized wave by an angular break of a thin dielectric slab," *J. Math. Sci.*, vol. 155, no. 3, pp. 390–396, 2008.

4.2 R. F. Harrington and J. R. Mautz, "An impedance sheet approximation for thin dielectric shells," *IEEE Trans. Antennas Propag.*, vol. 23, no. 4, pp. 531–534, 1975.

4.3 T. B. A. Senior and J. L. Volakis, *Approximate Boundary Conditions in Electromagnetics* (IEE Electromagnetic Wave Series 41). London, UK: Institution of Elect. Engineers, 1995.

4.4 V. M. Babich, M. A. Lyalinov and V. E. Grikurov, *Diffraction Theory: the Sommerfeld-Malyuzhinets Technique* (Alpha Science Series on Wave Phenomena). Oxford, UK: Alpha Science, 2008.

4.5 A. Sommerfeld, "Mathematische Theorie der Diffraction," *Math. Ann.*, vol. 47, pp. 317–374, 1896 (Transl.: *Mathematical Theory of Diffraction* (Progress in Mathematical Physics 35). Boston, MA: Birkhäuser, 2004).

4.6 M. A. Lyalinov, "Scattering of an acoustic axially symmetric surface wave propagating to the vertex of a right-circular impedance cone," *Wave Motion*, vol. 47, no. 4, pp. 241–252, 2010.

Chapter 5

5.1 M. Abramowitz and I. Stegun, Eds., *Handbook of Mathematical Functions*. New York, NY: Dover, 1972.

5.2 A. Sommerfeld, "Mathematische Theorie der Diffraction," *Math. Ann.*, vol. 47, pp. 317–374, 1896 (Transl.: *Mathematical Theory of Diffraction* (Progress in Mathematical Physics 35). Boston, MA: Birkhäuser, 2004).

5.3 Y. A. Antipov, "Diffraction of a plane wave by a circular cone with an impedance boundary condition," *SIAM J. Appl. Math.*, vol. 62, no. 4, pp. 1122–1152, 2002.

5.4 V. M. Babich, D. B. Dement'ev and B.A. Samokish, "On diffraction of high frequency waves by a cone of arbitrary shape," *Wave Motion*, vol. 21, no. 3, pp. 203–207, May 1995.

5.5 V. M. Babich, "The diffraction of a high-frequency acoustic wave by a narrow-angle absolutely rigid cone of arbitrary shape," *J. Appl. Math. Mech.*, vol. 60, no. 1, pp. 72–78, 1996.

5.6 V. M. Babich, "On PC-Ansätze," *J. Math. Sci.*, vol. 132, no. 11, pp. 2–10, 2006.

5.7 V. M. Babich, D. B. Dement'ev, B. A. Samokish and V. P. Smyshlyaev, "Scattering of a high-frequency wave by the vertex of an arbitrary cone (singular directions)," *J. Math. Sci.*, vol. 111, no. 4, pp. 3623–3631, 2002.

5.8 V. M. Babich and V. S. Buldyrev, *Asymptotic Methods in Short-Wavelength Diffraction Theory* (Alpha Science Series on Wave Phenomena). Oxford, UK: Alpha Science, 2009.

5.9 V. M. Babich, V. P. Smyshlyaev, D. B. Dement'ev and B. A. Samokish, "Numerical calculation of the diffraction coefficients for an arbitrary shaped perfectly conducting cone," *IEEE Trans. Antennas Propag.*, vol. 44, no. 5, pp. 740–747, 1996.

5.10 V. M. Babich, D. B. Dement'ev , B. A. Samokish and V. P. Smyshlyaev, "On evaluation of the diffraction coefficients for arbitrary 'non-singular' directions of a smooth convex cone," *SIAM J. Appl. Math.*, vol. 60, no. 2, pp. 536–573, 2000.

5.11 V. M. Babich and A.V. Kuznetsov, "Propagation of surface electromagnetic waves similar to Rayleigh waves in the case of Leontovich boundary conditions", *J. Math. Sci.*, vol. 138, no. 2, pp. 5483–5490, 2006.

5.12 R. Grimshaw, "Propagation of surface waves at high frequencies," *IMA J. Appl. Math.*, vo. 4, no. 2, pp. 174–193, Jun. 1968.

5.13 V. M. Babich, V. S. Buldyrev and I. A. Molotkov, *Space-time Ray Method: Linear and Non-linear Waves*. Leningrad, USSR: Leningrad Univ. Press, 1985.

5.14 V. M. Babich, M. A. Lyalinov and V. E. Grikurov, *Diffraction Theory: the Sommerfeld-Malyuzhinets Technique* (Alpha Science Series on Wave Phenomena). Oxford, UK: Alpha Science, 2008.

5.15 S. Blume and U. Uschkerat, "The radar cross section of the semi-infinite elliptic cone: numerical evaluation," *Wave Motion*, vol. 22, no. 3, pp. 311–326, Nov. 1995.

5.16 S. Blume and V. Krebs, "Numerical evaluation of dyadic diffraction coefficients and bistatic radar cross sections for a perfectly conducting semi-infinite elliptic cone," *IEEE Trans. Antennas Propag.*, vol. 46, no. 3, pp. 414–424, 1998.

5.17 V. A. Borovikov, *Diffraction by Polygons and Polyhedrons*. Moscow, USSR: Nauka, 1966.

5.18 J. J. Bowman, T. B. A. Senior and P. L. E. Uslenghi, Eds., *Electromagnetic and Acoustic Scattering by Simple Shapes*, rev. printing. New Yok, NY: Hemisphere, 1987.

5.19 J. Cheeger and M. E. Taylor, "On the diffraction of waves by conical singularities. I," *Commun. Pure Appl. Math.*, vol. 35, no. 3, pp. 275–331, 1982.

5.20 J. Cheeger and M. E. Taylor, "On the diffraction of waves by conical singularities. II," *Common. Pure Appl. Math.*, vol. 35, no. 4, pp. 487–529, 1982.

5.21 H. Bateman, *Tables of Integral Transforms*. New York, NY: McGraw-Hill, 1954.

5.22 L. B. Felsen, "Plane wave scattering by small-angle cones," *IRE Trans. Antennas Propag.*, vol. 5, no. 1, pp. 121–129, 1957.

5.23 L. B. Felsen and N. Marcuvitz, *Radiation and Scattering of Waves*. Englewood Cliffs, NJ: Prentice Hall, 1973.

5.24 I. S. Gradshteyn and I. M. Ryzhik, *Table of Integrals, Series and Products*, 4th ed. New York, NY: Academic Press, 1980.

5.25 D. S. Jones, "Scattering by a cone," *Quart. J. Mech. Appl. Math.*, vol. 50, no. 4, pp. 499–523, Nov. 1997.

5.26 D. S. Jones, *The Theory of Electromagnetism*. Oxford, UK: Pergamon, 1964.

5.27 J.-M. L. Bernard, "Méthode analytique et transformées fonctionnelles pour la diffraction d'ondes par une singularité conique: équation intégrale de noyau non oscillant pour le cas d'impédance constante," CEA, Saclay, France, Tech. Rep. CEA-R-5764, Sept. 1997. (erratum in J. Phys. A, vol. 32, p. L45)

5.28 J. M. L. Bernard and M. A. Lyalinov, "The leading asymptotic term for the scattering diagram in the problem of diffraction by a narrow circular impedance cone," *J. Phys. A: Math. Gen.*, vo. 32, no. 4, pp. L43–L48, Jan. 1999. (replace $\psi(\nu - 1/2)$ by $\psi(\nu + 1/2)$ in (12))

5.29 J.-M. L. Bernard and M. A. Lyalinov, "Diffraction of acoustic waves by an impedance cone of an arbitrary cross-section," *Wave Motion*, vol. 33, no. 2, pp. 155–181, 2001. (erratum: p. 177 replace $O(1/\cos(\pi(\nu - b)))$ by $O(\nu^d \sin(\pi\nu)/\cos(\pi(\nu - b)))$)

5.30 J.M.L. Bernard and M.A. Lyalinov, "The leading asymptotic term for the scattering diagram by a narrow impedance cone," in *Proc. of the IEEE AP-S Conf.*, Salt Lake City, 2000, pp. 398–401.

5.31 J.M.L. Bernard and M.A. Lyalinov, "Spectral domain solution and asymptotics for the diffraction by an impedance cone," *IEEE Trans. Antennas Propag.*, vol. 49, no. 12, pp. 1633–1637, 2001.

5.32 J.-M. L. Bernard, M. A. Lyalinov and N. Y. Zhu, "Analytical-numerical calculation of diffraction coefficients for a circular impedance cone," *IEEE Trans. Antennas Propag.*, vol. 56, no. 6, pp. 1616–1622, 2008.

5.33 B. D. Bonner, I. G. Graham and V. P. Smyshlyaev, "The computation of the conical diffraction coefficients in high-frequency acoustic wave scattering," *SIAM J. Numer. Anal.*, vol. 43, no. 3, pp. 1202–1230, 2005.

5.34 A. L. Brodskaya, A. V. Popov and S. A. Hoziossky, "Asymptotics of the wave reflected from a cone in the penumbral domain," in *Proc. of 6th All Union Symp. on Diffraction and Propagation of Waves*, Moscow-Erevan, 1976, pp. 227–231.
A. Popov, A. Ladyzhensky (Brodskaya) and S. Khozioski, "Uniform asymptotics of the wave diffracted by a cone of arbitrary cross section," *Russ. J. Math. Phys.*, vol. 16, no. 2, pp. 296–299, 2009.

5.35 V. S. Buldyrev and M. A. Lyalinov, *Mathematical Methods in Modern Electromagnetic Diffraction Theory* (Int. Series of Monographs on Advanced Electromagnetics 1). Tokyo, Japan: Science House, 2001.

5.36 V. V. Kamotskii, "Calculation of some integrals describing wave fields," *J. Math. Sci.*, vol. 108, no. 5, pp. 665–673, 2002.

5.37 L. Klinkenbusch, "Electromagnetic scattering by semi-infinite circular and elliptic cones," *Radio Sci.*, vol. 42, no. 6, RS6S10, 2007.

5.38 Y. A. Kravtsov and N. Y. Zhu, *Theory of Diffraction. Heuristic Approaches* (Alpha Science Series on Wave Phenomena). Oxford, UK: Alpha Science, 2010.

5.39 M. A. Lyalinov and N. Y. Zhu, "Acoustic scattering by a circular semi-transparent conical surface," *J. Eng. Math.*, vol. 59, no. 4, pp. 385–398, 2007.

5.40 M. A. Lyalinov, "The far-field asymptotics in the problem of diffraction of an acoustic plane wave by an impedance cone," *Russ. J. Math. Phys.*, vol. 16, no. 2, pp. 277–286, 2009.

5.41 M. A. Lyalinov, "Diffraction of a plane acoustic wave by an impedance cone. Surface waves," *J. Math. Sci.*, vol. 167, no. 5, pp. 651–659, 2010.

5.42 M. A. Lyalinov, "Scattering of an acoustic axially symmetric surface wave propagating to the vertex of a right-circular impedance cone," *Wave Motion*, vol. 47, no. 4, pp. 241–252, 2010.

5.43 M. A. Lyalinov, N. Y. Zhu and V. P. Smyshlyaev, "Scattering of a plane electromagnetic wave by a hollow circular cone with thin semi-transparent walls," *IMA J. Appl. Math.*, vol. 75, no. 5, pp. 676–719, Oct. 2010.

5.44 M.A. Lyalinov, "Acoustic scattering of a plane wave by a circular penetrable cone," *Wave Motion*, vol. 48, no. 1, pp. 62–82, 2011.

5.45 V. P. Smyshlyaev, "Diffraction by conical surfaces at high frequencies," *Wave Motion*, vol. 12, no. 4, pp. 329–339, 1990.

5.46 V. P. Smyshlyaev, "On the diffraction of waves by cones at high frequences," LOMI, Leningrad, USSR, Preprint E-9-89, 1989.

5.47 V. P. Smyshlyaev, "High-frequency asymptotic behavior of the Green's function for the Helmholtz equation in a region with a conical boundary," *Soviet Phys.: Doklady*, vol. 34, no. 11, pp. 970–972, 1989.

5.48 V. S. Vladimirov, *Equations of Mathematical Physics*. Moscow, USSR: Nauka, 1967.

5.49 K. E. Atkinson, *The Numerical Solution of Integral Equations of the Second Kind* (Cambridge Monographs on Appl. and Computational Math.). Cambridge, UK: Cambridge Univ. Press, 1997.

5.50 N. Y. Zhu and M. A. Lyalinov, "Diffraction by a wedge or by a cone with impedance-type boundary conditions and second-order functional difference equations," *Progress in Electromagnetcs Research B*, vol. 6, pp. 239–256, 2008.

5.51 G. D. Malyuzhinets, "Development in our concepts of diffraction phenomena (on the 130th anniversary of the death of Thomas Young)," *Soviet Phys. Uspekhi*, vol. 69 (ser. 2), no. 5, pp. 749–758, Sept.-Oct. 1959.

Chapter 6

6.1 J. J. Bowman, T. B. A. Senior and P. L. E. Uslenghi, Eds., *Electromagnetic and Acoustic Scattering by Simple Shapes*, rev. printing. New Yok, NY: Hemisphere, 1987.

6.2 M. A. Lyalinov, N. Y. Zhu and V. P. Smyshlyaev, "Scattering of a plane electromagnetic wave by a hollow circular cone with thin semi-transparent walls," *IMA J. Appl. Math.*, vol. 75, no. 5, pp. 676–719, Oct. 2010.

6.3 V. P. Smyshlyaev, "On the diffraction of waves by cones at high frequences," LOMI, Leningrad, USSR, Preprint E-9–89, 1989.

6.4 V. P. Smyshlyaev, "The high frequency diffraction of electromagnetic waves by cones of arbitrary cross section," *SIAM J. Appl. Math.*, vol. 53, no. 3, pp. 670–688, 1993.

6.5 V. M. Babich, D. B. Dement'ev , B. A. Samokish and V. P. Smyshlyaev, "On evaluation of the diffraction coefficients for arbitrary 'non-singular' directions of a smooth convex cone," *SIAM J. Appl. Math.*, vol. 60, no. 2, pp. 536–573, 2000.

6.6 O. V. Klyubina, "Diffraction of plane waves by bodies of different types," *J. Math. Sci.*, vol. 111, no. 4, pp. 3708–3716, 2002.

6.7 J.-M. L. Bernard and M. A. Lyalinov, "Electromagnetic scattering by a smooth convex impedance cone," *IMA J. Appl. Math.*, vol. 69, no. 3, pp. 285–333, Jun. 2004. (multiply $\sin(\zeta)$ by $n/|n|$ in (D.20) of appendix D)

6.8 M. A. Lyalinov, "Diffraction of a plane acoustic wave by an impedance cone. Surface waves," *J. Math. Sci.*, vol. 167, no. 5, pp. 651–659, 2010.

6.9 V. M. Babich, "On PC-Ansätze," *J. Math. Sci.*, vol. 132, no. 11, pp. 2–10, 2006.

6.10 J.-M. L. Bernard, "Méthode analytique et transformées fonctionnelles pour la diffraction d'ondes par une singularité conique: équation intégrale de noyau non oscillant pour le cas d'impédance constante," CEA, Saclay, France, Tech. Rep. CEA-R-5764, Sept. 1997. (erratum in J. Phys. A, vol. 32, p. L45)

6.11 D. S. Jones, *The Theory of Electromagnetism*. Oxford, UK: Pergamon, 1964.

6.12 M. Abramowitz and I. Stegun, Eds., *Handbook of Mathematical Functions*. New York, NY: Dover, 1972.

6.13 I. S. Gradshteyn and I. M. Ryzhik, *Table of Integrals, Series and Products*, 4th ed. New York, NY: Academic Press, 1980.

6.14 A. A. Tuzhilin, "The theory of Malyuzhinets' inhomogeneous functional difference equations," (in Russian), *Different. Uravn.*, vol. 9, no. 11, pp. 2058–2064, Nov. 1973.

6.15 V. M. Babich, M. A. Lyalinov and V. E. Grikurov, *Diffraction Theory: the Sommerfeld-Malyuzhinets Technique* (Alpha Science Series on Wave Phenomena). Oxford, UK: Alpha Science, 2008.

6.16 V. M. Babich, V. P. Smyshlyaev, D. B. Dement'ev and B. A. Samokish, "Numerical calculation of the diffraction coefficients for an arbitrary shaped perfectly conducting cone," *IEEE Trans. Antennas Propag.*, vol. 44, no. 5, pp. 740–747, 1996.

6.17 D. S. Jones, "Scattering by a cone," *Quart. J. Mech. Appl. Math.*, vol. 50, no. 4, pp. 499–523, Nov. 1997.

6.18 S. Blume and V. Krebs, "Numerical evaluation of dyadic diffraction coefficients and bistatic radar cross sections for a perfectly conducting semi-infinite elliptic cone," *IEEE Trans. Antennas Propag.*, vol. 46, no. 3, pp. 414–424, 1998.

6.19 L. Klinkenbusch, "Electromagnetic scattering by semi-infinite circular and elliptic cones," *Radio Sci.*, vol. 42, no. 6, RS6S10, 2007.

6.20 L. B. Felsen, "Electromagnetic properties of wedge and cone surfaces with a linearly varying surface impedance," *IRE Trans. Antennas Propag.*, vol. 7, no. 5, pp. 231–243, 1959.

6.21 L. B. Felsen, "Plane wave scattering by small-angle cones," *IRE Trans. Antennas Propag.*, vol. 5, no. 1, pp. 121–129, 1957.

6.22 S. Blume and U. Uschkerat, "The radar cross section of the semi-infinite elliptic cone: numerical evaluation," *Wave Motion*, vol. 22, no. 3, pp. 311–326, Nov. 1995.

6.23 V. A. Borovikov, *Diffraction by Polygons and Polyhedrons*. Moscow, USSR: Nauka, 1966.

6.24 J. Cheeger and M. E. Taylor, "On the diffraction of waves by conical singularities. I," *Commun. Pure Appl. Math.*, vol. 35, no. 3, pp. 275–331, 1982.

6.25 J. Cheeger and M. E. Taylor, "On the diffraction of waves by conical singularities. II," *Common. Pure Appl. Math.*, vol. 35, no. 4, pp. 487–529, 1982.

6.26 L. B. Felsen and N. Marcuvitz, *Radiation and Scattering of Waves*. Englewood Cliffs, NJ: Prentice Hall, 1973.

6.27 B. D. Bonner, I. G. Graham and V. P. Smyshlyaev, "The computation of the conical diffraction coefficients in high-frequency acoustic wave scattering," *SIAM J. Numer. Anal.*, vol. 43, no. 3, pp. 1202–1230, 2005.

6.28 K. E. Atkinson, *The Numerical Solution of Integral Equations of the Second Kind* (Cambridge Monographs on Appl. and Computational Math.). Cambridge, UK: Cambridge Univ. Press, 1997.

6.29 V. H. Weston, "Theory of absorbers in scattering," *IEEE Trans. Anennas Propag.*, vol. 11, no. 5, pp. 578–584, Sept. 1963.

6.30 V. M. Babich and A.V. Kuznetsov, "Propagation of surface electromagnetic waves similar to Rayleigh waves in the case of Leontovich boundary conditions", *J. Math. Sci.*, vol. 138, no. 2, pp. 5483–5490, 2006.

6.31 R. Grimshaw, "Propagation of surface waves at high frequencies," *IMA J. Appl. Math.*, vo. 4, no. 2, pp. 174–193, Jun. 1968.

6.32 J.-M. L. Bernard, M. A. Lyalinov and N. Y. Zhu, "Analytical-numerical calculation of diffraction coefficients for a circular impedance cone," *IEEE Trans. Antennas Propag.*, vol. 56, no. 6, pp. 1616–1622, 2008.

6.33 Y. A. Kravtsov and N. Y. Zhu, *Theory of Diffraction. Heuristic Approaches* (Alpha Science Series on Wave Phenomena). Oxford, UK: Alpha Science, 2010.

6.34 A. L. Brodskaya, A. V. Popov and S. A. Hoziossky, "Asymptotics of the wave reflected from a cone in the penumbral domain," in *Proc. of 6th All Union Symp. on Diffraction and Propagation of Waves*, Moscow-Erevan, 1976, pp. 227–231.
A. Popov, A. Ladyzhensky (Brodskaya) and S. Khozioski, "Uniform asymptotics of the wave diffracted by a cone of arbitrary cross section," *Russ. J. Math. Phys.*, vol. 16, no. 2, pp. 296–299, 2009.

6.35 M.A. Lyalinov, "Acoustic scattering of a plane wave by a circular penetrable cone," *Wave Motion*, vol. 48, no. 1, pp. 62–82, 2011.

6.36 J.-M. L. Bernard, M. A. Lyalinov and N. Y. Zhu, "Chapter 5. Diffraction of acoustic and electromagnetic waves by impedance cones," in *Electromagnetics and Network Theory and Their Microwave Technology Applications*, S. Lindenmeier and R. Weigel, Eds. Berlin, Germany: Springer, 2011, pp. 65–73.

Index

Printed in the USA
CPSIA information can be obtained
at www.ICGtesting.com
JSHW051410221024
72173JS00006B/1331

9 781613 53003